THE STANLEY CATALOG COLLECTION
Volume II

THE STANLEY CATALOG COLLECTION
VOLUME II

A Supplemental Collection of 19th Century
Stanley and Leonard Bailey Catalogs

THE ASTRAGAL PRESS
Mendham, New Jersey

Introduction copyright © 1998 by
The Astragal Press

Library of Congress Catalog Card Number: 97-73827
International Standard Book Number 1-879335-78-6

Printed in the United States of America

Published by
THE ASTRAGAL PRESS
5 Cold Hill Road, Suite 12
Mendham NJ 07945-0239

TABLE OF CONTENTS

INTRODUCTION 1

STANLEY RULE & LEVEL CO.
   1872 Catalog and Price List. 7
   1874 Catalog and Price List with 1876 Supplement. . . . 57
   1877 Catalog and Price List 135
   1884 Catalog and Price List 203
   1892 Catalog and Price List
      with Abridged Revision to 1897. 269

LEONARD BAILEY & CO.
   1876 Catalog and Price List 361
   1883 Catalog and Price List 393

MILESTONES in the GROWTH of STANLEY TOOLS 407

INTRODUCTION

Stanley catalogs have long served as primary references for Stanley tool enthusiasts. Studying these vintage catalogs is one of the best ways to develop a knowledge of the older tools, when they were introduced, their features, and how they changed over time. Ken Roberts, pioneering Stanley researcher, reproduced many of the company's catalogs in the mid-1970s and early '80s, all of which, to collectors' dismay, have been out of print for some time. Now, with this Volume II of the *Stanley Catalog Collection*, Astragal Press has provided collectors with these essential publications, along with much of Ken Roberts' accompanying commentary. Together with Volume I of the collection, this new reprinting of Stanley's and Leonard Bailey's major catalogs truly represents an important body of information on this admired company and its early products.

Selling hardware and tools in the "old" days was very different from the national advertising campaigns and direct marketing of today. Before 1900, Stanley's catalogs were distributed primarily through its salesmen who traveled across the country, personally visiting each hardware store along the way. What Stanley learned early on was to add descriptive information for many of its tools and other products. Even cattle ties and screw drivers had enthusiastically written selling points. Stanley's sales force surely had an easier job of it armed with these well designed catalogs.

The 1872 catalog was reprinted only once before, by Ken Roberts, in 1981. Its unusually large size, $8\frac{1}{2}$ by 11 inches (reduced for this volume), and beautifully engraved illustrations reflect the success the 19-year old Stanley Rule & Level Company experienced from its introduction of Leonard Bailey's and Charles Miller's metallic planes just two years earlier. Both the 1870 and 1872 catalogs can be said to represent the turning point and beginning development into the company known today as "Stanley Tools."

Stanley's attention to the needs of the architectural and building professions, as well as those of the cabinetmaker, shipwright and many other trades, is evident throughout each of these catalogs. During the years between 1855 and 1900, Stanley produced an average of 100 styles of rules annually.

Many were specialty rules with unique scales and calculating tables. Several highly complex styles were marked in "octagonal" and "edge and middle" scales used in brace and frame construction and by ship mast and spar makers. Other models had the Gunter's scale, an early form of slide rule, and still others were designed for engineers with tables for calculating volume, weight and pressure. A study of the special rules manufactured by Stanley provides wonderful insight into 19th century trades. Stanley made cordage and cotton gauges, rules for cog wheel gear calculation, printer's scales, milk can and barrel measures, patternmakers' shrinkage rules, and rules for blacksmiths, carpenters, even brewers. Stanley also made rules for board measure, button and hat manufacture, watch glass gauging, and shoe sizing.

The Stanley Rule & Level Company very gradually introduced new tools. Deliberately and cautiously, the company either acquired or developed products with patented features on which it secured sole production rights. Company employees Frederick and Justus Traut, and Leonard Bailey and Charles Miller, who worked as contractors, developed most of the patented tools offered in the 1872 catalog with the exception of Curtiss' screw driver, Wheeler's countersink and Clement's "Excelsior" tool handle. By 1874, Stanley's product line and catalog grew to include Winterbottom's patented mitre squares, and Traut's N° 60 and 60½ patented marking gauges, his patent adjustable plumb bob and N° 46 patented dado, filletster and plow plane.

Stanley's 1877 catalog demonstrates the company's resolve to offer a unique and competitive line of carpenters' tools. Its 1876 line of "Liberty Bell" planes, shown in this catalog, provides an interesting bit of history. Designed by employees Justus Traut and Henry Richards to commemorate the centennial of the nation's Declaration of Independence, these planes also allowed Stanley to offer a lower priced alternative to its premium Bailey line, a move which involved the company in litigation with Mr. Bailey for several years. The 1877 catalog also includes Stanley's first offering of its N° 101, 102 and 103 block planes, advertised as "likely to be wanted in every household," and a number of patented tools, including Traut's N° 62 "reversible spoke shave," the N° 70 box scraper, and the N° 80 and 90 steel cased rabbeting planes. Price reductions from the 1874 catalog can be seen throughout.

It is important to note that Stanley produced its 1874 and 1877 catalogs during one of the worst economic periods in modern world history. During the five year depression following the financial panic of 1873, both the Stanley Works and The Stanley Rule & Level Company faced extreme difficulties in acquiring materials and in shipping goods. Remarkably, however, sales of the Bailey planes grew from the 6,500 announced in 1871 to over 125,000 by 1877. This strengthened Stanley's determination to develop the most complete line of carpenters' tools available anywhere.

Stanley's 1884 catalog reveals that cattle ties remained in the product line; however, pinking irons did not. Handles for planes and saws continued, as did steak hammers, ice picks and sash cord irons; staples of the 19th century tool and household markets. Stanley's line of specialized rules continued to address technical applications as architectural design and mechanical engineering flourished. The carpenters' line had grown considerably to include Traut's N° 45 and 50 combination planes, his N° 14 adjustable try square, a wider and heavier Bailey plane, the N° 4½, the N° 130 double end block plane, and Stanley's version of Bailey's N13 circular plane, the N° 113. Improvements included Traut's development of the lateral adjustment for plane irons, a widely imitated feature, comparable in popularity to the classic features of the original Bailey plane.

By 1892, prices had stabilized on Stanley's traditional rules and levels, somewhat below those of 1877, but almost constant with the 1884 list. Prices on Stearn's rules, however, had gone up about 7% from the 1879 catalog. The Bailey line was reduced an average of 10%. A comparison of these catalogs demonstrates Stanley's continuous refinement of the product line and its presentation. The 1892 catalog introduced Traut's patented "Hand-y" groove level, the N° 16. Interestingly, Stanley protected this simple feature on levels, planes and squares with a total of seven patents. The N° 44 bit and square level was new that year, listed as 30¢ and one of the first products to be priced "each." Traut's N° 66 and 69 hand headers were relatively new, having been in production since 1886, and introduced about the same time as his N° 72 chamfer plane, his N° 74 floor plane, and his N° 1 level sights.

While essentially a reissue of the 1892 list, Stanley's 1897 catalog introduced a number of important patented features and tools. It is reprinted in this volume as an abridgement showing just the newer pages. Stanley's improved version of the N° 20 circular plane first appears here, as does Edmund Schade's 1892 patented cutter and Traut and Bodmer's patent throat adjustment for block planes. Justus Traut's endless flow of ideas fill the pages of the 1897 catalog. Shown here are his N° 98 and 99 side rabbet planes, his patent eclipse levels, N° 71½ router and N° 83 cabinet scraper. Also new was his N° 140 detachable-side rabbet and block plane and his N° 67 universal spoke shave. It was about this time that Stanley implemented Traut's patented machinery for improved manufacture of level vials and his device for adjusting spirit levels. Traut's work during this period included development of the N° 55 universal plane and the N° 57 corebox plane, both introduced in 1896–97. And although not offered until a few years later, in the nine months between April 1895 and January of 1896, Traut received patents on the basic features for the Bed Rock planes and for the N° 51/52 chuteboard and plane.

Justus Traut's name would surely have been forgotten were it not for the work of tool collectors and their research into the development and history of Stanley's fine vintage tools. While Leonard Bailey today still receives the honor of having his name cast into Stanley's Bailey line of planes, Traut and many other inventors and employees at The Stanley Rule & Level Company deserve much credit for the company's success. Their work, some of it dating back 150 years, could be called the "bed rock" of Stanley's heritage, a foundation so strong that it has allowed the company to build a future reaching into the 21st century.

Finally, Volume II of the *Stanley Catalog Collection* includes two of Leonard Bailey's catalogs. The first, originally published in 1876, represents Bailey's renewed efforts to manufacture and market his own products after leaving Stanley the previous year. Working from his new location in Hartford, Connecticut, Bailey issued this beautifully illustrated catalog, proudly introducing his new Victor line, ornamented by his spectacular, patented trademark. Bailey displayed the Victor tools at the 1876 Centennial exhibition in Philadelphia, winning the highest award in the category, and also at the 1876 Connecticut State Fair, where he received the gold medal.

Such a promising fresh start hardly predicted the trouble that followed. Because of his 1869 license agreement with Stanley, Bailey could not use his original, highly successful patents and so had developed new mechanisms for his Victor line. But even before his 1876 catalog was printed, Bailey was dragged into court by Justus Traut who claimed prior invention of the adjustment feature on Bailey's Victor planes. Although neither Bailey nor Traut had a patent on the issue, Traut argued successfully and won the case in December of 1876. Bailey had already reworked the mechanism and received a patent for a new design on Dec. 12, 1876, and again on April 10, 1977, for a refinement of the adjustment. Within a few months, however, The Stanley Rule & Level Company challenged Bailey's new mechanism and sued, claiming infringement on the 1869 license. Bailey countersued on the point that Stanley was itself infringing on the license by producing lower-priced planes in competition with his original line. Stanley won the case in June of 1878, but Bailey continued to assert the Victor line as unique and legitimate, crying out in his July 1878 letter to Stanley, "They who seek equity must do equity!"

Doggedly, Bailey continued to work, and before the year was out, joined forces with the Bailey Wringing Machine Company in its production of planes under the condition that he would continue manufacture of the Victor line at its Rhode Island factory. Seventeen months later the relationship ended and in 1879, Bailey returned to Hartford. It was at this point that Bailey, beleaguered and tired, came to Stanley with the proposal that they become his sales agent for the Victor line. Stanley accepted, and in the company's supplement to the 1879 catalog, the Victor tools were offered as part of Stanley's product line. It is likely that Bailey was compelled to mend his bridges as Stanley's ability to market and distribute his products had far exceeded his own efforts.

Bailey published his own catalog just once more. His 1883 issue is reprinted in this collection, a diminished offering, minus the various other manufacturers' products included with his 1876 list. Stanley is listed as general agent and customers were directed to the New Britain and New York offices for placing orders. By this time, Leonard Bailey had started work on a new idea and in 1882, patented his first hand crank letter copying machine, apparently deciding to leave the tool business to Stanley. In 1884, The Stanley Rule & Level Company purchased Leonard Bailey's entire tool operation, including goods, machinery and patents. Stanley offered the Victor tools for the last time in the 1888 catalog, selling out the remaining stock and discontinuing the line. Over the next 20 years, Bailey received 21 additional patents, mostly pertaining to his letter press machinery. He died in 1905 at the age of 79. His last patent was granted just 21 months before his death.

Millions of Bailey's planes exist throughout the world, many of which are still in use, a substantial testament to the craftsmanship of Bailey and of Stanley. It was especially fitting when in 1902, Stanley began to cast Bailey's name on the beds of Bailey iron planes, and in 1909, added his name to the castings on the Bailey wood-bottom planes. This tribute continues today.

John Walter
Randa Walter
Marietta, Ohio

1872 STANLEY RULE & LEVEL Co. CATALOG and PRICE LIST OF TOOLS

This catalog first came to my attention during a study of American hardware at the Library of the Connecticut Historical Society. Two unusual features were immediately apparent: (1) its size was $8\frac{1}{2}$" x 11", rather than the customary $5\frac{3}{4}$" x 9"; (2) it was printed by a rather obscure firm in Wallingford, Connecticut. The successes of the Bailey Patent Iron Planes were noted with an announcement that sales now totaled over 20,000. The 1872 Catalog marked the first appearance of the $9\frac{1}{2}$ Excelsior Block Plane.

Page 2 of this Catalog notes that Chas. L. Mead, Treasurer of the Stanley Rule and Level Co. was in charge of the New York City Warehouse at 55 Chambers St. Mead was born in Chesterfield, N.H. Soon after his birth, his parents moved to Brattleborough, Vt. His father was a successful lawyer and founded the first savings bank in Vermont. After being educated in local schools, he worked first in Providence, R.I., and after a few years removed to Springfield, Mass. He then came back to Brattleborough and bought the rule business of E.A. Stearns.

A short time thereafter the factory was burned out. Mead was about to rebuild there when Stanley Rule & Level Co. purchased the business in 1862. Soon after, Mead moved to New Britain, but still supervised the rule business at Brattleborough. In 1871 he removed to New York City to assume charge of the SR&L Co. Warehouse. The machinery at Vermont was then moved to New Britain. Soon after this, Mead was made Treasurer of SR&L Co. Upon the death of Henry Stanley in 1884, he was made president of the firm, continuing to hold both offices until the time of his death on August 21, 1899, at the age of 66.

Ken Roberts
1981

PRICE LIST

OF

U. S. STANDARD

Boxwood and Ivory Rules,

LEVELS, TRY SQUARES, GAUGES, IRON AND WOOD BENCH
PLANES, MALLETS, HAND SCREWS, SPOKE
SHAVES, SCREW DRIVERS, ETC.

MANUFACTURED BY THE

STANLEY RULE AND LEVEL CO.,
NEW BRITAIN, CONN.

WAREHOUSE, NO. 55 CHAMBERS ST. NEW YORK.

ORDERS FILLED AT THE WAREHOUSE OR AT NEW BRITAIN.

January, 1872.

Office of

THE STANLEY RULE & LEVEL CO.

New Britain, Conn. January 1, 1872.

Gentlemen:

We have pleasure in calling your attention to the extensive assortment of first-class Tools, etc., manufactured by us, more perfectly illustrated and classified in the Catalogue now presented, than heretofore. Our purpose will still be to excel in the uniform standard of all Goods produced by us, and to offer superior inducements and facilities for their general distribution amongst the Trade.

We shall be represented at our Warehouse **NO. 55 CHAMBERS ST. NEW YORK,** by **Mr. CHARLES L. MEAD,** *Treasurer of the Company,* and orders will receive prompt attention at the Warehouse, or at the Factory. *Respectfully,*

STANLEY RULE & LEVEL CO.

WALLINGFORD PRINTING COMPANY,
WALLINGFORD, CONN.

STANLEY'S BOXWOOD RULES. 3

EXPLANATION.

All Rules embraced in the following Lists, bear the Invoice Number by which they are sold, and are graduated to correspond with the description given of each.

BOXWOOD RULES.

ONE FOOT, FOUR FOLD, NARROW.

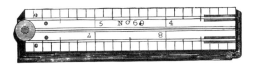

No. 69.

No.		Per Doz.
69.	Round Joint, Middle Plates, 8ths and 16ths of inches, ⅝ inch wide, .	$3.00

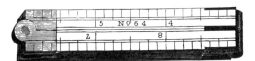

No. 64.

65.	Square Joint, Middle Plates, 8ths and 16ths of inches, ⅝ inch wide,								.	.	3.50
64.	"	"	Edge	"	"	"	"	⅝	"	"	5.00
65½.	Square Joint, Bound, 8ths and 16ths of inches,					⅝	"	"	.	.	11.00

No. 57.

55.	Arch Joint, Middle Plates, 8ths and 16ths of inches, ⅝ inch wide,								.	.	4.00
56.	"	"	Edge	"	"	"	"	⅝	"	"	6.00
57.	Arch Joint, Bound, 8ths and 16ths of inches,					⅝	"	"	.	.	12.00

STANLEY'S BOXWOOD RULES.

TWO FEET, FOUR FOLD, NARROW.

No. 68.

No.				Per Doz.
68.	Round Joint, Middle Plates, 8ths and 16ths of inches,		1 inch wide,	$ 3.50
8.	Round Joint, Middle Plates, Extra Thick, 8ths and 16ths of inches,		1 " "	4.00

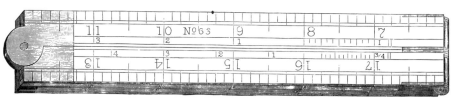

No. 63.

61.	Square Joint, Middle Plates, 8ths and 16ths of inches,	1 inch wide,	4.50
61½.	Square Joint, Middle Plates, 8ths and 16ths of inches,	¾ " "	5.50
63.	Square Joint, Edge Plates, 8ths, 10ths and 16ths of inches, Drafting Scale,	1 " "	7.00
63½.	Square Joint, Edge Plates, 8ths, 10ths and 16ths of inches, Extra quality,	¾ " "	8.00
84.	Square Joint, Half Bound, 8ths and 16ths of inches, Drafting Scale,	1 " "	12.00
62.	Square Joint, Bound, 8ths and 16ths of inches, Drafting Scale,	1 " "	15.00

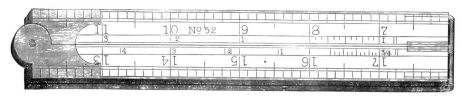

No. 52.

51.	Arch Joint, Middle Plates, 8ths and 16ths of inches, Drafting Scale,	1 inch wide,	6.00
53.	Arch Joint, Edge Plates, 8ths, 10ths and 16ths of inches, Drafting Scale,	1 " "	8.00
52.	Arch Joint, Half Bound, 8ths and 16ths of inches, Drafting Scale,	1 " "	13.00
54.	Arch Joint, Bound, 8ths and 16ths of inches, Drafting Scale,	1 " "	16.00
59.	Double Arch Joint, Bitted, 8ths and 16ths of inches, Drafting Scale,	1 " "	9.00
60.	Double Arch Joint, Bound, 8ths and 16ths of inches, Drafting Scale,	1 " "	21.00

TWO FEET, FOUR FOLD, BROAD.

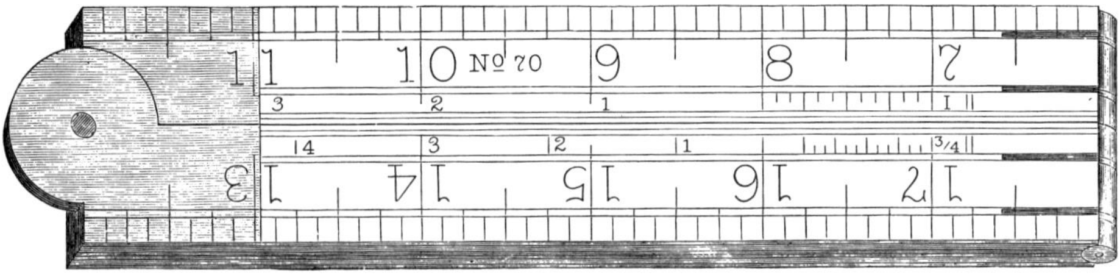

No. 70.

No.			Per Doz.
67.	Round Joint, Middle Plates, 8ths and 16ths of inches,	1⅜ inch wide,	$5.00
70.	Square Joint, Middle Plates, 8ths and 16ths of inches, Drafting Scale,	1⅜ " "	7.00
72.	Square Joint, Edge Plates, 8ths, 10ths and 16ths of inches, Drafting Scale,	1⅜ " "	9.00
72½.	Square Joint, Bound, 8ths, 10ths and 16ths of inches, Drafting Scale,	1⅜ " "	18.00

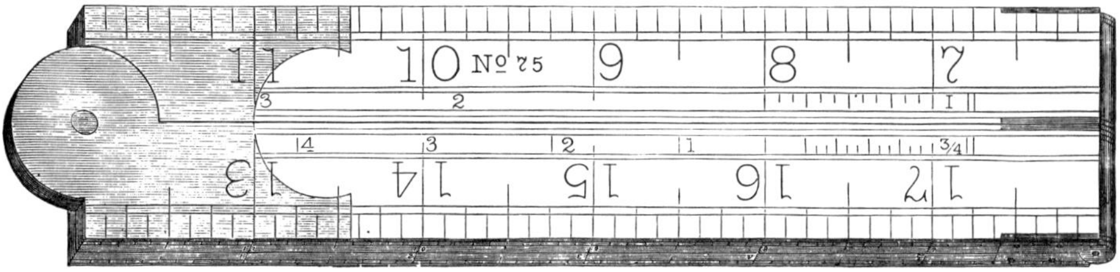

No. 75.

73.	Arch Joint, Middle Plates, 8ths and 16ths of inches, Drafting Scale,	1⅜ inch wide,	9.00
75.	Arch Joint, Edge Plates, 8ths, 10ths and 16ths of inches, Drafting Scale,	1⅜ " "	11.00

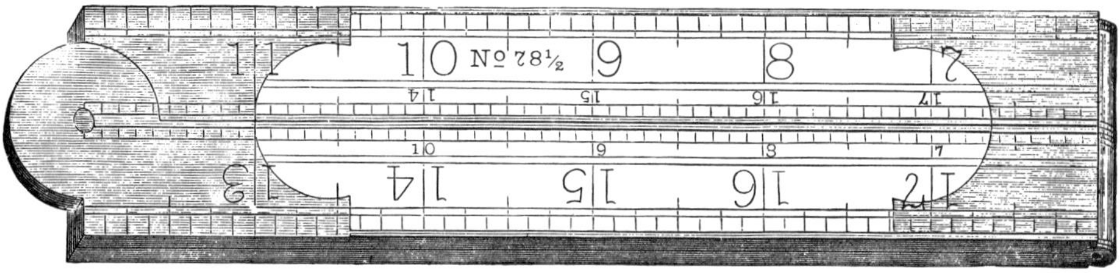

No. 78½.

76.	Arch Joint, Bound, 8ths, 10ths and 16ths of inches, Drafting Scale,	1⅜ inch wide,	20.00
77.	Double Arch Joint, Bitted, 8ths, 10ths and 16ths of inches, Drafting Scale,	1⅜ " "	12.00
78.	Double Arch Joint, Half Bound, 8ths, 10ths and 16ths of inches, Dr. Scale,	1⅜ " "	20.00
78½.	Double Arch Joint, Bound, 8ths, 10ths and 16ths of inches, Drafting Scale,	1⅜ " "	24.00
83.	Arch Joint, Edge Plates, with Gunter's Slide, 8ths, 10ths, 12ths and 16ths of inches, 100ths of a foot, and Octagonal Scales,	1⅜ " "	14.00

TWO FEET, SIX FOLD.

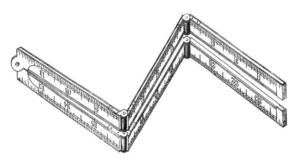

No. 58.

No.		Per Doz.
58.	Arch Joint, Edge Plates, 8ths and 16ths of inches, ¾ inch wide,	$13.00

BOXWOOD CALIPER RULES.

No. 32.

36.	Square Joint, 2 Fold, 6 inch, 8ths and 16ths of inches,	⅞ inch wide,	7.00
32.	Arch Joint, Edge Plates, 4 Fold, 12 inch, 8ths, 10ths and 16ths of inches,	1 " "	12.00

THREE FEET, FOUR FOLD RULES.

66.	Arch Joint, Middle Plates, 4 Fold, 16ths of inches, and 8ths of a yard,	1 inch wide,	8.00
66½.	" " " " 4 " 8ths and 16ths of inches,	1 " "	8.00

SHIP CARPENTERS' BEVELS.

42.	Boxwood, Double Tongue, 8ths and 16ths of inches,	6.00
43.	" Single " " " " "	6.00

BOARD MEASURE, TWO FEET, FOUR FOLD.

79.	Square Joint, Edge Plates, 16ths of inches, Drafting Scale, 1⅜ inch wide,	11.00
81.	Arch " " " " " " 1⅜ " "	13.00
82.	" " Bound, " " " " 1⅜ " "	22.00

BOXWOOD RULES.   7

TWO FEET, TWO FOLD.

No.			Per Doz.
29.	Round Joint, 8ths and 16ths of inches, . . .	1⅜ inch wide,	$3.50
18.	Square " " " " " . .	1½ " "	5.00
22.	" " Bitted, *Board Measure*, 16ths of inches, and Octagonal Scales, ...	1½ " "	8.00
1.	Arch Joint, 8ths and 16ths of inches, Octagonal Scales, .	1½ " "	7.00
2.	" " Bitted, 8ths, 10ths and 16ths of inches, Octagonal Scales,	1½ " "	8.00
4.	" " " Extra Thin, 8ths and 16ths of inches, Drafting and Octagonal Scales, ...	1½ " "	10.00
5.	Arch Joint, Bound, 8ths, 10ths and 16ths of inches, Drafting and Octagonal Scales, ...	1½ " "	16.00

TWO FEET, TWO FOLD, SLIDE.

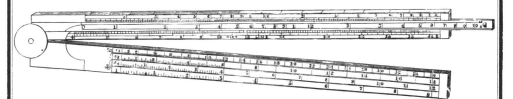

No. 12.

We have just had prepared an improved Treatise on the Gunter's Slide and Engineers' Rules, showing their utility and containing full and complete instructions, enabling mechanics to make their own calculations. It is also particularly adapted to the use of persons having charge of cotton or woolen machinery, surveyors and others. 200 pages, bound in cloth. Price $1.00, net. *Sent by mail, postpaid, on receipt of the price.*

No.			Per Doz.
26.	Square Joint, Plain Slide, 8ths, 10ths and 16ths of inches, Octagonal Scales, ...	1½ inch wide,	$9.00
27.	Square Joint, Bitted, Gunter's Slide, 8ths, 10ths and 16ths of inches, 100ths of a foot, Drafting and Octagonal Scales, .	1½ " "	12.00
12.	Arch Joint, Bitted, Gunter's Slide, 8ths, 10ths and 16ths of inches, 100ths of a foot, Drafting and Octagonal Scales, .	1½ " "	14.00
15.	Arch Joint, Bound, Gunter's Slide, 8ths, 10ths and 16ths of inches, Drafting and Octagonal Scales, . . .	1½ " "	24.00
6.	Arch Joint, Bitted, Gunter's Slide, Engineering, 8ths, 10ths and 16ths of inches, 100ths of a foot, Octagonal Scales, . .	1½ " "	18.00
16.	Arch Joint, Bound, Gunter's Slide, Engineering, 8ths, 10ths and 16ths of inches, Octagonal Scales, . . .	1½ " "	28.00

☞ With recently constructed machinery, we can furnish, to order, Rules marked with French or Spanish graduations.

IVORY RULES.

SIX INCH, TWO FOLD, CALIPER.

No. 38.

No.		Per Doz.
38.	Square Joint, German Silver, 2 Fold, 6 inch, 8ths and 16ths of inches, ⅞ inch wide,	$15.00

IVORY, ONE FOOT, FOUR FOLD, CALIPER.

39.	Square Joint, Edge Plates, German Silver, 4 Fold, 12 inch, 8ths, 10ths and 16ths of inches, ⅞ inch wide,	38.00
40.	Square Joint, German Silver, Bound, 4 Fold, 12 in. 8ths and 16ths of inches, ⅝ " "	44.00

IVORY, ONE FOOT, FOUR FOLD.

90.	Round Joint, Brass, Middle Plates, 8ths and 16ths of inches,	10.00
92½.	Square Joint, German Silver, Middle Plates, 8ths and 16ths of inches, ⅝ inch wide,	14.00
92.	Square Joint, German Silver, Edge Plates, 8ths and 16ths of inches, ⅝ " "	17.00
88½.	Arch Joint, German Silver, Edge Plates, 8ths and 16ths of inches, ⅝ " "	21.00
88.	Arch Joint, German Silver, Bound, 8ths and 16ths of inches, ⅝ " "	32.00
91.	Square Joint, German Silver, Edge Plates, 8ths, 10ths and 16ths of inches, ¾ " "	23.00

IVORY, SIX INCH, TWO FOLD.

93.	Round Joint, Brass, 8ths and 16ths of inches,	4.50

IVORY, TWO FEET, FOUR FOLD.

85.	Square Joint, German Silver, Edge Plates, 8ths, 10ths and 16ths of inches, ⅞ inch wide,	54.00
86.	Arch Joint, German Silver, Edge Plates, 8ths, 10ths and 16ths of inches, 100ths of a foot, Drafting Scale, 1 " "	64.00
87.	Arch Joint, German Silver, Bound, 8ths, 10ths and 16ths of inches, Drafting Scale, 1 " "	80.00
89.	Double Arch Joint, German Silver, Bound, 8ths, 10ths and 16ths of inches, Drafting Scale, 1 " "	92.00
95.	Arch Joint, German Silver, Bound, 8ths, 10ths and 16ths of inches, Drafting Scale, 1⅜ " "	102.00
97.	Double Arch Joint, German Silver, Bound, 8ths, 10ths and 16ths of inches, Drafting Scale, 1⅜ " "	116.00

Bench Rules.

No.		Per Doz.
34.	Bench Rules, ... 2 feet,	$3.00
35.	" " Board Measure, . . . 2 "	6.00

Board and Log Measures.

NOTE.—The following Numbers give the contents in Board Measure of 1 inch Boards. Nos. 43¼, 46, 46½, 47, 47½, 48, 49.

DIRECTIONS.—Place the stick across the flat surface of the Board, bringing the inside of the Cap snugly to the edge of the same; then follow with the eye the column of figures in which the length of the Board is given as the first figure under the Cap, and at the mark nearest the opposite edge of the Board will be found the contents of the Board in feet.

No. 46.

46. Board Stick, Octagon, Brass Cap, 16 lines, 8 to 23 feet, . . 2 feet, $8.00

No. 46½.

46½.	Board Stick, Square,	16 lines, 8 to 23 feet,	. . .	2 feet,	8.00
47.	" "	Octagon, " " " " . . .		3 "	12.00
47½.	" "	Square, " " " " . . .		3 "	12.00

No. 43¼.

43¼. Board Stick, Flat, Hickory, Cast Brass Head and Tip, 6 lines, 12 to 22 feet, 3 feet, 12.00

No. 49.

49. Board Stick, Flat, Hickory, Steel Head, Brazed, Extra Strong, 6 lines, 12 to 22 feet, 3 feet, 26.00

No. 48.

48. Walking Cane, Board Measure, Octagon, Hickory, Solid Cast Brass Head and Tip, 8 lines, 9 to 16 feet, ... 3 feet, 12.00
48½. Walking Cane, Log Measure, Octagon, Hickory, Solid Cast Brass Head and Tip, 3 feet, 15.00

NOTE.—This Log Measure gives the number of feet of one inch, square edged boards, which can be sawed from a log of any size, from 12 to 36 inches diameter, and of any length. The figures immediately under the head of the Cane, are for the length of Logs in feet. Under these figures, on the same line, at the mark nearest the diameter of the Log, will be found the number of feet the Log will make. If the Log to be measured is not over 15 feet long, the diameter should be taken at the small end; if over 15 feet, at the middle of the Log.

Yard Sticks.

33.	Yard Stick, Polished, ...	2.00
41.	" " Brass Tip, Polished, ...	3.50
50.	" " Hickory, Brass Capped Ends, Polished, . . .	4.50

Wantage and Gauging Rods.

44.	Wantage Rods,	8 lines, ...		5.00
37.	" "	" 12 "		7.00
45.	Gauging	" 120 gallons, . . .	3 feet,	7.00
45½.	" "	" Wantage Tables, . . .	4 "	18.00
49½.	Forwarding Sticks, ...		5 "	24.00

School Rules.

98. Boxwood, 12 inch, ¾ inch wide, Beveled Edge, 8ths and 16ths of inches, . 1.25
99. " " " " " " " " 10ths " " " . 1.50

STEARNS' RULES.

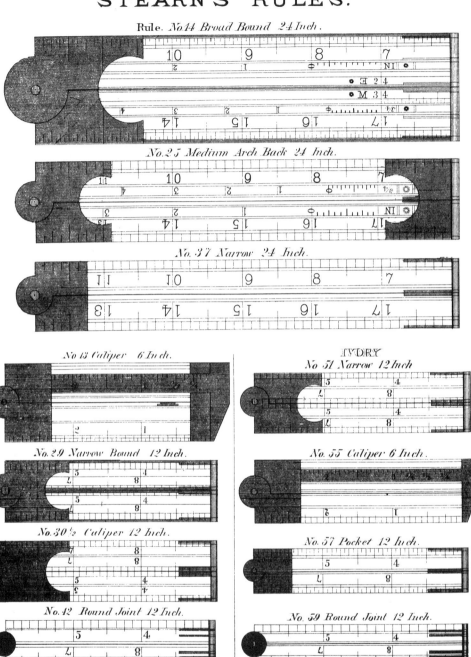

BOXWOOD RULES.

TWO FEET, TWO FOLD.

No.			Per. Doz.
1.	Arch Joint, Bound, Slide, Engineering, 10ths and 16ths of inches, 100ths of a foot, Drafting Scales,	1½ in. wide,	$28.00
2.	Arch Joint, Bitted, Slide, Engineering, 10ths, 12ths and 16ths of inches, 100ths of a foot, Drafting Scales,	1½ in. wide,	18.00
3.	Arch Joint, Bound, Gunter's Slide, 10ths, 12ths, and 16ths inches, 100ths of a foot, Drafting and Octagonal Scales,	1½ in. wide,	24.00
5.	Arch Joint, Bitted, Gunter's Slide, 10ths, 12ths and 16ths inches, 100ths of a foot, Drafting and Octagonal Scales,	1½ in. wide,	14.00
7.	Arch Joint, Bitted, 8ths and 16ths of inches, Drafting and Octagonal Scales,	1½ in. wide,	8.00
9.	Square Joint, Slide (graduating 8ths of inches), 8ths and 16ths of inches, Drafting and Octagonal Scales,	1½ in. wide,	9.00
11.	Square Joint, 8ths and 16ths of inches,	1½ in. wide,	5.00

TWO FEET, FOUR FOLD, BROAD.

14.	Arch Joint, Bound, Board Measure, 10ths, 12ths and 16ths of inches, Drafting and Octagonal Scales, [see Engraving],	1⅜ in. wide,	24.00
15.	Arch Joint, Bound, 8ths, 10ths, 12ths and 16ths of inches, Drafting and Octagonal Scales,	1⅜ in. wide,	22.00
17.	Arch Joint, Arch Back, Edge Plates, 8ths, 10ths, 12ths and 16ths of inches, 100ths of a foot, Drafting and Octagonal Scales,	1⅜ in. wide,	15.00
18.	Arch Joint, Triple Plated Edge Plates, Board Measure, 10ths, 12ths and 16ths of inches, 100ths of a foot, Drafting and Octagonal Scales,	1⅜ in. wide,	14.00
19.	Arch Joint, Triple Plated Edge Plates, Slide, 8ths, 10ths, 12ths and 16ths of inches, 100ths of a foot, Drafting and Octagonal Scales,	1⅜ in. wide,	15.00
20.	Arch Joint, Triple Plated Edge Plates, 8ths, 10ths, 12ths, and 16ths of inches, 100ths of a foot, Drafting and Octagonal Scales,	1⅜ in. wide,	12.00
21.	Arch Joint, Middle Plates, Bitted, 8ths and 16ths of inches, Drafting and Octagonal Scales,	1⅜ in. wide,	10.00
32.	Square Joint, Middle Plates, Bitted, 8ths and 16ths of inches, Drafting and Octagonal Scales,	1⅜ in. wide,	8.00

TWO FEET, FOUR FOLD, MEDIUM.

22.	Arch Joint, Bound, Board Measure, 16ths of inches, Drafting Scales,	1⅛ in. wide,	20.00
23.	Arch Joint, Bound, 8ths, 10ths, 12ths and 16ths of inches, Drafting Scales,	1⅛ in. wide,	18.00
25.	Arch Joint, Arch Back, Edge Plates, 8ths, 10ths and 16ths of inches, 100ths of a foot, Drafting and Octagonal Scales, [see Engraving],	1⅛ in. wide,	13.00
26.	Arch Joint, Edge Plates, Board Measure, 16ths of inches, Drafting Scales,	1⅛ in. wide,	11.00
27.	Arch Joint, Middle Plates, Bitted, 8ths and 16ths of inches, Drafting and Octagonal Scales,	1⅛ in. wide,	8.00

TWO FEET, FOUR FOLD, NARROW.

No.			Per Doz.
45.	Arch Joint, Edge Plates, 8ths and 16ths of inches,	1 in. wide,	$ 8.00
46.	Arch Joint, Middle Plates, 8ths and 16ths of inches,	1 " "	6.00
34.	Square Joint, Square Back Plates, (Edge Plates), 8ths and 16ths of inches,	1 " "	9.00
35.	Square Joint, Bound, 8ths and 16ths of inches, Drafting and Octagonal Scales,	1 " "	15.00
37.	Square Joint, Edge Plates, 8ths and 16ths of inches, [see Engraving],	1 " "	7.00
38.	Square Joint, Middle Plates, 8ths and 16ths of inches,	1 " "	4.50
41.	Round Joint, Middle Plates, 8ths and 16ths of inches,	1 " "	3.50

TWO FEET, FOUR FOLD, EXTRA NARROW.

31.	Arch Joint, Edge Plates, 8ths, 10ths and 16ths of inches,	$\frac{3}{4}$ " "	9.00
72.	Arch Joint, Bound, 8ths, 10ths, 12ths and 16ths of inches,	$\frac{3}{4}$ " "	17.00
33.	Square Joint, Edge Plates, 8ths, 10ths and 16ths of inches,	$\frac{3}{4}$ " "	8.00
36.	Square Joint, Middle Plates, 8ths and 16ths of inches,	$\frac{3}{4}$ " "	5.50
44.	Round Joint, Middle Plates, 8ths and 16ths of inches,	$\frac{3}{4}$ " "	4.50

TWO FEET, SIX FOLD.

$28\frac{1}{2}$.	Arch Joint, Edge Plates, 8ths, 10ths, 12ths and 16ths of inches, 100ths of a foot,	$\frac{3}{4}$ " "	13.50

ONE FOOT, FOUR FOLD.

29.	Arch Joint, Bound, 8ths, 10ths, 12ths and 16ths of inches, [see Engraving],	13-16ths " "	13.00
30.	Arch Joint, Edge Plates, 8ths, 10ths, 12ths and 16ths of inches, 100ths of a foot,	13-16ths " "	8.00
39.	Square Joint, Edge Plates, 8ths and 16ths of inches,	13-16ths " "	6.00
40.	Round Joint, Middle Plates, 8ths and 16ths of inches,	13-16ths " "	3.25
74.	Square Joint, Bound, 8ths and 16ths of inches,	5-8ths " "	12.00
75.	Square Joint, Middle Plates, 8ths and 16ths of inches,	5-8ths " "	4.00
42.	Round Joint, Middle Plates, (extra), 8ths and 16ths inches, [see Engraving],	5-8ths " "	3.25
43.	Round Joint, Middle Plates, 8ths and 16ths of inches,	5-8ths " "	3.00

ONE FOOT, FOUR FOLD, CALIPER.

$29\frac{1}{4}$.	Arch Joint, Bound, 8ths, 10ths, 12ths and 16ths of inches,	13-16ths " "	18.00
$30\frac{1}{2}$.	Arch Joint, Edge Plates, 8ths, 10ths, 12ths and 16ths of inches, 100ths of a foot, [see Engraving],	13-16ths " "	12.50

SIX INCH, TWO FOLD, CALIPER.

12.	Square Joint, Brass Case, Spring Caliper, 8ths and 16ths of inches,	$1\frac{1}{8}$ " "	12.00
13.	Square Joint, 8ths and 16ths of inches, [see Engraving],	$1\frac{1}{8}$ " "	8.00
$13\frac{1}{2}$.	Square Joint, 8ths and 16ths of inches,	13-16ths " "	7.00

MISCELLANEOUS ARTICLES.

No.			Per Doz.
$63\frac{1}{4}$.	Bench Rule, Boxwood, Bound, 8ths and 16ths of inches,	24 inches long,	$ 18.00
64.	Bench Rule, Maple, Capped Ends, 8ths and 16ths of inches,	24 " "	4.00
71.	Yardstick, Maple, Capped Ends,	36 " "	4.00
80.	Saddler's Rule, Maple, Capped Ends, 8ths and 16ths of inches, $1\frac{1}{2}$ in. wide,	36 " "	9.00
81.	Pattern Maker's Shrinkage Rule, Boxwood, 8ths and 16ths of inches,	$24\frac{1}{4}$ " "	12.00
82.	Pattern Maker's Shrinkage Rule, Two Fold, Boxwood, Triple Plated Edge Plates, 8ths and 16ths of inches,	$24\frac{1}{4}$ " "	18.00

IVORY RULES.

TWO FEET, FOUR FOLD, BROAD.

No.		Unbound.	Bound.
47.	Arch Joint, Triple Plated Edge Plates, German Silver, 8ths, 10ths, 12ths and 16ths of inches, 100ths of a foot on edges of unbound, Drafting and Octagonal Scales, 1½ in. wide,	$90.00	$110.00

IVORY, TWO FEET, FOUR FOLD, MEDIUM.

48.	Arch Joint, Edge Plates, German Silver, 8ths, 10ths, 12ths and 16ths of inches, (100ths of a foot on edges of unbound), Drafting Scales, 1¼ in. wide,	70.00	88.00

IVORY, TWO FEET, FOUR FOLD, NARROW.

50.	Square Joint, Edge Plates, German Silver, 8ths, 10ths, 12ths and 16ths of inches, (100ths of a foot on edges of unbound), Drafting Scales, 1 inch wide,	60.00	76.00

IVORY, TWO FEET, FOUR FOLD, EXTRA NARROW.

56.	Arch Joint, Edge Plates, German Silver, 8ths, 10ths, 12ths and 16ths of inches, (100ths of a foot on edges of unbound), ¾ inch wide,	54.00	68.00

IVORY, TWO FEET, SIX FOLD.

60.	Arch Joint, Edge Plates, German Silver, 8ths, 10ths, 12ths and 16ths of inches, (100ths of a foot on edges of unbound), ¾ inch wide,	80.00	100.00

IVORY, ONE FOOT, FOUR FOLD.

51.	Arch Joint, Edge Plates, German Silver, 8ths, 10ths, 12ths and 16ths of inches, (100ths of a foot on edges of unbound), [see Engraving], 13-16ths in. wide,	30.00	42.00
52.	Square Joint, Edge Plates, German Silver, 8ths, 10ths, 12ths and 16ths of inches, (100ths of a foot on edges of unbound), 13-16ths in. wide,	26.00	38.00
57.	Square Joint, Edge Plates, German Silver, 8ths and 16ths of inches, [see Engraving], 5-8ths in. wide,	18.00	30.00
58.	Square Joint, Edge Plates, Brass, 8ths and 16ths of inches, 5-8ths " "	15.00	
59.	Round Joint, Middle Plates, Brass, 8ths and 16ths of inches, [see Engraving], 9-16ths in. wide,	10.00	

IVORY, ONE FOOT, FOUR FOLD, CALIPER.

53.	Arch Joint, Edge Plates, German Silver, 8ths, 10ths, 12ths and 16ths of inches, (100ths of a foot on edges of unbound), 13-16ths in. wide,	44.00	58.00
54.	Square Joint, Edge Plates, German Silver, 8ths, 10ths, 12ths and 16ths of inches, (100ths of a foot on edges of unbound), 13-16ths in. wide,	40.00	54.00

IVORY, SIX INCH, TWO FOLD, CALIPER.

55.	Square Joint, German Silver, 8ths and 16ths, of inches, [see Engraving], 13-16ths in. wide,		15.00
55½.	Square Joint, German Silver Case, Spring Caliper, 8ths and 16ths of inches, 13-16ths in. wide,		18.00

PLUMBS AND LEVELS.

No.										Per Doz.
102.	Levels, Two Side views, Polished, Assorted,					10 to 16 Inch,	.	.	.	$ 9.00
103.	"	"	"	"	"	18 to 24 "		.	.	12.00

No. 0.

00.	Plumb and Level, Arch Top Plate, Two Side Views, Polished, Assorted,	20 to 24 Inches,	$ 16.00
0.	" " " " " " " " " "	26 to 30 "	18.00
1½.	Mahogany Plumb and Level, two Side Views, Polished, Assorted,	16 to 24 "	16.50

DESCRIPTION OF THE
Patent Improved Adjusting Plumb and Level,

MANUFACTURED BY THE

STANLEY RULE & LEVEL CO.

[Sectional Drawing].

    The Spirit-glass, (or bubble tube), in the Level, is set in a Metallic Case, which is attached to the Brass Top-plate above it,—at one end by a substantial hinge, and at the opposite end by an Adjusting Screw, which passes down through a flange on the Metallic Case. Between this flange and the Top-plate above, is inserted a stiff spiral spring; and by driving, or slacking the Adjusting Screw, should occasion require, the Spirit-glass can be instantly adjusted to a position parallel with the base of the Level.

    The Spirit-glass in the Plumb, is likewise set in a Metallic Case attached to the Brass Top-plate at its outer end. By the use of the Adjusting Screw at the lower end of the Top-plate, the Plumb-glass can be as readily adjusted to a right angle with the base of the Level, if occasion requires, and by the same method as adopted for the Level-glass.

    The simplicity of this improved method of adjusting the Spirit-glasses will commend itself to every Mechanic. But one screw is used in the operation, and the action of the Brass Spring is perfectly reliable under all circumstances.

Patent Adjustable Plumbs and Levels.

No. 3.

Per Doz.

2. Patent Adjustable Plumb and Level, Arch Top Plate, two Brass Lipped Side Views, Polished, Assorted, ... 26 to 30 inches, $27.00
3. Patent Adjustable Plumb and Level, Arch Top Plate, two Side Views, Polished and Tipped, Assorted, ... 26 to 30 " 30.00
4. Patent Adjustable Plumb and Level, Arch Top Plate, two Brass Lipped Side Views, Polished and Tipped, Assorted, . . . 26 to 30 " 39.00
5. Patent Adjustable Plumb and Level, Triple Stock, Arch Top Plate, two Brass Lipped Side Views, Polished and Tipped, Assorted, . . 26 to 30 " 48.00

No. 9.

1. Patent Adjustable Mahogany Plumb and Level, Arch Top Plate, two Side Views, Polished, Assorted, ... 26 to 30 inches, $22.50
6. Patent Adjustable Mahogany Plumb and Level, Arch Top Plate, two Brass Lipped Side Views, Polished, Assorted, . . . 26 to 30 " 33.00
9. Patent Adjustable Mahogany Plumb and Level, Arch Top Plate, two Brass Lipped Side Views, Polished and Tipped, Assorted, . . 26 to 30 " 48.00
10. Patent Adjustable Mahogany Plumb and Level, Triple Stock, two Extra Heavy Brass Lipped Side Views, and Arch Top Plate, Polished and Tipped, Assorted, ... 26 to 30 " 60.00
25. Patent Adjustable Mahogany Plumb and Level, Improved Double Adjusting Side Views, and Arch Top Plate, Polished and Tipped, . 30 " 54.00
11. Patent Adjustable Rosewood Plumb and Level, Arch Top Plate, two Extra Brass Lipped Side Views, Polished and Tipped, Assorted, . 28 to 30 " 90.00
12. Machinsts' Brass Bound Rosewood Plumb and Level, two Brass Side Views, Polished, 20 in. 125.00

PLUMBS AND LEVELS. Continued.

No.			Per Doz.
32.	Patent Adjustable and Graduating Plumb and Level, Mahogany,	28 inch,	$100.00
35.	Patent Adjustable Masons' Plumb and Level, 3¾ inch wide,	42 "	36.00
43.	Iron Plumb and Level, two side Views, Brass Top Plates,	9 "	20.00
45.	Machinists' Level, all Iron, Brass Top Plate, extra finish,	9 "	28.00

POCKET LEVELS.

[Improved Patterns.]

No. 41.

No.		Per Doz.
40.	Iron, Japanned, Pocket Levels,	$2.50
41.	Brass Top-plate, " "	3.00
42.	All Brass, " "	8.00

46.	Iron Pocket Levels, Brass Top-Plate, a superior article,	$3.00

Level Glasses.

		Per Gro.
Level Glasses, packed in ¼ gross boxes,	1¾ inch,	$12.00
" " " " " "	2 "	12.00
" " " " " "	2½ "	12.75
" " " " " "	3 "	13.50
" " " " " "	3½ "	15.00
" " " " " "	4 "	16.50
" " " " " "	4½ "	18.00
Assorted, 1¾, 2, 3 and 3½ inch,		14.50

Compass Dividers.

With Cast-Steel Spring, and Set-Screw.

5	6	7	8	9	10	11	12 inch.
$5.25	5.50	6.00	6.75	7.50	8.50	9.50	10.50

No. 32.

No. 35.

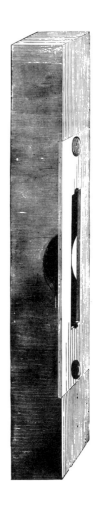

No. 45.

Nicholson's Patent Metallic Plumb and Level.

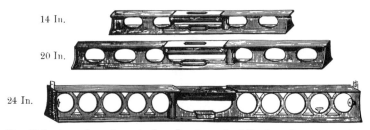

14 In.

20 In.

24 In.

The chief points of excellence in these Levels are the following: Once accurately constructed, they always remain so; the sides and edges are perfectly true, thus combining with a Level a convenient and reliable *straight-edge* ; shafting, or other work overhead, may be lined with accuracy by sighting the *bubble* from below, through an aperture in the base of the Level. Other points are their lightness, strength, and convenience of handling. All Levels, being thoroughly tested, are warranted reliable.

No.			Per Doz.
13.	14 Inches long,		$ 21.00
14.	20 " "		25.00
15.	24 " "		30.00

Patent Iron Frame Levels.

These Levels have a Cast Iron Frame, thoroughly braced, and combine the greatest strength with the least possible weight of metal. They are made absolutely correct in their form, and are subject to no change from warping or shrinking.

When finished, the sides are inlaid with Black Walnut, (as represented above), secured by the screws which pass from one to the other, through the Iron braces, and the sides may thus be easily removed, if occasion requires.

Per Doz.

No. 48, Patent Adjustable Plumb and Level, Iron Frame, with Black **Walnut** (inlaid) sides, 12 inch, ... $ 30.00

No. 49, Patent Adjustable Plumb and Level, Iron Frame, with Black Walnut (inlaid) sides, 18 inch, ... 40.00

PATENT COMBINATION TRY SQUARE AND BEVEL.

The Patent Combination Try Square and Bevel will be found a valuable acquisition to Carpenters, Cabinet Makers, Stone Cutters and others, as it combines within itself a perfect Square and Bevel. The Blade being made by improved machinery, can be relied upon as correct, and can be used either as an inside or outside Square. In using as a Bevel, adjust the blade to any given angle required, and secure it by means of the screw.

	Per Doz.
4 Inch,	$ 6.00
6 "	7.00
8 "	9.00
10 "	11.00
12 "	14.00

IMPROVED MITRE SQUARES.

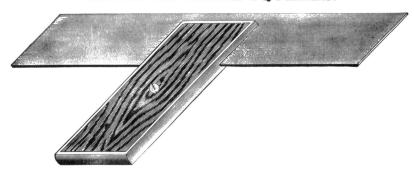

Iron Frame Handle, with Black Walnut (inlaid) Sides.

Inches,	8	10	12
Per dozen,	$9.00	10.50	12.00

Improved Try Squares, No. 1.

Iron Frame Handle, with Rosewood (inlaid) Sides.

Inches,	3	4	6	8	9	10	12
Per dozen,	$9.00	10.50	12.00	16.00	18.00	21.00	28.00

Patent Improved Try Squares, No. 2.

One-half Iron Frame Handle, Rosewood Back and Sides.

Inches,	3	4½	6	7½	9	12
Per dozen,	$4.75	5.75	6.25	7.50	9.00	12.50

Plated Try Squares.

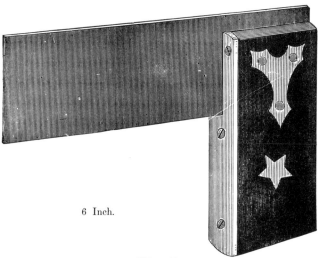

6 Inch.

No. 2.

		Per Doz
Rosewood,	3 Inch,	$ 3.00
"	4½ "	3.75
"	6 "	5.00
"	7½ "	5.75
"	9 "	6.50
"	12 "	8.50
" with Rest,	15 "	12.50
" " "	18 "	15.50

Plumb and Level Try Squares.

A very convenient Tool for Mechanics. A spirit glass being set in the inner edge of the Try Square handle, constitutes the handle a Level; and when the handle is brought to an exact level, the blade of the square will be upright, and become a perfect Plumb.

	Per Doz.
No. 1. 12 Inch, Rosewood, with Plumb and Level in Handle, and extra heavy Brass Mountings,	$ 36.00
No. 1. 7½ " " " " " " " "	18.00
No. 2. 7½ " " " " " " " "	12.00

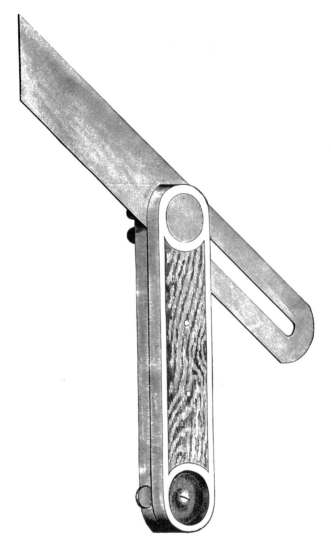

PATENT IMPROVED SLIDING T BEVELS.

The patent method of fastening the Blade of the Bevel at any desired angle, is a valuable feature in this Tool.

Iron Frame Handle, with Black Walnut (inlaid) Sides.

Inches,	8	10
Per dozen,	$9.00	11.00

PLATED SLIDING T BEVELS.

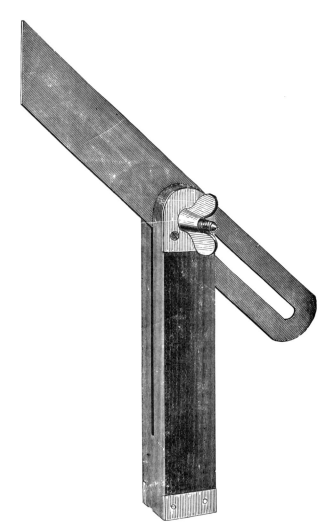

6 Inch.

							Per Doz.
6 Inch, Rosewood, with Brass Thumb Screw,	.	.	.	.	.	$ 5.50
8 " " " " " "	.	.	.	.	.	6.00
10 " " " " " "	.	.	.	.	.	6.50
12 " " " " " "	.	.	.	.	.	7.00
14 " " " " " "	.	.	.	.	.	7.50

GAUGES.

No. 68.

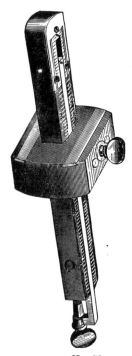
No. 79.

No.		Per Doz.
61.	Marking Gauge, Beechwood, Boxwood Thumb Screw, Oval Bar, Marked, Steel Points, 1 dozen in a box,	$1.00
61½.	Marking Gauge, Beechwood, Boxwood Thumb Screw, Oval Head and Bar, Marked, Steel Points, 1 dozen in a box,	1.50
62.	Marking Gauge, Beechwood, Boxwood Thumb Screw, Oval Bar, Marked, Steel Points, Polished, 1 dozen in a box,	2.00
64.	Marking Gauge, Polished, Plated Head, Boxwood Thumb Screw, Oval Bar, Marked, Steel Points, 1 dozen in a box,	2.75
65.	Marking Gauge, Boxwood, Polished, Plated Head, Boxwood Thumb Screw, Oval Bar, Marked, Steel Points, 1 dozen in a box,	4.00
66.	Marking Gauge, Rosewood, Plated Head and Bar, Brass Thumb Screw, Oval Bar, Marked, Steel Points, ½ dozen in a box,	5.50
70.	Cutting Gauge, Mahogany, Polished, Plated Head, Boxwood Thumb Screw, Oval Bar, Marked, Steel Cutter, 1 dozen in a box.	4.00

GAUGES.

No. 77.

No.		Per Doz.
67.	Mortise Gauge (Improved), Adjustable Wood Slide, Boxwood Thumb Screw, Oval Bar, Marked, Steel Points,	$ 5.00
68.	Mortise Gauge (Improved), Plated Head, Adjustable Wood Slide, Brass Thumb Screw, Oval Bar, Marked, Steel Points (see Engraving, page 24),	7.00
73.	Mortise Gauge, Boxwood or Satinwood, Polished, Plated Head, Brass Slide, Brass Thumb Screw, Oval Bar, Marked, Steel Points, ½ dozen in a box,	8.00
76.	Mortise Gauge, Boxwood, Plated Head, Screw Slide, Brass Thumb Screw, Oval Bar, Marked, Steel Points, ½ dozen in a box,	12.00
77.	Mortise Gauge, Rosewood, Screw Slide, Brass Thumb Screw, Oval Bar, Steel Points, ½ dozen in a box (see Engraving on this page),	10.00
78.	Mortise Gauge, Rosewood, Plated Head, Screw Slide, Brass Thumb Screw, Oval Bar, Marked, Steel Points, ½ dozen in a box,	11.00
79.	Mortise Gauge, Rosewood, Plated Head, Screw Slide, Brass Thumb Screw, Plated Bar, Marked, Steel Points (see Engraving, page 24), ½ dozen in a box,	13.00
80.	Mortise Gauge, Boxwood, Full Plated Head, Screw Slide, Brass Thumb Screw, Plated Bar, Marked, Steel Points, ½ dozen in a box,	18.00

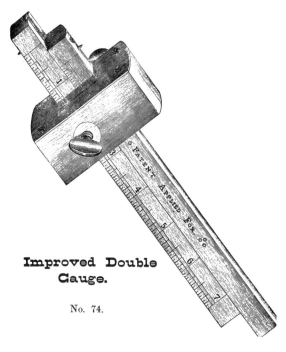

Improved Double Gauge.

No. 74.

No.		Per Doz.
71.	Double Gauge, (Marking and Mortise Gauge combined), Beechwood, Polished, Plated Head and Bars, Brass Thumb Screws, Oval Bars, Marked, Steel Points running through the Bars,	$9.00
72.	Double Gauge, (Marking and Mortise Gauge combined), Beechwood, Polished, Boxwood Thumb Screws, Oval Bars, Marked, Steel Points running through the Bars,	4.50
74.	Double Gauge, (Marking and Mortise Gauge combined), Boxwood, Polished, Full Plated Head and Bars, Brass Thumb Screws, Oval Bars, Marked, Steel Points running through the Bars, (see Engraving on this page),	14.00
83.	Handled Slitting Gauge, 17 inch Bar, Graduated, ½ dozen in a box,	8.00
84.	Handled Slitting Gauge, 17 inch Bar, Graduated, with Roller, ½ dozen in a box,	11.00
85.	Panel Gauge, Beechwood, Boxwood Thumb Screw, Oval Bar, Steel Points, 1 dozen in a box,	3.20
85½.	Panel Gauge, Rosewood, Brass Thumb Screw, Plated Oval Bar, Steel Points, ½ dozen in a box,	18.00
90.	Williams' Patent Combination Gauge, ½ dozen in a box,	22.00

PATENT IMPROVED SOLID CAST STEEL
SCREW DRIVERS.
WARRANTED.

By the aid of improved machinery we are now producing a Screw Driver which is superior to any other in the market—and at the prices paid for the ordinary Tools.

The Blades are made from the best quality of CAST STEEL, and are tempered with great care. They are ground down to a correct taper, and pointed at the end, by special machinery; thus procuring perfect uniformity in size, form and strength, while the peculiar shape of the point gives it unequaled firmness in the screw-head, when in use.

The shank of the Blade is properly slotted to receive a Patent metallic fastening, which secures it permanently in the handle, on all sizes above 3 inch. The Handles are of the most approved pattern, the Brass Ferules of the thimble form, *extra heavy*, and closely fitted.

Every objection which has existed in the ordinary Screw Drivers, is overcome in our new method of manufacturing; and Mechanics will appreciate the improvements now offered in this tool, for which they have so many uses.

PRICES.

Sizes,	1½	2	3	4	5	6	7	8	10	Inch.
	$1.00	1.75	2.00	2.50	3.00	3.50	4.00	4.75	6.25	Per Doz.

WHEELER'S PATENT COUNTERSINK,
FOR WOOD,
With Patent Gauge Attachment.

This Countersink cuts rapidly, will not clog with shavings, makes a perfectly smooth and round hole, and may be used equally well for any sized Screw. An ingenious adaptation of a Gauge, as shown in the above Engraving, gives great facility and accuracy where many Screws of the same size are to be used; as the Gauge can be easily adjusted so that the Countersink will cease to cut at any required depth.

Countersinks, 1 Dozen in a box, ... $4.00 Per Doz.
" 1 " " " With Gauge, . . . 5.50 " "

BAILEY'S PATENT PLANES.

No tools have ever been accorded more general praise than "Bailey's Patent Planes" have received from skilled Wood-workers throughout this country, and in foreign countries. The sale of them has already exceeded **20.000** Planes, and is rapidly increasing. The strictest care is pledged for the continued excellence of these Planes over all others. Every Dealer should introduce them to the notice of all good Mechanics: the Tools will commend themselves when brought into actual service.

☞ Each Plane is fitted in working order, when sent into Market.

Iron Planes.

No.									
1,	Smooth	Plane,	5 1-2	Inches in Length,	$1\frac{1}{4}$	In.	Cutter,	$	4.00
2,	"	"	7	" " "	$1\frac{5}{8}$	"	"		4.50
3,	"	"	8	" " "	$1\frac{3}{4}$	"	"		5.00
4,	"	"	9	" " "	2	"	"		5.50

5,	Jack Plane,	14	Inches in Length,	2	In.	Cutter,	6.00
6,	Fore "	18	" " "	$2\frac{3}{8}$	"	"	7.00
7,	Jointer "	22	" " "	$2\frac{3}{8}$	"	"	8.00
8,	" "	24	" " "	$2\frac{5}{8}$	"	"	9.00

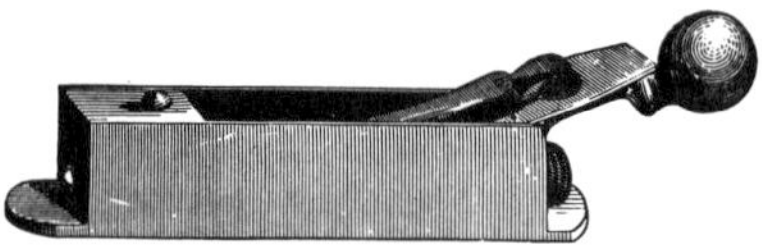

9,	Block Plane, 10 Inches in Length, 2 In. Cutter,	8.00
9 1-2,	Excelsior Block Plane, 6 In. in Length, $1\frac{3}{4}$ In. Cutter,	2.00
10,	Carriage Makers' Rabbet Plane, 14 Inches in Length, $2\frac{1}{8}$ In. Cutter,	6.50
11,	Belt Makers' Plane, $2\frac{3}{8}$ In. Cutter,	4.00

Bailey's Patent Adjustable Circular Plane.

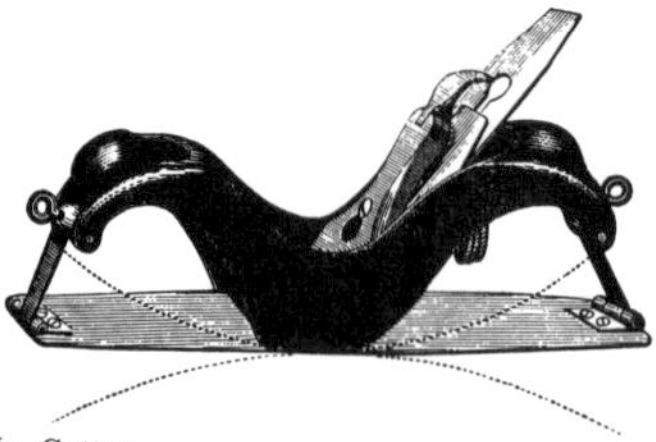

13, $1\frac{3}{4}$ In. Cutter, ... 5.00

This Plane has a *Flexible Steel Face*, and by means of the Thumb-Screws at each end of the Stock, can be easily adapted to plane circular work, either concave or convex.

Wood Planes.

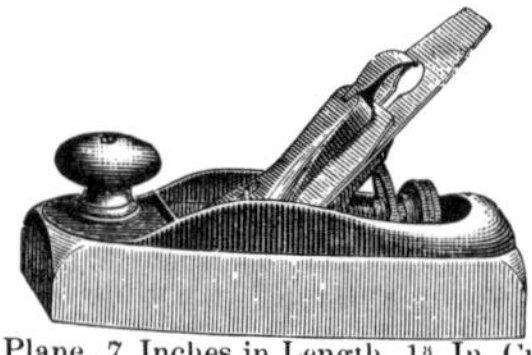

No.									
21,	Smooth	Plane,	7	Inches in Length,	$1\frac{3}{4}$	In.	Cutter,	$	2.50
22,	"	"	8	" " "	$1\frac{3}{4}$	"	"		2.50
23,	"	"	9	" " "	$1\frac{3}{4}$	"	"		2.50
24,	"	"	8	" " "	2	"	"		2.75
25,	Block,	"	9 1-2	" " "	$1\frac{3}{4}$	"	"		2.75

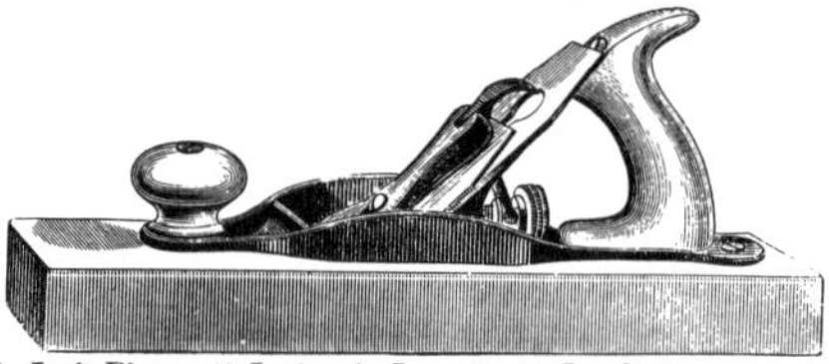

26,	Jack Plane,	15	Inches in Length,	2	In.	Cutter,	3.25
27,	" "	15	" " "	$2\frac{1}{8}$	"	"	3.50
28,	Fore "	18	" " "	$2\frac{3}{8}$	"	"	4.00
29,	" "	20	" " "	$2\frac{3}{8}$	"	"	4.00
30,	Jointer "	22	" " "	$2\frac{3}{8}$	"	"	4.25
31,	" "	24	" " "	$2\frac{3}{8}$	"	"	4.[illegible]5
32,	" "	26	" " "	$2\frac{5}{8}$	"	"	4.75
33,	" "	28	" " "	$2\frac{5}{8}$	"	"	4.75
34,	" "	30	" " "	$2\frac{5}{8}$	"	"	5.00

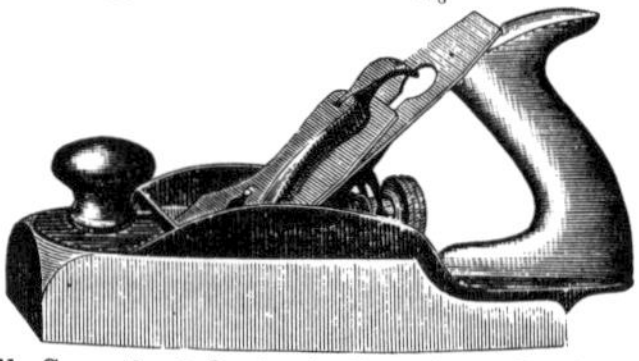

35,	Handle Smooth,	9	Inches in Length,	2	In.	Cutter,	3.50
36,	" "	10	" " "	$2\frac{3}{8}$	"	"	4.00
37,	Jenny "	13	" " "	$2\frac{3}{8}$	"	"	4.00

☞ Extra Plane-woods, of every style, can be supplied cheaply.

Bailey's Patent Adjustable Veneer Scraper.

12, 3 In. Cutter, ... 5.00

CAST STEEL, Hand, VENEER SCRAPERS, 3 x 5 In. per doz. 4.00

Bailey's Patent Plane Irons.

☞ These Plane-Irons have stood the severest tests applied to them by practical workmen, and are by the manufacturers fully WARRANTED.

PRICES.

SINGLE IRONS.

Inches,	1¼	1⅝	1¾	2	2⅛	2⅜	2⅝
Per Doz.	$3.50	4.25	4.75	5.50	5.75	6.00	6.50

DOUBLE IRONS.

Per Doz.	$6.50	7.50	8.50	9.00	9.50	10.00	10.50

☞ Each Plane-Iron is sharpened and put in perfect working order before leaving the Factory.

Bailey's Patent Adjustable Iron and Wood Bench Planes.

EXPLANATION.

The Plane-Iron is secured in its position by means of the Iron Lever, with a Cam and Thumb-piece at its upper end. A Screw passing down into the Iron bed-piece below, serves as a fulcrum upon which the Lever acts in clamping down the Plane-Iron.

The Lever may be put in position, or removed at pleasure, without the use of any tool, it being properly slotted for this purpose; and the pressure required for the best working of the Plane can be obtained at any time, by driving or slacking the central Screw upon which the Lever operates.

The Thumb-screw located under the Iron bed-piece and just in front of the Handle to the Plane, operates a simple device by means of which the Plane-Iron can be easily set forward or withdrawn, while it is still clamped down to the bed-piece; and without removing the hands from the Plane, or the Plane from the work, any desired thickness of shaving may be obtained with perfect accuracy.

The Bed-piece upon which the Plane-Iron rests, is attached to the stock of the Plane by two Screws; and can be moved forward or backward, sufficiently to open or close the mouth of the Plane, as the owner may desire.

TESTIMONIALS.

"The BEST PLANE now in use."—CHICKERING & SONS, *Piano Forte Manufacturers, Boston, Mass.*

"Superior to any others used in this Shop."—W. S. TOWN, *Master of Car Repairs, Hudson River R. R. Co.*

"Works splendidly on our Yellow Pine."—H. C. HALL, *Jacksonville, Florida.*

"All that is wanted to sell your Planes here, is to show them."—E. WATERS, *Manufacturer of Sporting Boats, Troy, N. Y.*

"Far superior to any I have seen used."—JOHN ENGLISH, *North Milwaukie, Wis. Car Shops.*

"I have used your Planes, and have never found their equal."—B. F. FURRY, *Foreman McLear & Kendall's Coach Works, Wilmington, Del.*

"We know of no other Plane that would be a successful substitute for them."—R. BURDETT & Co. *Organ Manufacturers, Chicago.*

"The best Plane ever introduced in carriage-making." JAS. L. MORGAN, *at J. B. Brewster & Co's Carriage Manufactory, New York.*

"From the satisfaction they give, would like to introduce them here."—J. ALEXANDER, *at Bell's Melodeon Factory, Guelph, Canada.*

"Bailey's Planes are used in our Factory. We can get only one verdict, and that is, the men would not be without them." STEINWAY & SONS, *Piano Forte Manufacturers, New York.*

"I have owned a set of Bailey's Planes for six months, and can cheerfully say that I would not use any others, for I believe they are the best in use."—C. C. HARRIS, *Stair Builder, St. Louis, Mo.*

"We have Bailey's Planes in our Factory, and there is but one opinion about them. They are the best and cheapest tool we have ever used. Good tools are always the cheapest." Jos. BECKHAUS, *Carriage Builder, Philadelphia.*

"I have used a set of Bailey's Planes about six months, and would not use any other now. There is inquiry about them here every day. Send me your descriptive Catalogue." EDWARD CARVILLE, *Sacramento City, California.*

A Boston Mechanic says: "I always tell my shopmates, when they wish to use my Plane, not to borrow a BAILEY PLANE unless they intend to buy one, as they will never be satisfied with any other Plane after using this."

BAILEY'S IRON SPOKE SHAVES.

☞ The SPOKE SHAVES in the following List are superior in style, quality and finish, to any in market. The CUTTERS are made of the best English CAST STEEL, tempered and ground by an improved method, and are in perfect working order when sent from the Factory.

No.		Per Doz.
51.	Patent Double Iron, Raised Handle, 10 Inch, 2⅛ Inch Cutter,	$4.00
52.	Patent Double Iron, Straight Handle, 10 Inch, 2⅛ Inch Cutter,	4.00
53.	Patent Adjustable, Raised Handle, 10 Inch, 2⅛ Inch Cutter,	5.00
54.	Patent Adjustable, Straight Handle, 10 Inch, 2⅛ Inch Cutter,	5.00
55.	Model Double Iron, Hollow Face, 10 Inch, 2⅛ Inch Cutter,	4.00
56.	Cooper's Spoke Shave, (Heavy), 18 Inch, 2⅝ Inch Cutter,	8.00
56½.	Cooper's Spoke Shave, (Heavy), 19 Inch, 4 Inch Cutter,	10.00
57.	Cooper's Spoke Shave, (Light), 18 Inch, 2⅛ Inch Cutter,	5.00
58.	Model Double Iron Shave, 10 Inch, 2⅛ Inch Cutter,	3.50
59.	Single Iron Shave, (Pattern of No. 56), 10 Inch, 2⅛ Inch Cutter,	4.00
60.	Double Cutter, Hollow and Straight, 10 Inch, 1½ Inch each Cutter,	5.00

PRICE LIST OF SPOKE SHAVE CUTTERS.

51, 52, 53, 54, 55, 57, 58, 59,	1.50
60, (in pairs),	2.25
56,	2.00
56½,	2.25

MILLER'S PATENT ADJUSTABLE METALLIC PLOW, FILLETSTER AND MATCHING PLANE.

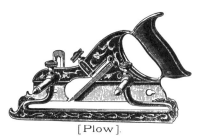

[Plow].

This Tool embraces in a most ingenious and successful combination, the common Carpenter's Plow, a Filletster and a Matching Plane. The entire assortment can be kept in smaller space, or made more portable, than an ordinary Carpenter's Plow.

Each Tool in this combination is complete in itself, and is capable of more perfect adjustment, for its specific uses, than the most improved form of the same Tool as manufactured separately by any other party.

The Bed-piece represented above can be fastened to the Stock instantly, by means of two screws projecting from the back of the same. The heads of the screws being passed through the front end of the slots, as shown in the drawing of the Plow, and then brought back to the rear end of the slots, a single turn will secure the Bed-piece to the Base-piece with perfect firmness; and it can be released at any time with equal facility.

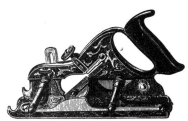

[Filletster].

The above engraving represents the Filletster, which may be readily adjusted to cut any required width, by regulating the horizontal Gauge which slides upon the two bars on the front side of the Stock. The depth to be cut can be adjusted by use of the upright Gauge, with a Thumb-screw, on the back side of the Stock.

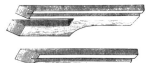

The Tonguing Tool and also the Plow-bits are made of a superior quality of STEEL, and have a V slot in the bottom surface which fits down upon a seat prepared accurately for the same. The Thumb-screw above will secure the tools in position with the greatest possible firmness.

☞ Each Plow is accompanied by a Tonguing tool (¼ in.) and eight Plow bits (1-8, 3-16, 1-4, 5-16, 3-8, 7-16, 1-2 and 5-8 inch), fully **WARRANTED**.

PRICES.

COMBINED PLOW, FILLETSTER, AND MATCHING PLANE, (including Plow-bits, Tonguing and Grooving Tools),
IRON STOCK and Gauge, ... $12.00
GUN METAL Stock and Gauge, ... 15.00

[Matching Plane].

PRICES.

COMBINED PLOW AND MATCHING PLANE,.(including Plow-bits, Tonguing and Grooving Tools):
IRON STOCK and Gauge, ... 8.00
GUN METAL Stock and Gauge, ... 12.00

CHALK-LINE REELS AND STEEL SCRATCH AWLS.

	Per Doz.
Chalk-Line Reels, 3 Dozen in a Box,	$ 0.50
" " " with 60 feet best quality Chalk-Line, 1 Dozen in a Box,	2.25
" " " " Steel Scratch Awls, 1 Dozen in a Box,	1.25
" " " " " " " and 60 feet best quality Chalk-Line, 1 Dozen in a Box,	3.00

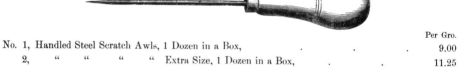

	Per Gro.
No. 1, Handled Steel Scratch Awls, 1 Dozen in a Box,	9.00
2, " " " " Extra Size, 1 Dozen in a Box,	11.25

HANDLED BRAD AWLS.

	Per Gro.
Handled Brad Awls, assorted, 1 Dozen in a Box,	6.75
" " " " large, " " "	8.50

Brad Awl Handles, Brass Ferrules, assorted, in Boxes,	3.50

HANDLES.

	Per Gro.
Polished Auger Handles, assorted,	$6.50
Polished Auger Handles, Extra Large,	8.00

In cases of about two gross.

Improved Auger Handles.

No.		Per Doz.
1.	Malleable Iron Shank, to fit Small Size Bits,	2.75
2.	" " " " Large " "	3.00

	Per Gro.
Polished Hickory Firmer Chisel Handles, assorted, in paper boxes,	5.25
" " " " " " large "	6.50

Polished Socket Firmer Chisel Handles, assorted, in paper boxes, . . 3.50

Polished Hickory Socket Framing Chisel Handles, Iron Ferrules, assorted, in paper boxes, 7.00

	13	14	15	Inch.
Polished Hickory Shingling Hatchet Handles,	$0.44	0.50	0.60	per doz.
	16	17	18	19	Inch.
Polished Hickory Bench Hatchet Handles,	$0.67	0.75	0.88	1.00	per doz.

	13	14	15	16	18	20	Inch.
Polished Hickory Carpenters' Hammer Handles,	$0.50	0.60	0.67	0.75	0.88	1.00	per doz.

MALLETS.

No.										Per Doz.
1,	Round Hickory Mallet,	Mortised,	5 In. Long,	3 In. Diameter,	.	.	.	$1.75
2,	" " "	"	5½ "	3½ "	"	.	.	.	2.25
3,	" " "	"	6 "	4 "	"	.	.	.	2.50
5,	" Lignumvitæ	"	5 "	3 "	"	.	.	.	3.00
6,	" "	"	5½ "	3½ "	"	.	.	.	4.00
7,	" "	"	6 "	4 "	"	.	.	.	5.00

8,	Square Hickory Mallet,	Mortised,	6 In. Long,	2½ by 3½,	.	.	2.25
9,	" " "	"	6½ "	2¾ " 3¾,	.	.	2.75
10,	" " "	"	7 "	3 " 4,	.	.	3.25
11,	" Lignumvitæ "	"	6 "	2¼ " 3½,	.	.	3.75
12,	" " "	"	6½ "	2¾ " 3¾,	.	.	4.75
13,	" " "	"	7 "	3 " 4,	.	.	5.75

| 14, | Round Mallet, Mortised, Iron King, 6 In. Long, 4 In. Diameter, | . | . | . | 6.50 |
| 14½, | " " " " " 5½ " 3½ " " | . | . | 4.50 |

4, Round Hickory Tinner's Mallet, 5½ In. Long, assorted, 2, 2¼ and 2½ In. Diameter, . 1.00

15, Iron Mallet, Mortised, Hickory Ends, 2½ In. Diameter, . . . 4.00

MALLETS, Continued.

No. 16, Mallets, Heavy, Malleable Iron Socket, Mortised, Hickory Ends, 3 inch diameter, $ 12.00

SAW HANDLES.

All full sizes, SPEAR & JACKSON's pattern, of perfect timber, well seasoned, and every way superior and reliable goods.

No.					Per Doz.
1.	Full size,	Cherry,	Polished Edges,		$ 1.65
2.	"	Beech,	" "		1.40
3.	"	"	Plain "		1.20
4.	Small panel,	"	Polished "	for 16 to 20 inch Saws,	1.35
5.	Meat Saws,	"	" "		1.35
6.	Compass Saw	"	" "		1.35
7.	Back Saw,	"	" "		1.35

Packed two gross in a case.

PLANE HANDLES.

	Per Doz.
Jack Plane Handles, (5 gross in a case),	$ 0.42
Fore, or Jointer Handles, (3¾ gross in a case),	0.75

PATENT EXCELSIOR TOOL HANDLE.

[Turkey Boxwood.]

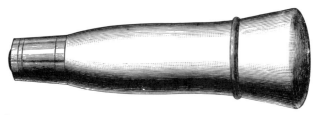

This Handle is the best article of the kind ever invented, it being simple in its construction, compact, and less liable to get out of order than any other in use, and dispenses entirely with the use of a wrench. In the interior of each Handle will be found twenty Brad-awls and Tools, adapted to the various kinds of work required of Mechanics, and useful in every household, or in ordinary branches of business. They are of the full size necessary for the several uses for which they are designed, as represented below.

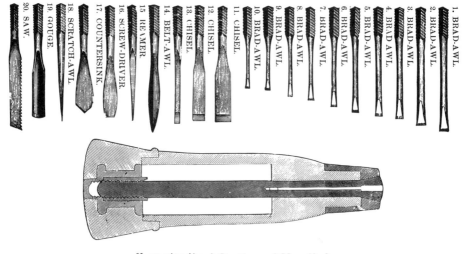

[Longitudinal Section of Handle.]

DIRECTIONS.—Unscrew the Cap of the Handle, and select the Tool needed; with the thumb the centre bolt may be thrust down sufficiently to open the clamp at small end of the Handle, into which the Tool may be inserted; then in replacing the cover and screwing it down to its place, the clamp will be closed, and the Tool firmly secured for use.

☞ Put up in boxes of one dozen, with one set of twenty Tools in each Handle, per dozen, $13.50

HANDLES, AWL HAFTS, ETC.

		Per Gross.
No. 8.	Carpenter's Tool Handle, Steel Screw and Nut, with Iron Wrench,	$ 14.00

No. 8½.   "   "   "   "   "   "   "   "   "   and 10 Brad Awls, assorted sizes, one Handle with Brad Awls, packed in a box, and 12

boxes in a package, ...   Per Doz. 4.75

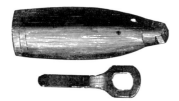

Per Gross.

No. 5. Hickory, Pegging Awl Handles, Plain Top, Steel Screw and Nut, with Iron Wrench, 1 dozen in a box, ... $ 11.00

No. 6. Hickory, Pegging Awl Handles, Leather Top, Steel Screw and Nut, with Iron Wrench, 1 dozen in a box, ... 12.00

No. 6¼. Appletree Sewing Awl Hafts, to hold any size Awl, Steel Screw and Nut, with Iron Wrench, 1 dozen in a box, ... $ 14.00

No. 7. Appletree Sewing Awl Hafts, to hold any size Awl, Steel Screw and Nut, with Iron Wrench, 1 dozen in a box, ... $ 26.00

No. 10. Common Sewing Awl Handle, with Brass Ferrule, . . . 3.50

Polished Cast Steel Awls.

Patent Pegging Awls, (short start, for Handles Nos. 5 and 6), assorted, Nos. 00, 0, 1, 2, 3, 4, 5 and 6. In boxes of one gross, ... $ 0.80

Patent Sewing Awls, (short start, for Handles Nos. 6½ and 7), assorted, from 2 to 3 inches. In boxes of one gross, ... 2.50

Sewing Awls, English Pattern, (for Handle No. 10), assorted, from 2½ to 3⅜ inches. In boxes of one gross, ... 2.25

HAND SCREWS.

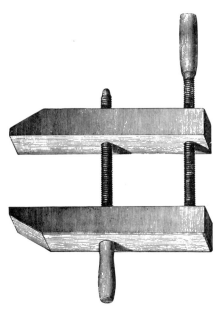

Diameter of Screws.	Length of Screws.	Length of Jaws.	Size of Jaws.	Per Doz.
$1\frac{1}{4}$ inch.	24 inch.	20 inch.	$2\frac{1}{4}$ by $2\frac{1}{4}$ inch.	$10.50
$1\frac{1}{8}$ "	20 "	18 "	$2\frac{5}{8}$ " $2\frac{5}{8}$ "	8.50
1 "	18 "	16 "	$2\frac{3}{8}$ " $2\frac{3}{8}$ "	6.50
$\frac{7}{8}$ "	16 "	14 "	2 " 2 "	4.75
$\frac{3}{4}$ "	12 "	10 "	$1\frac{5}{8}$ " $1\frac{5}{8}$ "	3.25
$\frac{5}{8}$ "	10 "	$8\frac{1}{2}$ "	$1\frac{3}{8}$ " $1\frac{3}{8}$ "	2.50
$\frac{1}{2}$ "	10 "	8 "	$1\frac{1}{4}$ " $1\frac{1}{4}$ "	2.25

Hand Screws, $\frac{1}{2}$, $\frac{5}{8}$ and $\frac{3}{4}$ inch, packed two dozen in a case; all other sizes, one dozen in a case.
Moulders' Flask Screws made to order.

IMPROVED CABINET MAKERS' CLAMP.

	Per Doz.
Cabinet Makers' Clamps, length inside the Jaws, 2 feet, ...	$9.50
" " " " " " 3 "	10.00
" " " " " " 4 "	11.00
" " " " " " 5 "	12.00

PLAIN DOOR STOPS, WITH IRON SCREWS.

No. 6.
Birch.

	Per Gross.
2½ Inch,	$4.50
3 "	5.00
3½ "	5.50

Black Walnut.

2½ Inch,	$5.00
3 "	5.50
3½ "	6.00

Packed in paper boxes of three dozen each.

IMPROVED ELASTIC DOOR STOPS.

No. 9.
Cherry, with Rosewood Cap.

2½ Inch,	$11.50
3 "	12.50
3½ "	14.00

Black Walnut, with Turkey Boxwood Cap.

2½ Inch,	$11.50
3 "	12.50
3½ "	14.00

Packed in paper boxes of three dozen each.

Improved Rubber Top Door Stops.

No. 8.

	Birch.	B. Walnut.	Mahogany.	Porcelain Enameled.
2½ Inch,	$10.00	$11.00	$12.00	$14.00 per gross.
3 "	11.00	12.00	13.00	15.00 " "
3½ "	12.00			

Packed in paper boxes of three dozen each.

Improved Sash Frame Pulleys.

We claim superiority for these Pulleys over any others in use, in these essential particulars:

First. In respect to wear and friction. We use Turkey Boxwood for the Wheels, and large smooth Iron Axles, instead of two grinding surfaces of rough cast iron; thus producing little friction, and no perceptible wear.

Second. The wheels being bored and turned *true*, the irregular motion and noise of the iron wheel is avoided.

Third. The groove of the wheel is turned out smooth, and of proper shape to receive the cord, which is thus prevented from wearing, and made to last much longer than in other pulleys.

		Per Doz.
Sash Frame Pulley, 1¾ inch,	. . . . . .	$ 0.65
" " " 2 "	. . . . . .	0.75
" " " 2¼ "	. . . . . .	1.25

Smith's Patent Sash Cord Iron.

These Irons can be more easily put into the edge of a Sash than any other, requiring only the boring of a round hole; and no screws are needed to keep them in position. When the windows are taken out for cleaning, or any other purpose, the Irons can be removed or replaced without the use of any Tool.

In boxes of one gross, ... $0.80

Trammel Points.

Bronze Metal, with Cast Steel Points and Pencil socket,

Large.	Medium.	Small.
$2.75	2.25	1.75 per **Pair**.

Patent Improved Cattle Tie.

It is believed that this cattle Tie, in both its parts, embraces more truly valuable features than any similar article heretofore sold. The snap is properly proportioned for the greatest strength required, and is made of Malleable Iron. The form of the hooked end prevents the snap becoming unfastened by any accidental means; while the *Improved Spiral Spring*, which is made of brass, and not liable to rust from being wet, or to break in cold weather, acts with certainty, and is protected in its position under an independent lug, or bar, upon which all shocks or pressure must first come.

The other part of the Cattle Tie consists of an iron socket, which may be secured at any desired position on a rope by means of a Malleable thumb-screw in one side of the socket; the thumb-screw having a perforated head, through which the snap is readily hooked. The socket may at once be changed from one position on the rope to another, and secured perfectly without the use of a screw-driver, or other tool, thus adapting the length of the rope to the size of the neck or horns of any animal. The extreme simplicity of both parts insures great durability.

	Per Doz.
Cattle Tie, with Rope, Japanned,	$4.50
" " " " X Plated,	5.00
" " without " Japanned,	1.75
" " " " X Plated,	2.25
Improved Spiral Spring Snaps, (only), Japanned,	0.80
" " " " " X Plated,	1.00

Pinking Irons.

Sizes,	$\tfrac{3}{8}$	$\tfrac{1}{2}$	$\tfrac{5}{8}$	$\tfrac{3}{4}$	$\tfrac{7}{8}$	1	$1\tfrac{1}{8}$	$1\tfrac{1}{4}$	$1\tfrac{1}{2}$	Inch.
Diamond Pattern,	$3.00	3.00	3.00	3.00	3.00	3.00	3.50	3.50	4.00	per doz.
Scolloped,	3.00	3.00	3.00	3.00	3.00	3.00	3.50	3.50	4.00	" "

Patent Improved Tack Hammer.

This Tack Hammer and Claw combined, is made of the best quality of Malleable Iron, polished, and the handle is inlaid with Black Walnut. In its appearance, it surpasses anything of the kind in market; and, being cast in a single piece, no liability exists of either the head or claw working loose. Perfect durability is thus secured.

Put up in paper boxes of one dozen each.

No. 4.   Malleable Iron, Polished, inlaid with Black Walnut, hardened face,   .   per doz. $2.75

Patent Improved Saddler's and Upholsterer's Hammer.

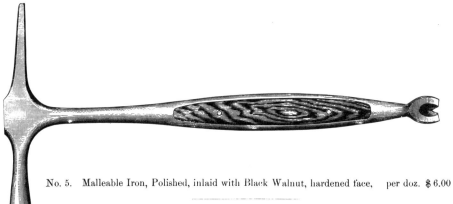

No. 5.   Malleable Iron, Polished, inlaid with Black Walnut, hardened face,   per doz. $6.00

Patent Improved Shoemaker's Hammer.

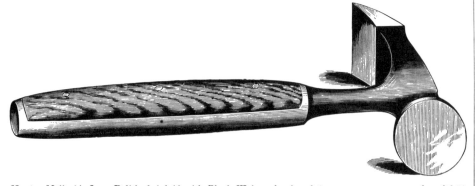

No. 6.   Malleable Iron, Polished, inlaid with Black Walnut, hardened face,   per doz. $6.50

Magnetic Tack Hammers.

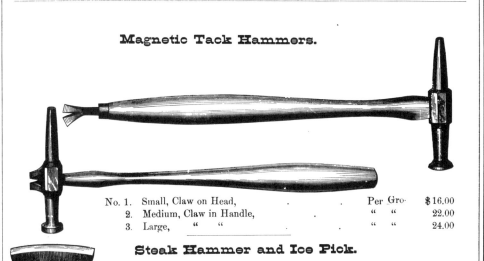

			Per Gro.	
No. 1.	Small, Claw on Head,	.	.	$16.00
2.	Medium, Claw in Handle,	.	" "	22.00
3.	Large, " "	.	" "	24.00

Steak Hammer and Ice Pick.

No.				Per Doz.
7.	Iron, Japanned,	.	.	$2.75
8.	X Plated, Polished Handle,	.	.	4.00

Improved Steak Hammer and Ice Pick.

No.			Per Doz.
9.	Iron, Japanned, Inlaid Handle,	. .	$4.00

TACKLE BLOCKS.

Rope Strapped Blocks, Hook and Thimble.

Length of Blocks,	4	5	6	7	8	9	10	11	12	Inch.
Diameter of Wheels,	2½	3	3½	4¼	5	5¾	6½	7¼	8	"
Diameter of Rope,	½	⅝	¾	⅞	1	1	1⅛	1⅛	1¼	"
Single, Iron Bushed,	$0.80	$1.00	$1.20	$1.40	$1.60	$1.80	$2.25	$2.75	$3.50	Each.
Double, Iron Bushed,	1.40	1.75	2.10	2.45	2.80	3.30	4.00	5.00	6.00	"
Triple, Iron Bushed,	1.75	2.10	2.45	2.80	3.50	4.50	5.50	6.50	7.50	"
Single, Roller Bushed,	1.35	1.70	2.00	2.35	2.70	3.00	3.40	4.00	4.50	"
Double, Roller Bushed,	2.50	3.12	3.75	4.35	4.95	5.60	6.50	7.50	8.50	"
Triple, Roller Bushed,	3.36	4.25	5.12	5.88	6.75	7.62	8.75	10.25	11.50	"

Rope Strapped Blocks, Hook and Thimble, Thick Mortise.

Length of Blocks,	9	10	11	12	Inch.
Diameter of Wheels,	5½	6¼	7	7½	"
Diameter of Rope,	1¼	1⅜	1½	1½	"
Single, Iron Bushed,	$3.50	$4.00	$4.50	$5.00	Each.
Double, Iron Bushed,	7.00	8.00	9.00	10.00	"
Single, Roller Bushed,	5.50	6.00	6.75	7.50	"
Double, Roller Bushed,	11.00	12.00	13.50	15.00	"

TACKLE BLOCKS.

Inside Iron Strapped Blocks, Loose Hook and Stiff Swivel.

Length of Blocks,	4	5	6	7	8	9	10	11	12 Inch.
Diameter of Wheels,	2½	3	3½	4¼	5	5¾	6½	7¼	8 "
Diameter of Rope,	½	⅝	¾	⅞	1	1	1⅛	1⅛	1¼ "
Single, Iron Bushed,	$1.04	$1.30	$1.60	$1.85	$2.14	$2.42	$2.66	$2.90	$3.25 Each.
Double, Iron Bushed,	1.95	2.42	2.91	3.42	3.89	4.39	4.86	5.50	6.00 "
Triple, Iron Bushed,	2.66	3.32	3.97	4.65	5.32	6.00	6.65	7.50	8.25 "
Single, Roller Bushed,	1.47	1.85	2.22	2.61	2.95	3.32	3.70	4.25	4.60 "
Double, Roller Bushed,	2.79	3.47	4.18	4.89	5.58	6.27	6.98	8.00	8.65 "
Triple, Roller Bushed,	3.91	4.86	5.89	6.89	7.84	8.85	9.83	11.25	12.35 "

Awning Blocks, Rope Strapped.

	1¼	1½	2	2½	3	3½	4	Inch.
Single,	14	14	14	16	18	20	22	cts. Each.
Double,	28	28	28	32	36	40	44	" "

Awning Blocks, Not Strapped.

	1¼	1½	2	2½	3	3½	4	Inch.
Single,	11	11	11	12	14	16	18	cts. Each.
Double,	22	22	22	24	28	32	36	" "

Plain Blocks, Not Strapped.

Iron Bushed.		Roller Bushed.	
4 to 9 Inch,	$0.10 per Inch.	4 to 9 Inch,	$0.18 per Inch.
10 to 11 "	0.14 " "	10 to 11 "	0.25 " "
12 to 13 "	0.16 " "	12 to 13 "	0.27 " "
14 to 15 "	0.22 " "	14 to 15 "	0.32 " "

INDEX.

	Page.
FRONTISPIECE, View of Factory at New Britain,	
AUGER HANDLES,	33
Awl Hafts and Handles,	37
Awls, Pegging and Sewing,	37
BRAD AWL HANDLES,	32
Brad Awls, Handled,	32
Bevels, Sliding T,	22 and 23
CABINET MAKERS' CLAMPS,	38
Chalkline Reels and Awls,	32
Carpenters' Tool Handles,	37
Cattle Ties,	41
Chisel Handles,	33
Compass Dividers,	16
Countersinks, Wheeler's Patent,	27
DOOR STOPS,	39
GAUGES,	24 to 26
HAMMER HANDLES,	33
Hatchet Handles,	33
Hand Screws,	38
LEVEL GLASSES,	16
MALLETS,	34 and 35
Mitre Squares,	19
Miscellaneous Articles, Stanley's,	9
" " Stearns',	12
PLANES, Bailey's Patent,	28
Plane-Irons, Bailey's Patent,	29
Plane Handles,	35
Pinking Irons,	41
Plumbs and Levels,	14 to 18
Pocket Levels,	16
Plow, Miller's Patent Adjustable,	31
RULES, Stanley's,	3 to 8
" Stearns',	10 to 13
SASH FRAME PULLEYS,	40
Sash Cord Irons,	40
Scratch Awls, Handled,	32
Saw Handles,	35
Screw Drivers,	27
Steak Hammers,	43
Shoemakers' Hammers,	42
Spoke Shaves, Bailey's Patent,	30
Spoke Shave Cutters,	30
TACKLE BLOCKS,	44 and 45
Tack Hammers,	42 and 43
Trammel Points,	40
Tool Handle and Tools, "Excelsior,"	36
Try Squares,	20 and 21
Try Square and Bevel, Patent Combination,	19
UPHOLSTERERS' HAMMERS,	42
VENEER SCRAPERS,	28

Attach to our Catalogue of January, 1872.

DISCOUNT SHEET.
STANLEY RULE & LEVEL COMPANY.
April 10, 1872.

Catalogue Pages.		Discount per cent.
33	Auger Handles,	20
37	Awl Hafts and Handles,	20
37	Awls, Pegging and Sewing,	20
32	Brad Awl Handles,	20
32	Brad Awls, Handled,	20
23	BEVELS, Sliding T,	40
22	" " Improved,	15
38	Cabinet Makers' Clamps,	20
32	Chalk-line Reels and Awls,	20
37	Carpenters' Tool Handles,	20
41	Cattle Ties,	20
33	Chisel Handles,	10
16	Compass Dividers	15
27	Countersinks, Wheeler's Pat.	20
39	Door Stops, No. 6,	10
39	" " No. 8 and 9,	25
24 to 26	GAUGES,	40
33	Hammer Handles,	10
33	Hatchet Handles,	10
38	HAND SCREWS,	20
16	Level Glasses,	50
34 and 35	MALLETS, Hickory,	10
	" Lignumvitæ,	Net
19	Mitre Squares, Improved,	15
28	PLANES, Bailey's Patent,	15
29	Plane Irons, " "	15
35	Plane Handles,	20
41	Pinking Irons,	50

Catalogue Pages.		Discount per cent.
14 to 16	PLUMBS AND LEVELS,	50
18	Plumbs and Levels, Iron Frame,	15
18	" " Nicholson's Pat.	15
16	Pocket Levels, Imp'd Patterns,	40
31	PLOWS, Miller's Pat. Adjustable, with Filletster,	15
31	‡PLOW and Matching Plane Combined,	15
3 to 8	*RULES, Stanley's,	25
9	Miscellaneous Articles, "	25
10 to 13	*RULES, Stearns',	25
12	Miscellaneous Articles, "	25
40	SASH FRAME PULLEYS,	20
40	Sash Cord Irons,	10
32	Scratch Awls, Handled,	20
35	Saw Handles	20
27	SCREW DRIVERS, Flat H'dl.	20
27	" " Oval Handle,	10
43	Steak Hammers,	10
42	Shoemakers' Hammers,	10
30	SPOKE SHAVES, Bailey's Pat.	15
30	Spoke Shave Cutters, " "	15
44 and 45	TACKLE BLOCKS,	15
42 and 43	Tack Hammers,	10
40	Trammel Points,	20
36	Tool Handles, "Excelsior,"	10
21	TRY SQUARES, No. 2,	40
20	" " Improved,	15
21	Try Squares, Plumb and Level,	40
19	" " and Bevel Combined,	15
42	Upholsterers' Hammers,	10
28	Veneer Scrapers, Bailey's Pat.	15

DISCOUNT for Cash (IF PAID WITHIN 30 DAYS), 10 per cent.

☞ TERMS CASH. Accounts unpaid Thirty Days from date of Invoice are subject to our Draft, payable at sight.

* The List price of Rules No. 68, (Stanley's) and No. 41 (Stearns') is changed from $3.50 to $4.00 Per Doz. Also, the List Price of Rules No. 61, (Stanley's,) and No. 38 (Stearns') is changed from $4.50 to $5.00 Per Doz.

‡ We desire to direct especial attention to the low price of this superior tool.

1874 STANLEY RULE & LEVEL Co. CATALOG and PRICE LIST, with 1876 Supplement

This Catalog, 1874 with 1876 Supplement, shows considerable progress and the addition of several new items since the catalogs of the early 1870s.

New products appearing in the 1874 Catalog include:

- Improved Mitre Try Square with Cast Brass Handle, p.30.
- Winterbottom's Patent Combined Try & Mitre Square, p.30½; probably first offered during 1875.
- Nos. 110 Improved Iron and 15½ Bailey's Patent Adjustable Block Planes, p.30½, probably offered 1875.
- Bailey's Patent Flush T Bevel, (US Pat.4145,715), p.33.
- Patent Eureka Flush T Bevel, Iron Handle, p.34
- Nos. 60 and 60½ Patent Iron Marking Gauges, p.35.
- Though previously offered in 1870 as a non-numbered item, Bailey's Patent Adjustable Veneer Scraper was first assigned No. 12 in the Plane Line, p.39.
- Bailey's Patent Adjustable Circular Plane, p.39 (US Pat.#113,003, Mar.14,1871).
- Wheeler's Patent Countersink, p.43. (US Pat.101,796, Apr.12,1870); with Gauge (US Pat.116,901, July 11,1871).
- Traut's Patent Adjustable Dado, Filletster, etc., No.46, p.45, (US Pat.#136,469, Mar.4,1873).
- Patent Improved Solid Cast Steel Screw Drivers, p.46, (J.P.Curtiss, US Pat.#115,582, June 6,1871).

- Improved Trammel Points, p.56.
- Adjustable Plumb Bobs, p.56, (J.A.Traut, US Pat.151,521, June 2,1874).

New products appearing in the 1876 Supplement include:

- $9\frac{1}{2}$ and $9\frac{3}{4}$ Excelsior Block Planes, p.66.
- Patent Tongueing and Grooving Plane, No. 48, p.71, (Chas. A. Miller, US Pat.165,355, July 9, 1875).
- Patent Improved Mitre Box, p. 71, (J.A.Traut, US Pat.192,139, June 16, 1877).
- Traut's Patent Adustable Dado, No.47, p.70, (US Pat.136,469, Mar. 4, 1873).

Ken Roberts
1978

Price List

OF

U. S. STANDARD

BOXWOOD AND IVORY

RULES,

LEVELS, TRY SQUARES, GAUGES, MALLETS,
SCREW DRIVERS, HAND SCREWS,
IRON AND WOOD PLANES,
SPOKE SHAVES, ETC.

MANUFACTURED BY THE

Stanley
Rule and Level Co.,

NEW BRITAIN, CONN.

---•••---

WAREROOMS:

No. 35 CHAMBERS STREET, NEW YORK.

ORDERS FILLED AT THE WAREROOMS, OR AT NEW BRITAIN.

---•••---

JANUARY, 1874.

OFFICE OF THE

STANLEY RULE AND LEVEL CO.,

New Britain, Ct., Jan. 1, 1874.

Gentlemen :

This Catalogue and Price List will be found to embrace the largest assortment of first-class Carpenters' Tools manufactured by any concern in this country.

It will be observed that we have increased several of the leading lines of our Tools, while some of the specialties made by us, are for the first time presented here, and are classified under their proper headings. Our aim is to adopt every genuine improvement in the various lines of Tools we manufacture ; and, with the extensive facilities enjoyed by us for their production, we claim for our Goods superiority in quality and finish.

Our customers will please note the removal of our Warerooms (formerly No. 55 Chambers Street) to <u>35 Chambers Street, New York.</u>

Respectfully,

STANLEY RULE & LEVEL CO.

PRICE LIST.

☞ All Rules embraced in the following Lists, bear the Invoice **Number** by which they are sold, and are graduated to correspond with the description given of each.

Stanley's Boxwood Rules.

One Foot, Four Fold, Narrow.

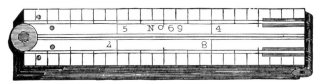

No.		Per Dozen.
69.	Round Joint, Middle Plates, 8ths and 16ths of inches, ⅝ inch wide...	$3 00

65.	Square Joint, Middle Plates, 8ths and 16ths of inches, ⅝ inch wide...	3 50
64.	Square Joint, Edge Plates, 8ths and 16ths of inches, ⅝ inch wide...	5 00
65½.	Square Joint, Bound, 8ths and 16ths of inches, ⅝ inch wide,	11 00

55.	Arch Joint, Middle Plates, 8ths and 16ths of inches, ⅝ inch wide...	4 00
56.	Arch Joint, Edge Plates, 8ths and 16ths of inches, ⅝ inch wide...	6 00
57.	Arch Joint, Bound, 8ths and 16ths of inches, ⅝ inch wide..	12 00

Two Feet, Four Fold, Narrow.

No.			Per Dozen.
68.	Round Joint, Middle Plates, 8ths and 16ths of inches....1	in. wide,	$4 00
8.	Round Joint, Middle Plates, Extra Thick, 8ths and 16ths of inches....1	"	4 50
61.	Square Joint, Middle Plates, 8ths and 16ths of inches....1	"	5 00
63.	Square Joint, Edge Plates, 8ths, 10ths, and 16ths of inches, Drafting Scales....1	"	7 00
84.	Square Joint, Half Bound, 8ths, 10ths, and 16ths of inches, Drafting Scales....1	"	12 00
62.	Square Joint, Bound, 8ths, 10ths, and 16ths of inches, Drafting Scales....1	"	15 00
51.	Arch Joint, Middle Plates, 8ths, 10ths, and 16ths of inches, Drafting Scales....1	"	6 00
53.	Arch Joint, Edge Plates, 8ths, 10ths, and 16ths of inches, Drafting Scales....1	"	8 00
52.	Arch Joint, Half Bound, 8ths, 10ths, and 16ths of inches, Drafting Scales....1	"	13 00
54.	Arch Joint, Bound, 8ths, 10ths, and 16ths of inches, Drafting Scales....1	"	16 00
59.	Double Arch Joint, Bitted, 8ths, 10ths, and 16ths of inches, Drafting Scales....1	"	9 00
60.	Double Arch Joint, Bound, 8ths, 10ths, and 16ths of inches, Drafting Scales....1	"	21 00

Two Feet, Four Fold, Extra Narrow.

61½.	Square Joint, Middle Plates, 8ths and 16ths of inches....¾	in. wide,	5 50
63½.	Square Joint, Edge Plates, 8ths, 10ths, and 16ths of inches....¾	"	8 00
62½.	Square Joint, Bound, 8ths, 10ths, 12ths, and 16ths of inches....¾	"	15 00

STANLEY'S BOXWOOD RULES. 7

Stanley's Two Feet, Four Fold, Narrow Rules.

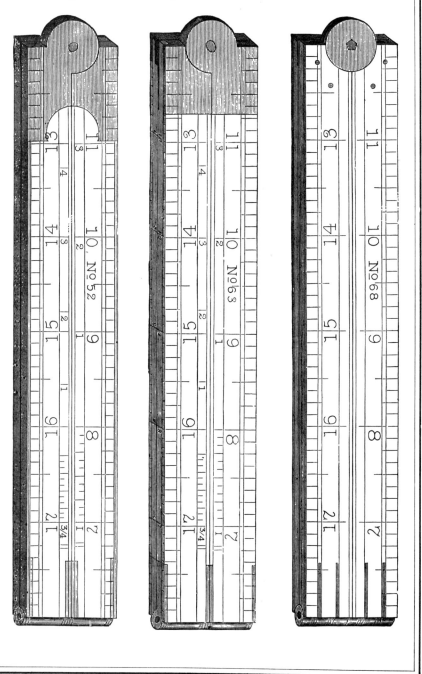

Two Feet, Four Fold, Broad.

No.			Per Dozen.
67.	Round Joint, Middle Plates, 8ths and 16ths of inches	$1\frac{3}{8}$ in. wide,	$5 00
70.	Square Joint, Middle Plates, 8ths and 16ths of inches, Drafting Scales	$1\frac{3}{8}$ "	7 00
72.	Square Joint, Edge Plates, 8ths, 10ths, and 16ths of inches, Drafting Scales	$1\frac{3}{8}$ "	9 00
$72\frac{1}{2}$	Square Joint, Bound, 8ths, 10ths, and 16ths of inches, Drafting Scales	$1\frac{3}{8}$ "	18 00
73.	Arch Joint, Middle Plates, 8ths, 10ths, and 16ths of inches, Drafting Scales	$1\frac{3}{8}$ "	9 00
75.	Arch Joint, Edge Plates, 8ths, 10ths, and 16ths of inches, Drafting Scales	$1\frac{3}{8}$ "	11 00
76.	Arch Joint, Bound, 8ths, 10ths, and 16ths of inches, Drafting Scales	$1\frac{3}{8}$ "	20 00
77.	Double Arch Joint, Bitted, 8ths, 10ths, and 16ths of inches, Drafting Scales	$1\frac{3}{8}$ "	12 00
78.	Double Arch Joint, Half Bound, 8ths, 10ths, and 16ths of inches, Drafting Scales	$1\frac{3}{8}$ "	20 00
$78\frac{1}{2}$	Double Arch Joint, Bound, 8ths, 10ths, and 16ths of inches, Drafting Scales	$1\frac{3}{8}$ "	24 00
83.	Arch Joint, Edge Plates, Slide, 8ths, 10ths, 12ths, and 16ths of inches, 100ths of a foot, and Octagonal Scales	$1\frac{3}{8}$ "	14 00

Board Measure, Two Feet, Four Fold.

79.	Square Joint, Edge Plates, 12ths and 16ths of inches, Drafting Scales	$1\frac{3}{8}$ in. wide,	11 00
81.	Arch Joint, Edge Plates, 12ths and 16ths of inches, Drafting Scales	$1\frac{3}{8}$ "	13 00
82.	Arch Joint, Bound, 12ths and 16ths of inches, Drafting Scales	$1\frac{3}{8}$ "	22 00

STANLEY'S BOXWOOD RULES.

Stanley's Two Feet, Four Fold, Broad Rules.

No. 78½

No. 75

No. 70

Two Feet, Two Fold.

No.			Per Dozen.
29.	Round Joint, 8ths and 16ths of inches....... 1⅜ in. wide,		$3 50
18.	Square Joint, 8ths and 16ths of inches....... 1½	"	5 00
22.	Square Joint, Bitted, *Board Measure*, 10ths and 16ths of inches, and Octagonal Scales..... 1½	"	8 00
1.	Arch Joint, 8ths and 16ths of inches, Octagonal Scales............................1½	"	7 00
2.	Arch Joint, Bitted, 8ths, 10ths, and 16ths of inches, Octagonal Scales................1½	"	8 00
4.	Arch Joint, Bitted, Extra Thin, 8ths and 16ths of inches, Drafting and Octagonal Scales..1½	"	10 00
5.	Arch Joint, Bound, 8ths, 10ths, and 16ths of inches, Drafting and Octagonal Scales.....1½	"	16 00

Two Feet, Two Fold, Slide.

26.	Square Joint, Slide, 8ths, 10ths, and 16ths of inches, Octagonal Scales................1½ in. wide,		9 00
27.	Square Joint, Bitted, Gunter's Slide, 8ths, 10ths, and 16ths of inches, 100ths of a foot, Drafting and Octagonal Scales...................1½	"	12 00
12.	Arch Joint, Bitted, Gunter's Slide, 8ths, 10ths, and 16ths of inches, 100ths of a foot, Drafting and Octagonal Scales...................1½	"	14 00
15.	Arch Joint, Bound, Gunter's Slide, 8ths, 10ths, and 16ths of inches, Drafting and Octagonal Scales................................1½	"	24 00
6.	Arch Joint, Bitted, Gunter's Slide, Engineering, 8ths, 10ths, and 16ths of inches, 100ths of a foot, Octagonal Scales..................1½	"	18 00
16.	Arch Joint, Bound, Gunter's Slide, Engineering, 8ths, 10ths, and 16ths of inches, Octagonal Scales................................1½	"	28 00

☞ With recently constructed machinery, we can furnish, to order, Rules marked with French or Spanish graduations.

STANLEY'S SLIDE RULES.

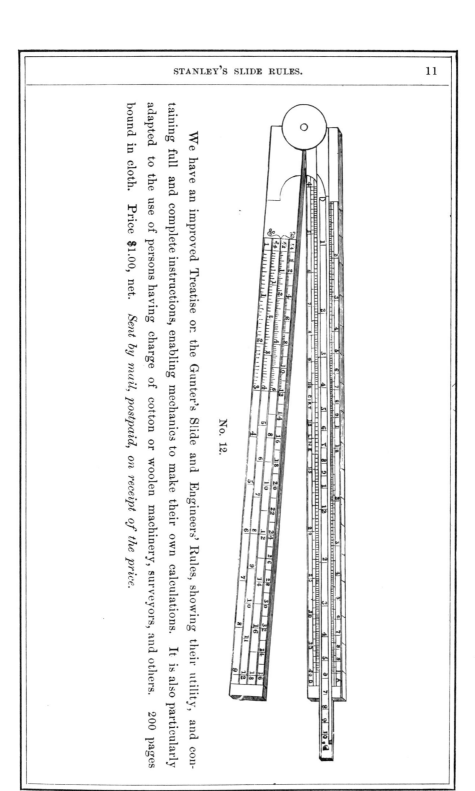

No. 12.

We have an improved Treatise or the Gunter's Slide and Engineers' Rules, showing their utility, and containing full and complete instructions, enabling mechanics to make their own calculations. It is also particularly adapted to the use of persons having charge of cotton or woolen machinery, surveyors, and others. 200 pages bound in cloth. Price $1.00, net. *Sent by mail, postpaid, on receipt of the price.*

Boxwood Caliper Rules.

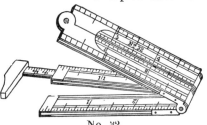

No. 32.

No.		Per Dozen.
36.	Square Joint, Two Fold, 6 inch, 8ths, 10ths, and 16ths of inches................ ⅞ in. wide,	$7 00
36½.	Square Joint, Two Fold, 12 inch, 8ths, 10ths, 12ths, and 16ths of inches............1⅜ "	12 00
32.	Arch Joint, Edge Plates, Four Fold, 12 inch, 8ths, 10ths, 12ths, and 16ths of inches......1 "	12 00
32½.	Arch Joint, Bound, Four Fold, 12 inch, 8ths, 10ths, 12ths, and 16ths of inches..........1 "	20 00

Two Feet, Six Fold Rules.

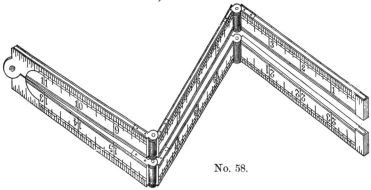

No. 58.

58.	Arch Joint, Edge Plates, 8ths and 16ths of inches........................... ¾ in. wide,	13 00

Three Feet, Four Fold Rules.

66.	Arch Joint, Middle Plates, Four Fold, 16ths of inches, and Yard Stick graduations on inside, 1 in. wide,	8 00
66½.	Arch Joint, Middle Plate, Four Fold, 8ths and 16ths of inches........................1 "	8 00

Ship Carpenters' Bevels.

42.	Boxwood, Double Tongue, 8ths and 16ths of inches,	6 00
43.	Boxwood, Single Tongue, 8ths and 16ths of inches,	6 00

Stanley's Ivory Rules.

Caliper.

No. 38.

No.		Per Dozen.
38.	Square Joint, German Silver, Two Fold, 6 inch, 8ths, 10ths, and 16ths of inches........ 7/8 in. wide,	$15 00
39.	Square Joint, Edge Plates, German Silver, Four Fold, 12 inch, 8ths, 10ths, 12ths, and 16ths of inches.................................. 7/8 "	38 00
40.	Square Joint, German Silver, Bound, Four Fold, 12 inch, 8ths and 16ths of inches......... 5/8 "	44 00

Ivory, One Foot, Four Fold.

90.	Round Joint, Brass, Middle Plates, 8ths and 16ths of inches.............................	10 00
92½.	Square Joint, German Silver, Middle Plates, 8ths and 16ths of inches................ 5/8 in. wide,	14 00
92.	Square Joint, German Silver, Edge Plates, 8ths and 16ths of inches..................... 5/8 "	17 00
88½.	Arch Joint, German Silver, Edge Plates, 8ths and 16ths of inches................... 5/8 "	21 00
88.	Arch Joint, German Silver, Bound, 8ths and 16ths of inches............................ 5/8 "	32 00
91.	Square Joint, German Silver, Edge Plates, 8ths, 10ths, 12ths, and 16ths of inches......... 3/4 "	23 00

Ivory, Two Feet, Four Fold.

85.	Square Joint, German Silver, Edge Plates, 8ths, 10ths, 12ths, and 16ths of inches......... 7/8 in. wide,	54 00
86.	Arch Joint, German Silver, Edge Plates, 8ths, 10ths, and 16ths of inches, 100ths of a foot, Drafting Scale............................1 "	64 00
87.	Arch Joint, German Silver, Bound, 8ths, 10ths, and 16ths of inches, Drafting Scale.......1 "	80 00
89.	Double Arch Joint, German Silver, Bound, 8ths, 10ths, and 16ths of inches, Drafting Scale, 1 "	92 00
95.	Arch Joint, German Silver, Bound, 8ths, 10ths, and 16ths of inches, Drafting Scale,....... 1 3/8 "	102 00
97.	Double Arch Joint, German Silver, Bound, 8ths, 10ths, and 16ths of inches, Drafting Scale...1 3/8 "	116 00

Miscellaneous Articles.

Bench Rules.

No.		Per Dozen.
34.	Bench Rule, Maple, Brass Tips..................2 feet,	$3 00
35.	" Board Measure, Brass Tips..........2 "	6 00
31.	" Satin Wood, Brass Capped Ends.....2 "	12 00

Board, Log, and Wood Measures.

46.	Board Stick, Octagon, Brass Cap, 8 to 23 feet......2 feet,	8 00
46½.	" " Square, " 8 " 23 " ......2 "	8 00
47.	" " Octagon, " 8 " 23 " ......3 "	12 00
47½.	" " Square, " 8 " 23 " ......3 "	12 00
43½.	" " Flat, Hickory, Cast Brass Head and Tip, 6 Lines, 12 to 22 feet....................3 "	12 00
49.	Board Stick, Flat, Hickory, Steel Head, Brazed, Extra Strong, 6 Lines, 12 to 22 feet.............3 "	26 00
48.	Walking Cane, Board Measure, Octagon, Hickory, Solid Cast Brass Head and Tip, 8 Lines, 9 to 16 feet..3 "	12 00
48½.	Walking Cane, Scribner's Log Measure, Octagon, Hickory, Solid Cast Brass Head and Tip..........3 "	15 00
71.	Wood Measure, Brass Caps, 8ths of inches and 10ths of a foot....................................4 "	8 00

Yard Sticks.

33.	Yard Stick, Polished...........................	2 00
41.	" " Brass Tip, Polished.................	3 50
50.	" " Hickory, Brass Cap'd Ends, Polished..	4 50

Wantage and Gauging Rods.

44.	Wantage Rod, 8 lines........................	5 00
37.	" " 12 lines.......................	7 00
45.	Gauging " 120 gallons...................3 feet,	7 00
45½.	" " Wantage Tables................4 "	18 00
49½.	Forwarding Stick..........................5 "	24 00

Scholars' Rules.

23.	Maple, 12 inch, ¾ inch wide, Beveled Edge, 16ths of inches...	1 00
98.	Boxwood, 12 inch, ¾ inch wide, Beveled Edge, 8ths and 16ths of inches.............................	1 25
99.	Boxwood, 12 inch, ¾ inch wide, Beveled Edge, 10ths and 16ths of inches............................	1 50

Board and Log Measures.

NOTE.—Nos. 43½, 46, 46½, 47, 47½, 48, and 49 give the contents in Board Measure of 1 inch Boards.

DIRECTIONS.—Place the stick across the flat surface of the Board, bringing the inside of the Cap snugly to the edge of the same; then follow with the eye the column of figures in which the length of the Board is given as the first figure under the Cap, and at the mark nearest the opposite edge of the Board will be found the contents of the Board in feet.

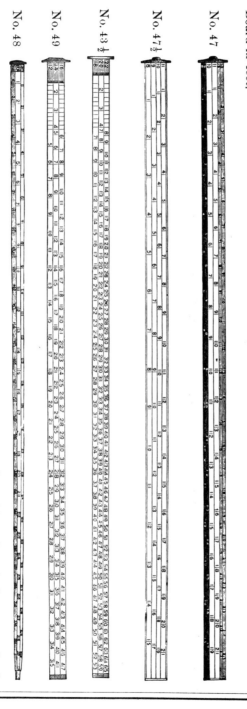

NOTE.—The Log Measure (No. 48½) has Scribner's Tables, and gives the number of feet of one inch square edged boards, which can be sawed from a log of any size, from 12 to 36 inches in diameter, and of any length. The figures immediately under the head of the Cane, are for the length of Logs in feet. Under these figures, on the same line, at the mark nearest the diameter of the Log, will be found the number of feet the Log will make. If the Log to be measured is not over 15 feet long, the diameter should be taken at the small end; if over 15 feet, at the middle of the Log.

Stearns' Boxwood Rules.

Two Feet, Two Fold.

No.		Per Dozen.
1.	Arch Joint, Bound, Slide, Engineering, 10ths and 16ths of inches, 100ths of a foot, Drafting Scales.................................1½ in. wide,	$28 00
2.	Arch Joint, Bitted, Slide, Engineering, 10ths, 12ths, and 16ths of inches, 100ths of a foot, Drafting Scales......................1½ "	18 00
3.	Arch Joint, Bound, Gunter's Slide, 10ths, 12ths, and 16ths of inches, 100ths of a foot, Drafting and Octagonal Scales...................1½ "	24 00
5.	Arch Joint, Bitted, Gunter's Slide, 10ths, 12ths, and 16ths of inches, 100ths of a foot, Drafting and Octagonal Scales...................1½ "	14 00
7.	Arch Joint, Bitted, 8ths and 16ths of inches, Drafting and Octagonal Scales..........1½ "	8 00
9.	Square Joint, Slide, 8ths and 16ths of inches, Drafting and Octagonal Scales..........1½ "	9 00
11.	Square Joint, 8ths and 16ths of inches.......1½ "	5 00

Two Feet, Four Fold, Broad.

14.	Arch Joint, Bound, Board Measure, 10ths, 12ths, and 16ths of inches, Drafting and Octagonal Scales [see Engraving, p. 17]..............1½ in. wide,	24 00
15.	Arch Joint, Bound, 8ths, 10ths, 12ths, and 16ths of inches, Drafting and Octagonal Scales..1½ "	22 00
17.	Arch Joint, Arch Back, Edge Plates, 8ths, 10ths, 12ths, and 16ths of inches, 100ths of a foot, Drafting and Octagonal Scales.............1½ "	15 00
18.	Arch Joint, Triple Plated Edge Plates, Board Measure, 10ths, 12ths, and 16ths of inches, 100ths of a foot, Drafting and Octagonal Scales................................1½ "	14 00
19.	Arch Joint, Triple Plated Edge Plates, Slide, 8ths, 10ths, 12ths, and 16ths of inches, 100ths of a foot, Drafting and Octagonal Scales...1½ "	15 00
20.	Arch Joint, Triple Plated Edge Plates, 8ths, 10ths, 12ths, and 16ths of inches, 100ths of a foot, Drafting and Octagonal Scales......1½ "	13 00
21.	Arch Joint, Middle Plates, Bitted, 8ths and 16ths of inches, Drafting and Octagonal Scales..1½ "	11 00
32.	Square Joint, Middle Plates, Bitted, 8ths and 16ths of inches, Drafting and Octagonal Scales...................................1½ "	9 00

Stearns' Two Feet, Four Fold Rules.

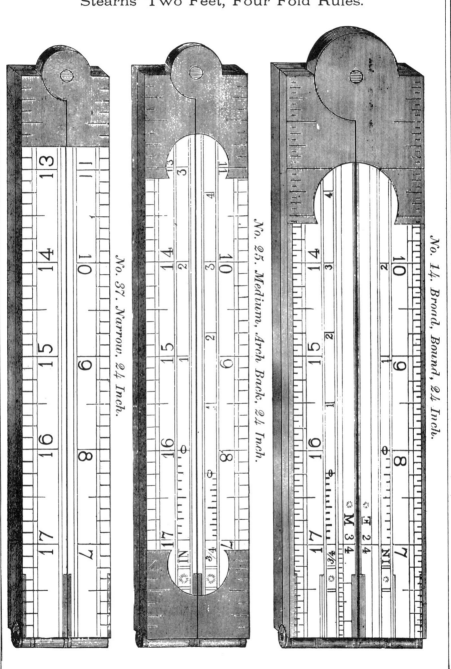

Two Feet, Four Fold, Medium.

No.   Per Dozen.

22. Arch Joint, Bound, Board Measure, 16ths of inches, Drafting Scales .... $1\frac{1}{8}$ in. wide, $20 00
23. Arch Joint, Bound, 8ths, 10ths, 12ths, and 16ths of inches, Drafting Scales .... $1\frac{1}{8}$ " 18 00
25. Arch Joint, Arch Back, Edge Plates, 8ths, 10ths, and 16ths of inches, 100ths of a foot, Drafting and Octagonal Scales [see Engraving, p. 17], $1\frac{1}{8}$ " 13 00
26. Arch Joint, Edge Plates, Board Measure, 16ths of inches, Drafting Scales .... $1\frac{1}{8}$ " 11 00
27. Arch Joint, Middle Plates, Bitted, 8ths and 16ths of inches, Drafting and Octagonal Scales .... $1\frac{1}{8}$ " 9 00

Two Feet, Four Fold, Narrow.

45. Arch Joint, Edge Plates, 8ths and 16ths of inches, 1 in. wide, 8 00
46. Arch Joint, Middle Plates, 8ths and 16ths of inches, 1 " 6 00
34. Square Joint, Square Back Plates, Edge Plates, 8ths and 16ths of inches .... 1 " 10 00
35. Square Joint, Bound, 8ths and 16ths of inches, Drafting and Octagonal Scales .... 1 " 15 00
37. Square Joint, Edge Plates, 8ths and 16ths of inches [see Engraving, p. 17] .... 1 " 7 00
38. Square Joint, Middle Plates, 8ths and 16ths of inches .... 1 " 5 00
41. Round Joint, Middle Plates, 8ths and 16ths of inches .... 1 " 4 00

Two Feet, Four Fold, Extra Narrow.

31. Arch Joint, Edge Plates, 8ths, 10ths, and 16ths of inches .... $\frac{3}{4}$ in. wide, 9 00
72. Arch Joint, Bound, 8ths, 10ths, 12ths, and 16ths of inches .... $\frac{3}{4}$ " 17 00
33. Square Joint, Edge Plates, 8ths, 10ths, and 16ths of inches .... $\frac{3}{4}$ " 8 00
36. Square Joint, Middle Plates, 8ths and 16ths of inches .... $\frac{3}{4}$ " 5 50

Two Feet, Six Fold, Extra Narrow.

$28\frac{1}{2}$. Arch Joint, Edge Plates, 8ths, 10ths, 12ths, and 16ths of inches, 100ths of a foot .... $\frac{3}{4}$ in. wide, 13 50

One Foot, Four Fold.

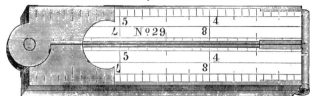

No.		Per Dozen.
29. Arch Joint, Bound, 8ths, 10ths, 12ths, and 16ths of inches..................13-16ths in. wide,		$13 00
30. Arch Joint, Edge Plates, 8ths, 10ths, 12ths, and 16ths of inches, 100ths of a foot.....13-16ths	"	8 00
39. Square Joint, Edge Plates, 8ths and 16ths of inches.............................13-16ths	"	6 00
40. Round Joint, Middle Plates, 8ths and 16ths of inches...............................13-16ths	"	3 25
74. Square Joint, Bound, 8ths and 16ths of inches................................5-8ths	"	12 00
75. Square Joint, Middle Plates, 8ths and 16ths of inches................................5-8ths	"	4 00
43. Round Joint, Middle Plates, 8ths and 16ths of inches................................5-8ths	"	3 00

One Foot, Four Fold, Caliper.

29½. Arch Joint, Bound, 8th's, 10ths, 12ths, and 16ths of inches................... 13-16ths in. wide,		18 00
30½. Arch Joint, Edge Plates, 8ths, 10ths, 12ths, and 16ths of inches, 100ths of a foot.....13-16ths	"	12 50

Six Inch, Two Fold, Caliper.

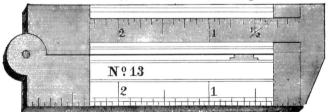

12. Square Joint, Brass Case, Spring Caliper, 8ths and 16ths of inches....................1⅛ in. wide,		12 00
13. Square Joint, 8ths and 16ths of inches.... .1⅛	"	8 00
13½. Square Joint, 8ths and 16ths of inches, 13-16ths	"	7 00

Stearns' Ivory Rules.

Ivory, One Foot, Four Fold.

No.		Per Dozen. Unbound.	Bound.
51.	Arch Joint, Edge Plates, German Silver, 8ths, 10ths, 12ths, and 16ths of inches (100ths of a foot on edges of unbound)..13-16ths in. wide,	$30 00	$40 00
52.	Square Joint, Edge Plates, German Silver, 8ths, 10ths, 12ths, and 16ths of inches (100ths of a foot on edges of unbound)..13-16ths in. wide,	27 00	37 00

Ivory, One Foot, Four Fold, Caliper.

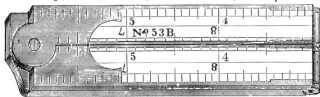

53.	Arch Joint, Edge Plates, German Silver, 8ths, 10ths, 12ths, and 16ths of inches (100ths of a foot on edges of unbound)...13-16ths in. wide,	35 00	45 00
54.	Square Joint, Edge Plates, German Silver, 8ths, 10th, 12ths, and 16ths of inches (100ths of a foot on edges of unbound)..13-16ths in. wide,	32 00	42 00

Ivory, Six Inch, Two Fold, Caliper.

55.	Square Joint, German Silver, 8ths and 16ths of inches................13-16ths in. wide,	15 00
55½.	Square Joint, German Silver Case, Spring Caliper, 8ths and 16ths of inches, 13-16ths in. wide,	18 00

Ivory, One Foot, Four Fold.

57.	Square Joint, Edge Plates, German Silver, 8ths and 16ths of inches......... 5-8ths in. wide,	18 00	28 00
58.	Square Joint, Edge Plates, Brass, 8ths and 16ths of inches..............5-8ths in. wide,	15 00	
59.	Round Joint, Middle Plates, Brass, 8ths and 16ths of inches...........................	12 00	

Ivory, Two Feet, Four Fold, Broad.

No.		Per Dozen.	
		Unbound.	Bound.
47.	Arch Joint, Triple Plated Edge Plates, German Silver, 8ths, 10ths, 12ths, and 16ths of inches, 100ths of a foot, Drafting and Octagonal Scales....................1½ in. wide,	$95 00	$110 00

Ivory, Two Feet, Four Fold, Medium.

48.	Arch Joint, Edge Plates, German Silver, 8ths, 10ths, 12ths, and 16ths of inches (100ths of a foot on edges of unbound), Drafting Scales......................1⅛ in. wide,	75 00	85 00

Ivory, Two Feet, Four Fold, Narrow.

50.	Square Joint, Edge Plates, German Silver, 8ths, 10ths, 12ths, and 16ths of inches (100ths of a foot on edges of unbound), Drafting Scales....................... 1 inch wide,	65 00	75 00

Ivory, Two Feet, Four Fold, Extra Narrow.

56.	Arch Joint, Edge Plates, German Silver, 8ths, 10ths, 12ths, and 16ths of inches (100ths of a foot on edges of unbound).......¾ inch wide,	55 00	65 00

Ivory, Two Feet, Six Fold, Extra Narrow.

60.	Arch Joint, Edge Plates, German Silver, 8ths, 10ths, 12ths, and 16ths of inches (100ths of a foot on edges of unbound) .....¾ inch wide,	80 00	90 00

Miscellaneous Articles.

No.			Per Dozen.
63½.	Bench Rule, Boxwood, Bound, 8ths and 16ths of inches ....24 in. long,		$18 00
64.	Bench Rule, Maple, Capped Ends, 8ths and 16ths of inches.......................24	"	4 00
71.	Yardstick, Maple, Capped Ends............36	"	4 00
80.	Saddlers' Rule, Maple, Capped Ends, 8ths and 16th of inches, 1½ inch wide. 36	"	9 00
81.	Pattern Makers' Shrinkage Rule, Boxwood, 8ths and 16ths of inches.... ....24¼	"	15 00
82.	Pattern Makers' Shrinkage Rule, Two Fold, Boxwood, Triple Plated Edge Plates, 8ths and 16ths of inches........................24¼	"	18 00

DESCRIPTION OF THE
Patent Improved Adjustable Plumbs and Levels,
MANUFACTURED BY THE
STANLEY RULE AND LEVEL CO.

[Sectional Drawing.]

The Spirit-glass (or bubble tube), in the Level, is set in a Metallic Case, which is attached to the Brass Top-plate above it—at one end by a substantial hinge, and at the opposite end by an Adjusting Screw, which passes down through a flange on the Metallic Case. Between this flange and the Top-plate above, is inserted a stiff spiral spring; and by driving, or slacking the Adjusting Screw, should occasion require, the Spirit-glass can be instantly adjusted to a position parallel with the base of the level.

The Spirit-glass in the Plumb, is likewise set in a Metallic Case attached to the Brass Top-plate at its outer end. By the use of the Adjusting Screw at the lower end of the Top-plate, the Plumb-glass can be as readily adjusted to a right angle with the base of the Level, if occasion requires, and by the same method as adopted for the Level-glass.

The simplicity of this improved method of adjusting the Spirit-glasses will commend itself to every Mechanic. But one screw is used in the operation, and the action of the Brass Spring is perfectly reliable under all circumstances.

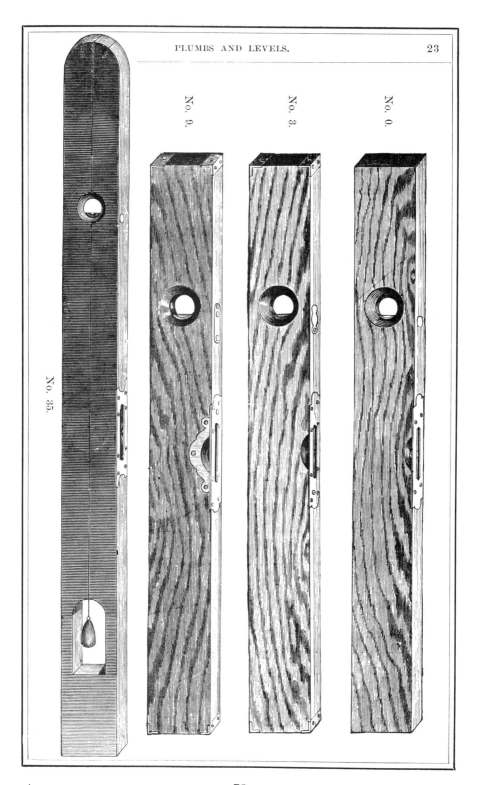

Plumbs and Levels.

No.		Per Dozen.
102.	Levels, Arch Top Plate, Two Side Views, Polished, Assorted...............10 to 16 inch,	$9 00
103.	Levels, Arch Top Plate, Two Side Views, Polished, Assorted...............18 to 24 "	12 00
00.	Plumb and Level, Arch Top Plate, Two Side Views, Polished, Assorted............18 to 24 "	16 00
0.	Plumb and Level, Arch Top Plate, Two Side Views, Polished, Assorted [see Engraving, p. 23]...............24 to 30 "	18 00
1½.	Mahogany Plumb and Level, Arch Top Plate, Two Side Views, Polished, Assorted....18 to 24 "	16 50
2½.	Plumb and Level, Arch Top Plate, Two Brass Lipped Side Views, Polished, Assorted...24 to 30 "	25 00
3½.	Plumb and Level, Arch Top Plate, Two Side Views, Polished and Tipped, Assorted..24 to 30 "	30 00
4½.	Plumb and Level, Arch Top Plate, Two Brass Lipped Side Views, Polished and Tipped, Assorted...............24 to 30 "	37 00
6½.	Mahogany Plumb and Level, Arch Top Plate, Two Brass Lipped Side Views, Polished, Assorted...............24 to 30 "	31 00
7.	Masons' Plumb and Level, Arch Top Plate, Two Side Views, Polished and Tipped.......36 "	36 00
12.	Machinists' Brass Bound Rosewood Plumb and Level, Two Brass Side Views, Polished........20 "	115 00
43.	Iron Plumb and Level, Two Side Views, Brass Top Plate...............9 "	20 00
45.	Machinists' Iron Level, Brass Top Plate, Extra Finish...............9 "	28 00

PATENT IMPROVED

Adjustable Plumbs and Levels.

No.   Per Dozen.

1. Patent Adjustable Mahogany Plumb and Level, Arch Top Plate, two Side Views, Polished, Assorted..........................26 to 30 inch, $27 00

2. Patent Adjustable Plumb and Level, Arch Top Plate, two Brass Lipped Side Views, Polished, Assorted..........................26 to 30   "   27 00

3. Patent Adjustable Plumb and Level, Arch Top Plate, two Side Views, Polished and Tipped, Assorted [see Engraving, p. 23]........26 to 30   "   32 00

4. Patent Adjustable Plumb and Level, Arch Top Plate, two Brass Lipped Side Views, Polished and Tipped, Assorted................26 to 30   "   39 00

5. Patent Adjustable Plumb and Level, Triple Stock, Arch Top Plate, two Ornamental Brass Lipped Side Views, Polished and Tipped, Assorted..........................26 to 30   "   48 00

6. Patent Adjustable Mahogany Plumb and Level, Arch Top Plate, two Brass Lipped Side Views, Polished, Assorted...................26 to 30   "   33 00

9. Patent Adjustable Mahogany Plumb and Level, Arch Top Plate, two Ornamental Brass Lipped Side Views, Polished and Tipped, Assorted [see Engraving, p. 23]............... 26 to 30   "   48 00

10. Patent Adjustable Mahogany Plumb and Level, Triple Stock, two Ornamental Brass Lipped Side Views, Arch Top Plate, Polished and Tipped, Assorted...... ....26 to 30   "   60 00

11. Patent Adjustable Rosewood Plumb and Level, Arch Top Plate, two Ornamental Brass Lipped Side Views, Polished and Tipped, Ass'd..26 to 30   "   90 00

25. Patent Adjustable Mahogany Plumb and Level, Arch Top Plate, Improved Double Adjusting Side Views, Polished and Tipped............30   "   54 00

32. Patent Adjustable Mahogany Graduated Plumb and Level (easily adjusted to work at any angle, or elevation, required)...............28   "   100 00

35. Patent Adjustable Masons' Plumb and Level [see Engraving, p. 23], $3\frac{3}{4}$ inch wide........42   "   36 00

NICHOLSON'S PATENT
Metallic Plumbs and Levels.

The chief points of excellence in these Levels are the following: Once accurately constructed, they always remain so; the sides and edges are perfectly true, thus combining with a Level a convenient and reliable *straight edge;* shafting, or other work overhead, may be lined with accuracy by sighting the *bubble* from below, through an aperture in the base of the Level. Other points are their lightness, strength, and convenience of handling. All Levels, being thoroughly tested, are warranted reliable.

No.		Per Dozen.
13.	14 inches long	$21 00
14.	20 " "	25 00
15.	24 " "	32 00

IMPROVED
Iron Frame Plumbs and Levels.

These Levels have a Cast-Iron Frame, thoroughly braced, and combine the greatest strength with the least possible weight of metal. They are made absolutely correct in their form, and are subject to no change from *warping* or *shrinking*. The sides are inlaid with Black Walnut, secured by the screws which pass from one to the other, through the Iron braces, and the sides may be easily removed, if occasion requires.

No.			Per Dozen.
48.	Patent Adjustable Plumb and Level, Iron Frame, with Black Walnut (Inlaid) Sides	12 inch,	$27 00
49.	Patent Adjustable Plumb and Level, Iron Frame, with Black Walnut (Inlaid) Sides	18 "	36 00

Pocket Levels.

IMPROVED PATTERNS.

No. 41½.

No.		Per Dozen.
40.	Cast Iron Top Plate, Japanned, 1 dozen in a box	$2 50
40½.	Cast Iron Top Plate, Extra Finish, 1 dozen in a box	3 00
41.	Brass Top Plate, 1 dozen in a box	3 00
41½.	Cast Brass Top Plate, Extra Thick, 1 dozen in a box	3 50
42.	All Brass, Pocket Level, 1 dozen in a box	8 00

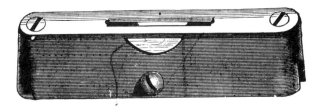

46. Iron, Polished, Brass Top Plate, a superior article 3 50

LEVEL GLASSES.

Per Gross.

Level Glasses, packed in ¼ gross boxes	1¾ inch,	$12 00
Level Glasses, " " "	2 "	12 00
Level Glasses, " " "	2½ "	12 75
Level Glasses, " " "	3 "	13 50
Level Glasses, " " "	3½ "	15 00
Level Glasses, " " "	4 "	16 50
Level Glasses, " " "	4½ "	18 00
Assorted, 1¾, 3, and 3½ inch, in ¼ gross boxes		14 50

Plated Try Squares, No. 2.

		Per Dozen.
Rosewood, 1 dozen in a box 3	inch,	$3 00
Rosewood, ½ " 4½	"	3 75
Rosewood, ½ " 6	"	5 00
Rosewood, ½ " 7½	"	5 75
Rosewood, ½ " 9	"	6 50
Rosewood, ½ " 12	"	8 50
Rosewood, with Rest, ½ dozen in a box............15	"	12 50
Rosewood, with Rest, ½ " 18	"	15 50

Plumb and Level Try Squares.

A very convenient Tool for Mechanics. A spirit glass being set in the inner edge of the Try Square Handle, constitutes the Handle a Level; and when the Handle is brought to an exact level, the blade of the square will be upright, and become a perfect Plumb.

No.		Per Dozen.
1. 7½ inch, Rosewood, with extra heavy Brass Mountings.....		$15 00
1. 12 inch, Rosewood, with extra heavy Brass Mountings.....		24 00
2. 7½ inch, Rosewood, with extra heavy Brass Mountings.....		7 50

½ dozen in a box.

Improved Try Squares, No. 1.

Iron Frame Handle, with Rosewood (Inlaid) Sides, Steel Blades, Square inside and out.

Inches,	3	4	6	8	9	10	12
Per Dozen,	$8 00	9 00	11 00	14 00	16 50	20 00	26 00

½ dozen in a box.

PATENT

Improved Try Squares, No. 2.

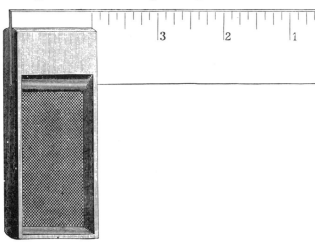

Iron Handle, Graduated Steel Blade, Square inside and out.

Inches,	4	6	8	10	12
Per Dozen,	$3 50	4 50	5 50	6 50	8 00

½ dozen in a box.

Improved Mitre Try Squares.

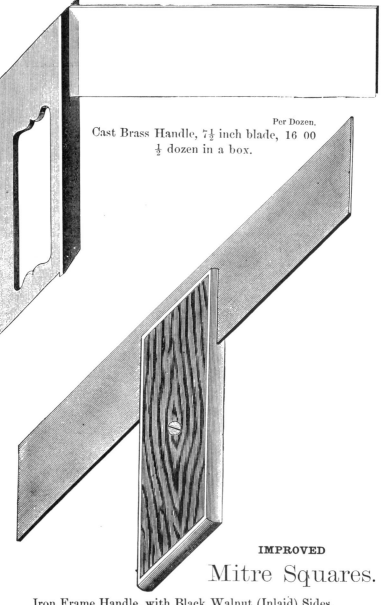

Per Dozen.
Cast Brass Handle, 7½ inch blade, 16 00
½ dozen in a box.

IMPROVED Mitre Squares.

Iron Frame Handle, with Black Walnut (Inlaid) Sides.

Inches,	8	10	12
Per Dozen,	$9 00	$10 00	$11 00

½ dozen in a box.

WINTERBOTTOM'S PATENT
Combined Try and Mitre Square.

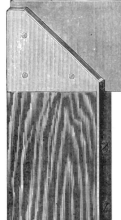

No. 2.

This Tool can be used with equal convenience and accuracy, as a Try Square, or a Mitre Square. By simply changing the position of the handle, and bringing the mitred face at the top of the handle against one edge of the work in hand, a perfect mitre, or angle of forty-five degrees, can be struck from either edge of the blade.

No. 1. Iron Frame Handle, with Black Walnut (Inlaid) Sides, Graduated Steel Blades.

Inches,	4	6	8
Per Dozen,	$6 00	7 50	9 00

No. 2. Rosewood Handle, with Steel Blades.

Inches,	$4\tfrac{1}{2}$	6	$7\tfrac{1}{2}$
Per Dozen,	$4 00	5 00	6 00

½ dozen in a box.

BAILEY'S
Patent Adjustable Planes.

No. 15½,

Two additional Block Planes may now be included in the List contained on page 39 of our Catalogue. They are adjusted by a Screw and Lever movement, as shown in the engraving. The handle to No. 15½ is secured by the use of a thumb-nut; and it may be easily attached to, or liberated from the Plane, as the convenience of the workman may require.

No.   (No. 15 is the same Plane, without the handle.)   Each.
15. Block Plane, 7 inches in Length, 1¾ inch Cutter $2 50
15½. "   "   7   "   "   1¾   "   " Rosewood H'dle, 3 00

(Nos.15 & 15-1/2 Block Planes are believed to have been introduced during 1875)

TRY AND MITRE SQUARE, ETC.   30½

WINTERBOTTOM'S PATENT

Combined Try and Mitre Square.

This Tool can be used with equal convenience and accuracy, as a Try Square, or a Mitre Square. By simply changing the position of the handle, and bringing the mitred face at the top of the handle against one edge of the work in hand, a perfect mitre, or angle of forty-five degrees, can be struck from either edge of the blade.

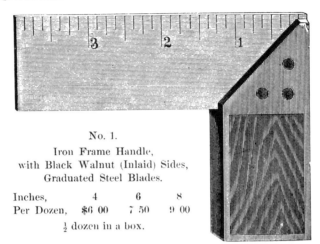

No. 1.
Iron Frame Handle,
with Black Walnut (Inlaid) Sides,
Graduated Steel Blades.

Inches,	4	6	8
Per Dozen,	$6 00	7 50	9 00

½ dozen in a box.

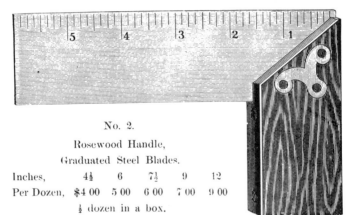

No. 2.
Rosewood Handle,
Graduated Steel Blades.

Inches,	4½	6	7½	9	12
Per Dozen,	$4 00	5 00	6 00	7 00	9 00

½ dozen in a box.

Improved Iron Block Plane.

No. 110   7½ inches long, 1¾ inch Cutter..................Per Dozen, $9 00
        Cutters, warranted Cast Steel......................   "   2 00

(The No.110 Block Plane is believed to have been introduced in 1875, after the No.15)

PATENT COMBINATION

Try Square and Bevel.

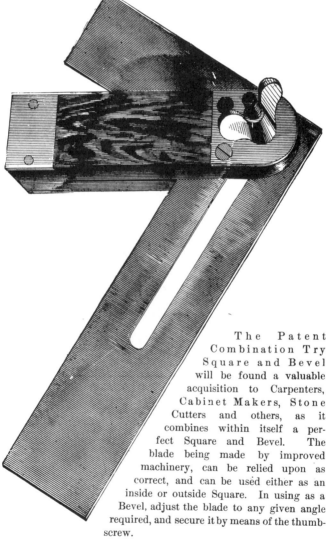

The Patent Combination Try Square and Bevel will be found a valuable acquisition to Carpenters, Cabinet Makers, Stone Cutters and others, as it combines within itself a perfect Square and Bevel. The blade being made by improved machinery, can be relied upon as correct, and can be used either as an inside or outside Square. In using as a Bevel, adjust the blade to any given angle required, and secure it by means of the thumb-screw.

Inches,	4	6	8	10	12
Per Dozen,	$6 50	7 50	9 00	10 50	12 00

½ dozen in a box.

Plated Sliding T Bevels.

ROSEWOOD, WITH BRASS THUMB SCREW.

	Per Dozen.
6 inch	$5 50
8 "	6 00
10 "	6 50
12 "	7 00
14 "	7 50

½ dozen in a box.

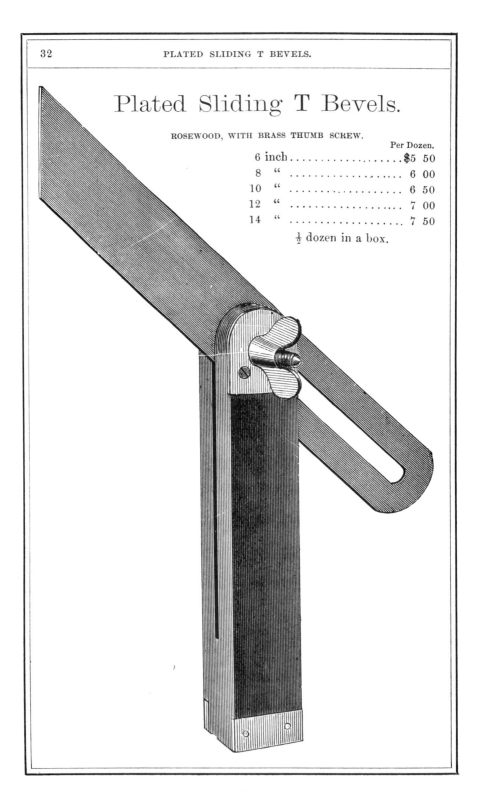

Bailey's Patent Flush T Bevel.

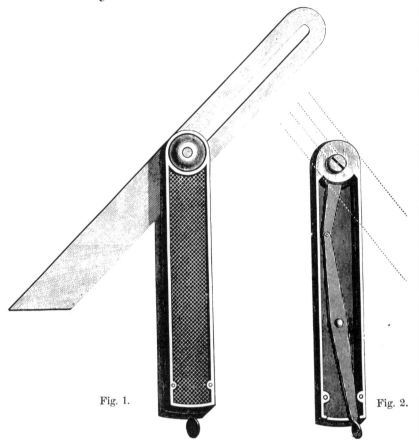

Fig. 1. Fig. 2.

The Handle is made of Cast Iron. The Blade is of fine quality Steel, spring temper, and with perfectly parallel edges.

The method of securing the Blade at any angle desired, is simple, yet very effective. Fig. 1 represents the Bevel complete. Fig. 2 shows the internal construction of the Handle. By moving the Thumb-piece at the lower end of the Handle, the long Lever acts upon the shorter one by means of strong pivots. The strength of a compound Lever is produced by this arrangement, and the short Lever being attached to a Nut inside the upper end of the Handle, operates as a wrench, to turn it upon the Screw, and thus to fasten, or release, the Blade at the pleasure of the owner.

PRICES:

Inches,	8	10
Per Dozen,	$15 00	$17 00

½ dozen in a box.

Patent Eureka Flush T Bevel.

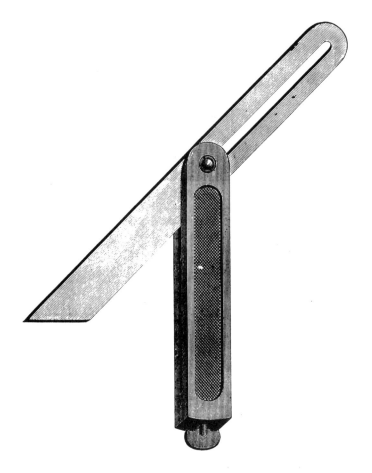

Iron Handle, Steel Blade, with Parallel Edges.

The Blade is easily secured at any angle, by turning the Thumb screw at the lower end of the Handle.

PRICES:

Inches,	6	8	10
Per Dozen,	$8 00	$9 00	$10 00

½ dozen in a box.

Gauges.

☞ Particular attention is invited to the improved features in our Gauges.

All Marking Gauges, excepting Nos. 61 and 61½, have an Adjusting Point of finely-tempered steel, which may be readily removed and replaced if it needs sharpening. The point can be thrust down as it wears away, or if by any means it be broken off, it can be easily repaired.

All Gauges with Brass Thumb Screws, excepting No. 68, have also a Brass Shoe inserted in the head, under the end of the Thumb Screw. This shoe protects the gauge-bar from being dented by the action of the screw; and the broad surface of the shoe being in contact with the bar, the head is held more firmly in position than by any other method, and with less wear of the screw threads.

The engraving below (No. 60) shows an Iron Marking and Cutting Gauge. No. 60½ is the same in form, but has a Reversible Brass Slide slotted into the face of the bar. When a Mortise Gauge is required, the brass slide may be turned over in the bar. The point in the brass slide may be moved to any position, and the slide will be secured by a single turn of the screw which fastens the head of the Gauge.

No. 60.

No.		Per Dozen.
60	Patent Iron Marking and Cutting Gauge, Oval Bar, Marked, Adjusting Steel Point, equally well adapted for use on Metals, or Wood, ½ dozen in a box..............................	$5 50
60½	Patent Iron Reversible Gauge, Mortise, Marking and Cutting combined, Brass Slide, Oval Bar, Marked, Adjusting Steel Point, equally well adapted for use on Metals, or Wood, ½ dozen in a box..............................	9 00

No. 61.

61.	Marking Gauge, Beechwood, Boxwood Thumb Screw, Oval Bar, Marked, Steel Point, 1 dozen in a box..................	1 00
61½	Marking Gauge, Beechwood, Boxwood Thumb Screw, Oval Head and Bar, Marked, Steel Point, 1 dozen in a box..........	1 50

GAUGES.

No. 65.

No.		Per Dozen.
62.	Patent Marking Gauge, Beechwood, Polished, Boxwood Thumb Screw, Oval Bar, Marked, Adjusting Steel Point, 1 dozen in a box....	$2 00
64.	Patent Marking Gauge, Polished, Plated Head, Boxwood Thumb Screw, Oval Bar, Marked, Adjusting Steel Point, 1 dozen in a box	2 75
64½	Patent Marking Gauge, Polished, Oval Plated Head, Brass Thumb Screw and Shoe, Oval Bar, Marked, Adjusting Steel Point, ½ dozen in a box	5 00
65.	Patent Marking Gauge, Boxwood, Polished, Plated Head, Brass Thumb Screw and Shoe, Oval Bar, Marked, Adjusting Steel Point, ¼ dozen in a box	5 00
66.	Patent Marking Gauge, Rosewood, Oval Plated Head and Bar, Brass Thumb Screw and Shoe, Oval Bar, Marked, Adjusting Steel Point, ½ dozen in a box	6 00

No. 71.

71.	Patent Double Gauge (Marking and Mortise Gauge combined), Beechwood, Polished, Plated Head and Bars, Brass Thumb Screws and Shoes, Oval Bars, Marked, Steel Points, ½ dozen in a box	9 00
72.	Patent Double Gauge (Marking and Mortise Gauge combined), Beechwood, Polished, Boxwood Thumb Screws, Oval Bars, Marked, Steel Points, ½ dozen in a box	4 00
74.	Patent Double Gauge (Marking and Mortise Gauge combined), Boxwood, Polished, Full Plated Head and Bars, Brass Thumb Screws and Shoes, Oval Bars, Marked, Steel Points, ½ dozen in a box	14 00
70.	Cutting Gauge, Mahogany, Polished, Plated Head, Boxwood Thumb Screw, Oval Bar, Marked, Steel Cutter, 1 dozen in a box	4 00
83.	Handled Slitting Gauge, 17-inch Bar, Marked, ½ dozen in a box	9 00
84.	Handled Slitting Gauge, with Roller, 17-inch Bar, Marked, ½ dozen in a box	11 00

GAUGES.

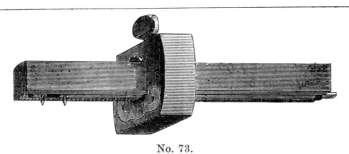

No. 73.

No.		Per Dozen.
73.	Patent Mortise Gauge, Boxwood, Polished, Plated Head, Brass Slide, Brass Thumb Screw and Shoe, Oval Bar, Marked, Steel Points, ½ dozen in a Box..............................	$8 00
76.	Patent Mortise Gauge, Boxwood, Polished, Plated Head, Screw Slide, Brass Thumb Screw and Shoe, Oval Bar, Marked, Steel Points, ½ dozen in Box..............................	11 00
80.	Patent Mortise Gauge, Boxwood, Full Plated Head, Plated Bar, Screw Slide, Brass Thumb Screw and Shoe, Marked, Steel Points, ½ dozen in a Box..............................	18 00
67.	Mortise Gauge, Adjustable Wood Slide, Boxwood Thumb Screw, Oval Bar, Marked, Steel Points, ½ dozen in a Box..........	4 50
68.	Mortise Gauge, Plated Head, Adjustable Wood Slide, Brass Thumb Screw, Oval Bar, Marked, Steel Points, ½ dozen in a Box..	6 50

No. 77.

77.	Patent Mortise Gauge, Rosewood, Plated Head, Improved Screw Slide, Brass Thumb Screw and Shoe, Oval Bar, Marked, Steel Points, ½ dozen in a Box..............................	10 00
78.	Patent Mortise Gauge, Rosewood, Plated Head, Screw Slide, Brass Thumb Screw and Shoe, Oval Bar, Marked, Steel Points, ½ dozen in a Box..............................	11 00
79.	Patent Mortise Gauge, Rosewood, Plated Head and Bar, Screw Slide, Brass Thumb Screw and Shoe, Marked, Steel Points, ½ dozen in a box......................	13 00
85.	Panel Gauge, Beechwood, Boxwood Thumb Screw, Oval Bar, Steel Point, 1 dozen in a box........	3 20
85½.	Panel Gauge, Rosewood, Plated Head and Bar, Brass Thumb Screw, Steel Point, ½ dozen in a box..........	18 00
90.	Williams' Patent Combination Gauge, ½ dozen in a Box...	22 00

Baileys' Patent Planes.

The sale of over **80,000** of these Planes already, and a constantly increasing demand for them, is the best evidence that all dealers in Mechanics' Tools can have of the estimation in which "Bailey's Patent Adjustable Planes" are held by the men who use them. A few of the numerous testimonials to the superiority of these Tools, from practical men in all branches of wood-working, will be found on page 41 of this Catalogue.

☞ The Planes are in perfect working order when sent into market; and a printed description of the method of adjustment accompanies each Plane.

Iron Planes.

No.							Each.	
1. Smooth Plane,	$5\frac{1}{2}$	inches in length,	$1\frac{1}{4}$	inch Cutter.			~~$3 50~~	3 00
2. "	"	7	"	"	$1\frac{5}{8}$	" "	~~4 00~~	3 50
3. "	"	8	"	"	$1\frac{3}{4}$	" "	~~4 50~~	4 00
4. "	"	9	"	"	2	" "	~~5 00~~	4 50

5. Jack Plane, 14 inches in length, 2 inch Cutter.	~~5 50~~	5 00
6. Fore " 18 " " $2\frac{3}{8}$ " "	~~6 50~~	6 00
7. Jointer " 22 " " $2\frac{3}{8}$ " "	~~7 50~~	7 00
8. " " 24 " " $2\frac{5}{8}$ " "	~~8 50~~	8 00

9. Block Plane, 10 inches in Length, 2 inch Cutter	8 00	7 50
10. Carriage Makers' Rabbet Plane, 14 inches in Length, $2\frac{1}{8}$ inch Cutter	6 00	6 00
11. Belt Makers' Plane, $2\frac{3}{8}$ inch Cutter	4 00	

(The lower prices inked in the right margin are believed to have been reduced during 1875)

BAILEY'S PATENT PLANES. 39

Iron Planes.
(CONTINUED.)

Each.
No. 9½. Excelsior Block Plane, 6 inches in Length, 1¾ in. Cutter, $2 00

No. 9¾. Excelsior Block Plane, with Rosewood handle, 6 inches in Length, 1¾ inch Cutter... .... ~~2 50~~ 2,25

Bailey's Patent Adjustable Veneer Scraper.

No. 12. 3 inch Cutter... ~~5 00~~ 4,50
CAST STEEL, Hand, VENEER SCRAPERS, 3x5 inch, per dozen.... 4 50

Bailey's Patent Adjustable Circular Plane.

No. 13. 1¾ inch Cutter.. ~~5 00~~ 4,50

This Plane has a *Flexible Steel Face*, and by means of the thumb screws at each end of the Stock, can be easily adapted to plane circular work—either concave or convex.

(The lower prices inked in the right margin are believed to have been reduced during 1875)

Wood Planes.

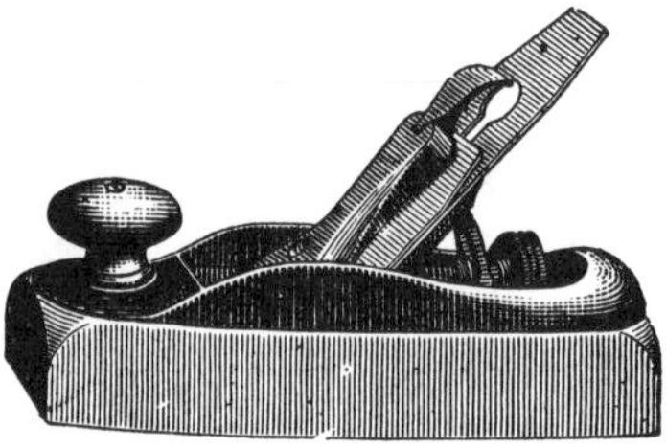

No.								Each.	
21.	Smooth Plane,	7	inches in Length,	$1\frac{3}{4}$	inch	Cutter	....	~~$3 00~~	2.50
22	"	"	8	"	"	$1\frac{3}{4}$	" "	...1.50.. ~~3 00~~	2.50
23.	"	"	9	"	"	$1\frac{3}{4}$	" "	.... ~~3 00~~	2.50
24.	"	"	8	"	"	2	" "	.... ~~3 25~~	2.75
25.	Block	"	$9\frac{1}{2}$	"	"	$1\frac{3}{4}$	" "	.... ~~3 25~~	2.75

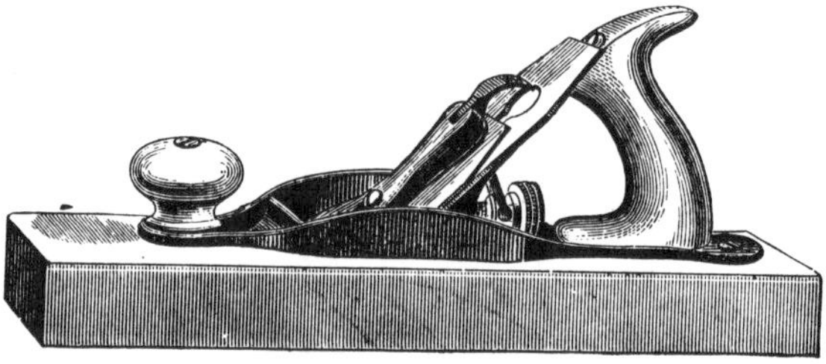

26.	Jack Plane,	15	inches in Length,	2	inch	Cutter	....	~~3 50~~	3.00
27.	"	"	15	"	"	$2\frac{1}{8}$	" "	....	~~3 75~~	3.25
28.	Fore	"	18	"	"	$2\frac{3}{8}$	" "	....	~~4 00~~	3.50
29.	"	"	20	"	"	$2\frac{3}{8}$	" "	....	~~4 00~~	3.50
30.	Jointer	"	22	"	"	$2\frac{3}{8}$	" "	....	~~4 25~~	3.75
31.	"	"	24	"	"	$2\frac{3}{8}$	" "	....	~~4 25~~	3.75
32.	"	"	26	"	"	$2\frac{5}{8}$	" "	....	~~4 75~~	4.25
33.	"	"	28	"	"	$2\frac{5}{8}$	" "	....	~~4 75~~	4.25
34.	"	"	30	"	"	$2\frac{5}{8}$	" "	....	~~5 00~~	4.50

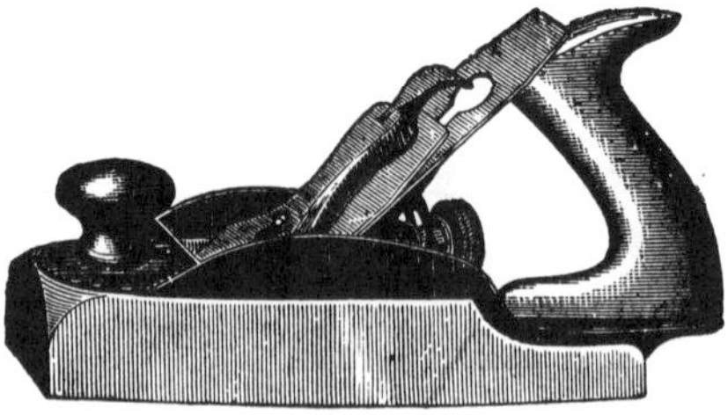

35.	Handle Smooth,	9	inches in Length,	2	inch	Cutter	....	4 00	3.50
36.	"	"	10	"	"	$2\frac{3}{8}$	" "	....	4 25	3.75
37.	Jenny	"	13	"	"	$2\frac{5}{8}$	" "	....	4 50	4.00

☞ Extra Plane-woods, of every style, can be supplied cheaply.

(The lower prices inked in the right margin are believed to have been reduced during 1875)

Bailey's Patent Plane Irons.

☞ These Plane-Irons have stood the severest tests applied to them by practical workmen, and are by the manufacturers fully WARRANTED.

PRICES:

SINGLE IRONS.

Inches,	1¼	1⅝	1¾	2	2⅛	2⅜	2⅝
Per Dozen,	$4.50	5.00	5.50	6.00	6.25	6.50	7.00

DOUBLE IRONS.

Per Dozen,	$7.75	8.25	9.00	9.50	10.00	10.50	11.00

☞ Each Plane-Iron is sharpened and put in perfect working order before leaving the Factory.

TESTIMONIALS.

"Works splendidly on our Yellow Pine."—H. C. HALL, *Jacksonville, Florida.*

"Far superior to any I have seen used."—JOHN ENGLISH, *North Milwaukie, Wis., Car Shops.*

"The BEST PLANE now in use."—CHICKERING & SONS, *Piano Forte Manufacturers, Boston, Mass.*

"Superior to any others used in this Shop."—W. S. TOWN, *Master of Car Repairs, Hudson River R. R. Co.*

"All that is wanted to sell your planes here is to show them."—E. WATERS, *Manufacturer of Sporting Boats, Troy, N. Y.*

"We know of no other plane that would be a successful substitute for them."—R. BURDETT & Co., *Organ Manufacturers, Erie, Pa.*

"The best plane ever introduced in carriage-making."—JAS. L. MORGAN, *at J. B. Brewster & Co.'s Carriage Manufactory, New York.*

"From the satisfaction they give would like to introduce them here."—J. ALEXANDER, *at Bell's Melodeon Factory, Guelph, Canada.*

"I have used your planes and have never found their equal."—B. F. FURRY, *Foreman McLear & Kendall's Coach Works, Wilmington, Del.*

"Bailey's Planes are used in our Factory. We can get only one verdict, and that is, the men would not be without them."—STEINWAY & SON, *Piano Forte Manufacturers, New York.*

"I have owned a set of Bailey's Planes for six months, and can cheerfully say that I would not use any others, for I believe they are the best in use."—C. C. HARRIS, *Stair Builder, St. Louis, Mo.*

"We have Bailey's Planes in our Factory, and there is but one opinion about them. They are the best and cheapest tool we have ever used. Good tools are always the cheapest."—JOS. BECKHAUS, *Carriage Builder, Philadelphia.*

"I have used a set of Bailey's Planes about six months, and would not use any other now. There is inquiry about them here every day. Send me your descriptive Catalogue.—EDWARD CARVILLE, *Sacramento City, California.*

A Boston Mechanic says : "I always tell my shopmates, when they wish to use my Plane, not to borrow a BAILEY PLANE unless they intend to buy one, as they will never be satisfied with any other Plane after using this."

Bailey's Iron Spoke Shaves.

☞ The SPOKE SHAVES in the following List, are superior in style, quality, and finish, to any in market. The CUTTERS are made of the best English CAST STEEL, tempered and ground by an improved method, and are in perfect working order when sent from the Factory.

No.   Per Dozen.
51. Patent Double Iron, Raised Handle, 10 inch, $2\frac{1}{8}$ inch Cutter, $4 00

52. Patent Double Iron, Straight Handle, 10 inch, $2\frac{1}{8}$ inch Cutter,   4 00

53. Patent Adjustable, Raised Handle, 10 inch, $2\frac{1}{8}$ inch Cutter,   5 00

54. Patent Adjustable, Straight Handle, 10 inch, $2\frac{1}{8}$ inch Cutter,   5 00

55. Model Double Iron, Hollow Face, 10 inch, $2\frac{1}{8}$ inch Cutter,   4 00

56. Coopers' Spoke Shave, 18 inch, $2\frac{5}{8}$ inch Cutter............8 00
$56\frac{1}{2}$ Coopers' Spoke Shave (Heavy), 19 inch, 4 inch Cutter, 10 00

57. Coopers' Spoke Shave (Light), 18 inch, $2\frac{1}{8}$ inch Cutter,   5 00

Bailey's Iron Spoke Shaves.

(CONTINUED.)

No.		Per Dozen.
58.	Model Double Iron, 10 inch, 2¼ inch Cutter	$3 50
59.	Single Iron (Pattern of No. 56), 10 in., 2⅛ in. Cutter	4 00
60.	Double Cutter, Hollow and Straight, 10 inch, 1½ in. Cutters,	5 00

Price List of Spoke Shave Cutters.

No. 51 52 53 54 55 57 58 59	1 50
No. 60 (in pairs)	2 25
No. 56	2 00
No. 56½	2 50

WHEELER'S

Patent Countersink.

FOR WOOD.

OVER 20,000 HAVE BEEN SOLD.

The Bit of this Countersink is in the shape of a hollow eccentric cone, thus securing a cutting edge of uniform draft from the point to the base of the tool, and obviating the tendency of such a tool, to lead off into the wood at its cutting edge, and to leave an angular line where it ceases to cut. The form of the tool between the cutter and the shank is that of a hollow half-cone, inverted, thus leaving ample space, just back of and above the cutter, for the free escape of shavings. This Countersink works equally well for every variety of screw, the pitch of the cone being the same as the taper given to the heads of all sizes of screws, thereby rendering only a single tool necessary for every variety of work. The Countersink cuts rapidly, and is easily sharpened by drawing a thin file lengthwise inside of the cutter. The ingenious method of attaching a gauge to a Countersink will be observed by reference to the engraving. By fastening the gauge at a given point, any number of screws may be driven so as to leave the heads flush with the surface, or at a uniform depth below it. The gauge can be easily moved, or detached entirely, by means of the set-screw; and the Countersinks are sold with or without the gauge.

PRICES:

	Per Dozen.
Countersinks, 1 dozen in a box	$4 00
Countersinks, 1 dozen in a box, with Gauge	5 00

MILLER'S PATENT COMBINED
Plow, Filletster and Matching Plane.

This Tool embraces, in a most ingenious and successful combination, the common Carpenter's PLOW, an adjustable FILLETSTER, and a perfect MATCHING PLANE. The entire assortment can be kept in smaller space, or made more portable, than an ordinary Carpenter's Plow.

☞ Each Plane is accompanied by a Tonguing tool (¼ inch), a Filletster Cutter, and eight Plow-bits (1-8, 3-16, 1-4, 5-16, 3-8, 7-16, 1-2, and 5-8 inch).

PRICES, Including Plow Bits, Tonguing and Grooving Tools.
No. 41. IRON Stock and Fence.........$10 50 | No. 42. GUN METAL Stock and Fence......$13 00

COMBINED
Plow and Matching Plane.

The above engraving represents the Tool, adjusted for use as a Plow. With each Plow eight Bits (1-8, 3-16, 1-4, 5-16, 3-8, 7-16, 1-2, and 5-8 inch) are furnished ; also a Tonguing Tool (1-4 inch), and by use of the latter, together with the 1-4 inch Plow Bit for grooving, a perfect Matching Plane is made.

PRICES, Including Plow Bits, Tonguing and Grooving Tools.
No. 43, IRON Stock and Fence..........$8 00 | No. 44, GUN METAL Stock and Fence......$10 50

☞ The Tool is packed in a box, and a printed description accompanies each one.

TRAUT'S PATENT ADJUSTABLE
Dado, Filletster, Plow, Etc.

The Tool here represented, consists of two sections:—A main stock, with two bars, or arms; and a sliding section, having its bottom, or face, level with that of the main stock.

It can be used as a Dado of any required width, by inserting the bit into the main stock, and bringing the sliding section snugly up to the edge of the bit. The two spurs, one on each section of the plane, will thus be brought exactly in front of the edges of the bit. The gauge on the sliding section will regulate the depth to which the tool will cut.

By attaching the Guard-plate, shown above, to the sliding section, the Tool may be readily converted into a Plow, a Filletster, or a Matching Plane—as explained in the printed instructions which go in every box.

The Tool is accompanied by eight Plow Bits, (1-8, 3-16, 1-4, 5-16, 3-8, 7-16, 1-2, and 5-8 inch), a Filletster Cutter, and a Tonguing Tool. All these tools are secured in the main stock on a *skew*.

PRICES, Including Plow Bits, Tonguing and Grooving Tools.

No. 46. IRON Stock and Fence.. $10 50

SOLID CAST STEEL SCREW DRIVERS.

PATENT IMPROVED

Solid Cast Steel Screw Drivers.

By the aid of improved machinery we are now producing Screw Drivers which are superior to any others in the market—and at the prices paid for the ordinary Tools.

The Blades are made from the best quality of Cast Steel, and are tempered with great care. They are ground down to a correct taper, and pointed at the end, by special machinery; thus procuring perfect uniformity in size, form, and strength, while the peculiar shape of the point gives it unequaled firmness in the screw-head when in use.

No. 1 Screw Drivers, all sizes above 3 inch, have the shank of the Blade properly slotted to receive a patent metallic fastening, which secures it permanently in the Handle. The Handles are of the most approved pattern, the Brass Ferrules, of the thimble form, *extra heavy*, and closely fitted.

No. 2 Screw Drivers, have the same Blade, without the metallic fastening, and are sold from the same list as No. 1. The difference in discount will be noted in our Discount Sheets.

PRICES:

Sizes	$1\frac{1}{2}$	2	3	4	5	6	7	8	10	Inch.
	$1.00	1.50	2.00	2.50	3.00	3.50	4.00	4.75	6.00	Per Dozen.

Sizes $1\frac{1}{2}$, 2, 3, and 4 inch, are packed 1 dozen in a box, all other sizes $\frac{1}{2}$ dozen in a box.

Sewing-Machine Screw Drivers.

Special Prices will be quoted for these in bulk.

Chalk-Line Reels, Etc.

	Per Dozen.
Chalk-Line Reels, 3 dozen in a box..........................	$0 45
Chalk-Line Reels, with 60 feet best quality Chalk-Line, 1 dozen in a box...	2 00
Chalk-Line Reels, with Steel Scratch Awls, 1 dozen in a box...	1 10
Chalk-Line Reels, with Steel Scratch Awls, and 60 feet best quality Chalk-Line, 1 dozen in a box..................	2 75

Handled Scratch Awls.

	Per Gross.
No. 1. Handled, Steel Scratch Awls, 1 dozen in a box........	$7 50
No. 2. Handled, Steel Scratch Awls, Large, 1 dozen in a box...	9 00

Handled Brad Awls, Etc.

	Per Gross.
Handled Brad Awls, assorted, 1 dozen in a box..	7 50
Handled Brad Awls, assorted, Large, 1 dozen in a box........	9 00

Brad Awl Handles, Brass Ferrules, assorted, 3 dozen in a box..	3 50

Chisel Handles, Etc.

Per Gross.
Polished Hickory Firmer Chisel Handles, asst'd, 1 doz. in a box, $5 25
Polished Hickory Firmer Chisel Handles, " Large, " " 6 25
Polished Apple Firmer Chisel Handles, " " " 6 00
Polished Apple Firmer Chisel Handles, " Large, " " 7 00

Polished Socket Firmer Chisel Handles, asst'd, 3 doz. in a box, 3 50
Polished Apple Socket Firmer Chisel Handles, assorted, 3 doz. in a box.................................... 4 50

Polished Hickory Socket Framing Chisel Handles, Iron Ferrules, assorted, 1 dozen in a box............................ 6 00

File Handles, Brass Ferrules, assorted, 3 dozen in a box....... 3 50
File Handles, Brass Ferrules, assorted, Large, 3 dozen in a box, 4 00

Polished Auger Handles, assorted........ .... 5 50
Polished Auger Handles, assorted, Large.................... 6 50
Packed 2 gross in a box.

Hickory Cross Cut Saw Handles, 15 inches long, 60 cents per dozen.

Saw Handles.

All full sizes, SPEAR & JACKSON's pattern, of perfect timber, well seasoned, and every way superior and reliable goods.

No.					Per Dozen.
1. Full size, Cherry,	Varnished Edges.				$1 65
2. " Beech,	"	"			1 40
3. " "	Plain	"			1 20
4. Small panel, "	Varnished	"	for 16 to 20 in. Saws		1 35
5. Meat Saw, "	"	"			1 35
6. Compass Saw "	"	"			1 25
7. Back Saw, "	"	"			1 35

Packed 2 gross in a case.

Plane Handles.

	Per Dozen.
Jack Plane Handles, 5 gross in a case	$0 42
Fore, or Jointer Handles, 3¾ gross in a case	0 75

Mallets.—Mortised Handles.

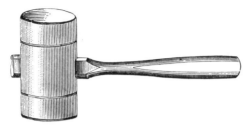

No.						Per Dozen.
1.	Round Hickory Mallet, Mortised,	5 in. long,	3 in. diameter,			$1 50
2.	" "	"	5½ "	3½ "		2 00
3.	" "	"	6 "	4 "		2 50
5.	" Lignumvitæ,	"	5 "	3 "		3 00
6.	" "	"	5½ "	3½ "		4 00
7.	" "	"	6 "	4 "		5 00

8.	Square Hickory Mallet, Mortised,	6 in. long,	2½ by 3½ inch,	2 00
9.	" "	"	6½ "	2¾ by 3¾ inch,	2 50
10.	" "	"	7 "	3 by 4 inch,	3 00
11.	" Lignumvitæ,	"	6 "	2½ by 3½ inch,	3 75
12.	" "	"	6½ "	2¾ by 3¾ inch,	4 75
13.	" "	"	7 "	3 by 4 inch,	5 75

14.	Round Mallet, Mortised, Iron Ring, 6 in. long, 4 in. diam...	6 00
14½.	" " " 5½ " 3½ " ...	4 00

Mallets.—Mortised Handles.

(CONTINUED.)

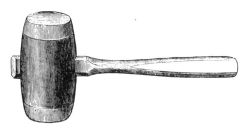

No.		Per Dozen.
15.	Round Iron Mallet, Mortised, Hickory Ends, 2½ in. diameter,	$4 00

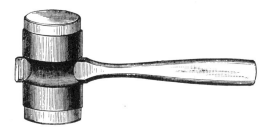

16. Round Mallet, Heavy Iron Socket, Mortised, Hickory Ends, 3 in. diameter.................................... 8 00

Tinners' Mallets.

No.		Per Dozen.
4.	Round Hickory Tinners' Mallet, 5½ in. long, assorted, 2, 2¼, and 2½ in. diameter....................................	$1 00

Patent Excelsior Tool Handle.
[TURKEY BOXWOOD.]

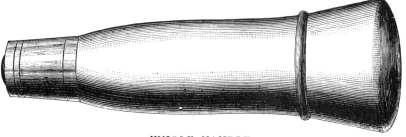

WHOLE HANDLE.

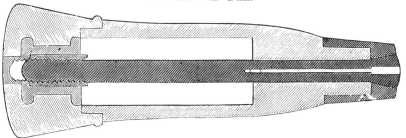

LONGITUDINAL SECTION.

This Handle is the best article of the kind ever invented, it being simple in its construction, compact, less liable to get out of order than any other, and dispensing entirely with the use of a wrench. In the interior of each Handle will be found twenty Brad Awls and Tools, adapted to the various kinds of work required of Mechanics, and useful in every household, or in ordinary branches of business. They are of the full size necessary for the several uses for which they are designed, as represented below.

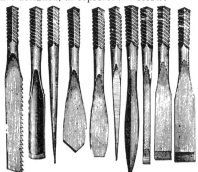 

DIRECTIONS.—Unscrew the Cap of the Handle, and select the Tool needed; with the thumb the center bolt may be thrust down sufficiently to open the clamp at the small end of the Handle, into which the Tool can be inserted; then in replacing the cover and screwing it down to its place, the clamp will be closed, and the Tool firmly secured for use.

☞ Put up in boxes of one dozen, with one set of twenty Tools in each Handle...Per Dozen, $13 50

HANDLES, AWL HAFTS, ETC.

Carpenters' Tool Handles.

No.   Per Dozen.

8. Carpenters' Tool Handle, Steel Screw, and Nut, with Iron Wrench, 1 dozen in a box............................ $1 00

8½. Carpenters' Tool Handle, Steel Screw, and Nut, with Iron Wrench, and 10 Brad Awls, assorted sizes. One handle with brad awls, packed in a box, and 12 boxes in a package, 4 00

Awl Hafts.

No.   Per Gross.

5. Hickory, Pegging Awl Haft, Plain Top, Steel Screw, and Nut, with Iron Wrench, 1 dozen in a box............$10 00

6. Hickory, Pegging Awl Haft, Leather Top, Steel Screw, and Nut, with Iron Wrench, 1 dozen in a box.......... 12 00

6½. Appletree Sewing All Haft, to hold any size Awl, Steel Screw, and Nut, with Iron Wrench, 1 dozen in a box...... 12 00

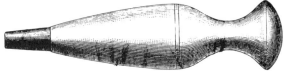

10. Common Sewing Awl Haft, Brass Ferrule, 3 dozen in a box, 3 50
11. Common Pegging Awl Haft, Brass Ferrule, 3 dozen in a box, 3 50

PATENT PEGGING AWLS.

PATENT PEGGING AWLS (short start, for Handles Nos. 5 and 6), assorted Nos. 00, 0, 1, 2, 3, 4, 5, and 6, 1 gross in a box.. 0 75

Hand Screws.

BEADED JAWS.

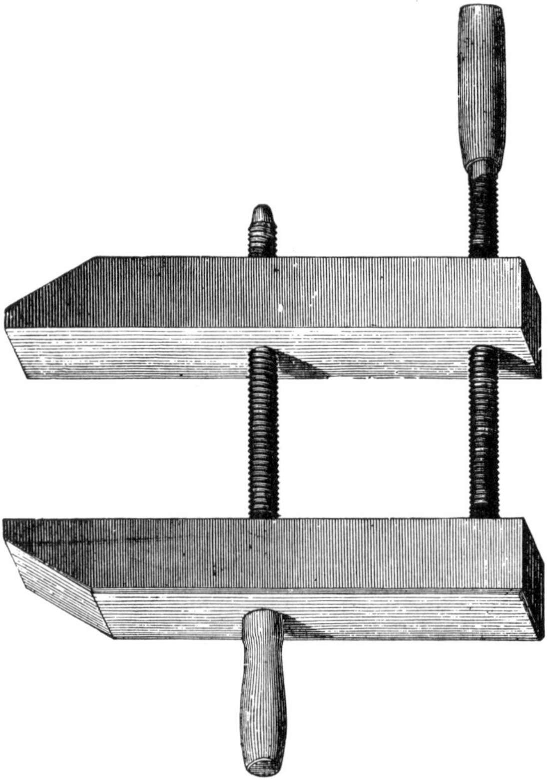

Diameter of Screws.	Length of Screws.	Length of Jaws.	Size of Jaws.	Per Dozen.
62 $1\frac{1}{4}$ inch.	24 inch.	20 inch.	$2\frac{7}{8}$ by $2\frac{7}{8}$ inch.	$10 50
77 $1\frac{1}{8}$ inch.	20 inch.	18 inch.	$2\frac{5}{8}$ by $2\frac{5}{8}$ inch	8 50
206 1 inch.	18 inch.	16 inch.	$2\frac{3}{8}$ by $2\frac{3}{8}$ inch.	6 50
307 $\frac{7}{8}$ inch.	16 inch.	14 inch.	2 by 2 inch.	4 75
382 $\frac{3}{4}$ inch.	12 inch.	10 inch.	$1\frac{5}{8}$ by $1\frac{5}{8}$ inch.	3 25
132 $\frac{5}{8}$ inch.	10 inch.	$8\frac{1}{2}$ inch.	$1\frac{3}{8}$ by $1\frac{3}{8}$ inch.	2 50
41 $\frac{1}{2}$ inch.	10 inch.	8 inch.	$1\frac{1}{4}$ by $1\frac{1}{4}$ inch.	2 25

Hand Screws, $\frac{1}{2}$, $\frac{5}{8}$, and $\frac{3}{4}$ inch, packed 2 dozen in a case; all other sizes, 1 dozen in a case.

$ 6.00 pr Dz net

Molders' Flask Screws, Rammers, and Mallets made to order.

Cabinet Makers' Clamps.

				Per Dozen.
6	2 feet, inside of Jaws			$9 00
14	3 "	"	"	10 00
12	4 "	"	"	11 00
9	5 "	"	"	12 00

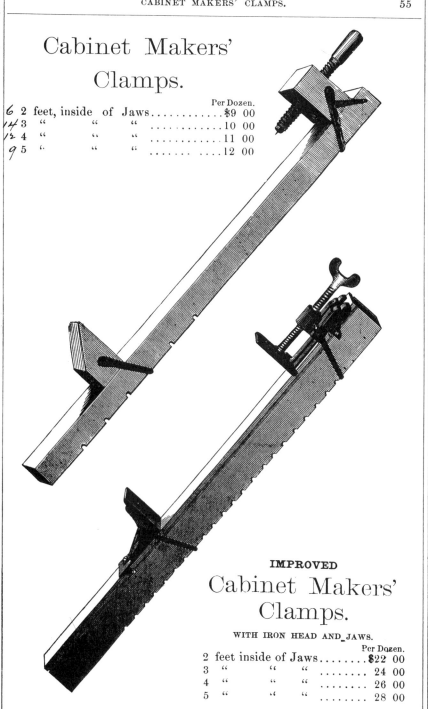

IMPROVED
Cabinet Makers' Clamps.

WITH IRON HEAD AND JAWS.

	Per Dozen.
2 feet inside of Jaws	$22 00
3 " " " "	24 00
4 " " " "	26 00
5 " " " "	28 00

Improved Trammel Points.

These Tools are used by Millwrights, Machinists, Carpenters, and all Mechanics having occasion to strike arcs, or circles, larger than can be conveniently done with ordinary Compass dividers. They may be used on a straight wooden bar of any length, and when secured in position by the thumb-screws, all circular work can be readily laid out by their use. They are made of Bronze Metal, and have Steel Points, either of which can be removed, and replaced by the pencil socket which accompanies each pair, should a pencil mark be preferred in laying out work.

PRICES:

No.		Per Pair.
1. (Small) Br'ze Metal, Steel Points,		$1 50
2. (Medium) " "		2 00
3. (Large) " "		2 75

Adjustable Plumb Bobs.

These Plumb Bobs are constructed with a reel at the upper end, upon which the line may be kept; and by dropping the bob with a slight jerk, while the ring is held in the hand, any desired length of line may be reeled off. A spring, which has its bearing on the reel, will check and hold the bob firmly at any point on the line. The pressure of the spring may be increased, or decreased, by means of the screw which passes through the reel. A suitable length of line comes already reeled on each Plumb Bob.

PRICES:

No.		Each.
1. (Small) Bronze Metal, with Steel Point		$1 75
2. (Large) " " " "		2 25
5. (Large) Iron " " "		1 00

Door Stops.

Plain Door Stops, with Iron Screws.

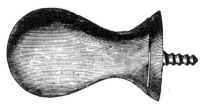

No. 6.

	Birch.	B. Walnut.	
2½ inch,	6 / $4 50	35 $5 50	Per Gross.
3 "	37 5 00	23 6 00	"

Packed 3 dozen in a box.

Improved Rubber Top Door Stops.

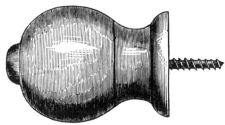

No. 8.

	Birch.	Chestnut.	B. Walnut.	Porcelain Enameled.	
2½ inch,	187 $10 00	$11 00	335 $11 00	5 $14 00	Per Gross.
3 "	99 11 00	12 00	122 12 00	/ 15 00	"

Packed 3 dozen in a box.

Floor Door Fenders.

	Chestnut.	B. Walnut.	
2¼ inch,	$12 00	$12 00	Per Gross.

Packed 3 dozen in a box.

Improved Sash Frame Pulleys.

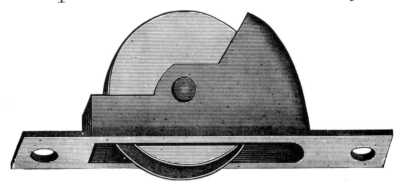

Superiority is claimed for these Pulleys over any others in use, in these essential particulars:

First. In respect to wear and friction. Turkey Boxwood is used for the Wheels, and large smooth Iron Axles, instead of two grinding surfaces of rough cast-iron; thus producing little friction, and no perceptible wear.

Second. The wheels being bored and turned *true*, the irregular motion and noise of the iron wheel is avoided.

¶Third. The groove of the wheel is turned out smooth, and of proper shape to receive the cord, which is thus prevented from wearing, and made to last much longer than in other pulleys.

	Per Dozen.
Sash Frame Pulley, 1¾ inch, 1 dozen in a box	$0 60
Sash Frame Pulley, 2 " " "	0 75
Sash Frame Pulley, 2¼ " " "	1 00

SMITH'S
Sash Cord Iron.

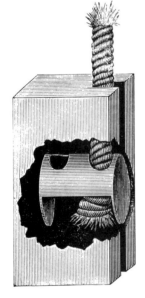

These Irons can be more easily put into the edge of a Sash than any other, requiring only the boring of a round hole; and no screws are needed to keep them in position. When the windows are taken out for cleaning, or any other purpose, the Irons can be removed or replaced without the use of any Tool.

In boxes of 1 gross..................75 cts.

Patent Improved Cattle Tie.

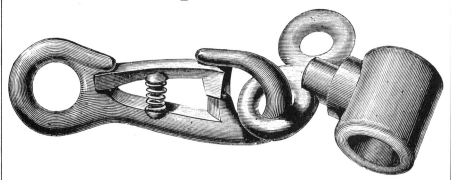

It is believed that this Cattle Tie, in both its parts, embraces more truly valuable features than any similar article heretofore sold. The snap is properly proportioned for the greatest strength required, and is made of Malleable iron. The form of the hooked end prevents the snap becoming unfastened by any accidental means; while the *Improved Spiral Spring*, which is made of brass, and not liable to rust from being wet, or to break in cold weather, acts with certainty, and is protected in its position under an independent lug, or bar, upon which all shocks or pressure must first come.

The other part of the Cattle Tie consists of an iron socket, which may be secured at any desired position on a rope, by means of a Malleable thumb screw in one side of the socket; the thumb screw having a perforated head, through which the snap is readily hooked. The socket may at once be changed from one position on the rope to another, and secured perfectly without the use of a screw-driver, or other tool, thus adapting the length of the rope to the size of the neck or horns of any animal. The extreme simplicity of both parts insures great durability.

Per Dozen.
Cattle Tie, with Rope, Japanned $4 00
Cattle Tie, without Rope, Japanned, 1 dozen in a box 1 50
Improved Spiral Spring Snaps (only), Japanned 75

HAMMERS.

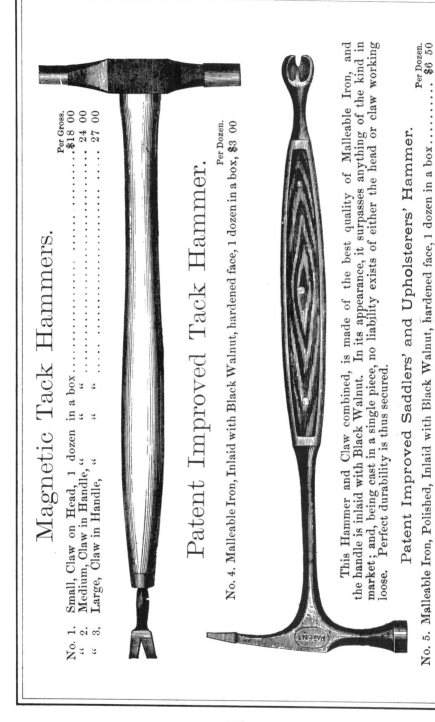

Magnetic Tack Hammers.

		Per Gross.
No. 1.	Small, Claw on Head, 1 dozen in a box	$18 00
" 2.	Medium, Claw in Handle, " " "	24 00
" 3.	Large, Claw in Handle, " " "	27 00

Patent Improved Tack Hammer.

Per Dozen.
No. 4. Malleable Iron, Inlaid with Black Walnut, hardened face, 1 dozen in a box, $3 00

Patent Improved Saddlers' and Upholsterers' Hammer.

This Hammer and Claw combined, is made of the best quality of Malleable Iron, and the handle is inlaid with Black Walnut. In its appearance, it surpasses anything of the kind in market; and, being cast in a single piece, no liability exists of either the head or claw working loose. Perfect durability is thus secured.

Per Dozen.
No. 5. Malleable Iron, Polished, Inlaid with Black Walnut, hardened face, 1 dozen in a box $6 50

STEAK HAMMERS, ETC. 61

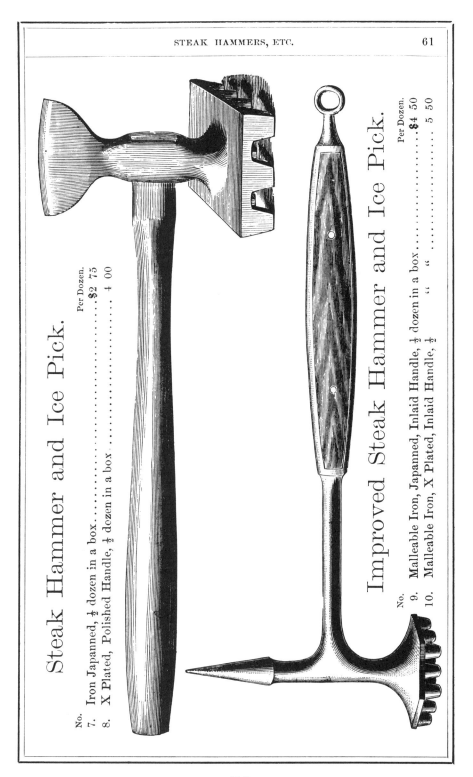

Steak Hammer and Ice Pick.

Per Dozen.
No.
7. Iron Japanned, ¼ dozen in a box................ $2 75
8. X Plated, Polished Handle, ½ dozen in a box........ 4 00

Improved Steak Hammer and Ice Pick.

Per Dozen.
No.
9. Malleable Iron, Japanned, Inlaid Handle, ½ dozen in a box......$4 50
10. Malleable Iron, X Plated, Inlaid Handle, ½   "   " 5 50

Tackle Blocks.

Rope Strapped Blocks,
HOOK AND THIMBLE.

Length of Blocks,	4	5	6	7	8	9	10	11	12 Inch.
Diameter of Wheels,	$2\frac{1}{2}$	3	$3\frac{1}{2}$	$4\frac{1}{4}$	5	$5\frac{3}{4}$	$6\frac{1}{2}$	$7\frac{1}{4}$	8 "
Diameter of Rope,	$\frac{1}{2}$	$\frac{5}{8}$	$\frac{3}{4}$	$\frac{7}{8}$	1	1	$1\frac{1}{8}$	$1\frac{1}{8}$	$1\frac{1}{4}$ "
Single, Iron Bushed,	$0 75	0 95	1 15	1 35	1 85	2 15	2 45	3 00	3 50 Each.
Double, Iron Bushed,	1 30	1 60	1 90	2 20	3 00	3 70	4 20	5 50	6 00 "
Triple, Iron Bushed,	1 65	2 05	2 45	2 85	3 95	4 85	5 40	7 50	8 25 "
Single, Roller Bushed,	1 40	1 75	2 10	2 45	2 95	3 25	3 55	4 35	4 85 "
Double, Roller Bushed,	2 60	3 25	3 90	4 55	5 20	5 85	6 50	8 00	8 85 "
Triple, Roller Bushed,	3 55	4 45	5 35	6 25	7 15	8 00	8 95	11 20	12 25 "

Tackle Blocks.

(CONTINUED.)

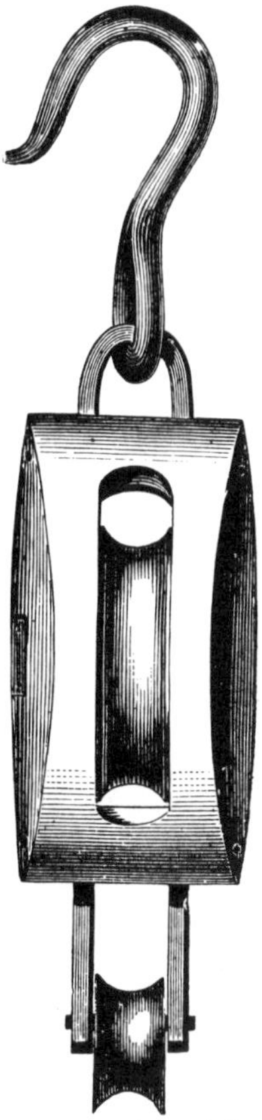

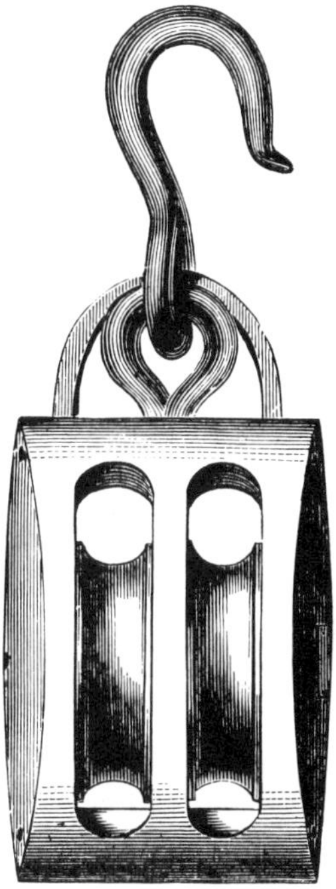

Inside Iron Strapped Blocks.

LOOSE HOOKS AND STIFF SWIVELS.

Length of Blocks,	4	5	6	7	8	9	10	11	12	Inch.
Diameter of Wheels,	$2\frac{1}{2}$	3	$3\frac{1}{2}$	$4\frac{1}{4}$	5	$5\frac{3}{4}$	$6\frac{1}{2}$	$7\frac{1}{4}$	8	"
Diameter of Rope,	$\frac{1}{2}$	$\frac{5}{8}$	$\frac{3}{4}$	$\frac{7}{8}$	1	1	$1\frac{1}{8}$	$1\frac{1}{8}$	$1\frac{1}{4}$	"
Single, Iron Bushed,	$1 10	1 40	1 65	2 00	2 25	2 55	2 80	3 35	3 65	Each.
Double, Iron Bushed,	2 00	2 50	3 00	3 50	4 05	4 55	5 05	6 15	6 75	"
Triple, Iron Bushed,	2 70	3 40	4 10	4 80	5 50	6 10	6 80	8 35	9 15	"
Single, Roller Bushed,	1 55	1 95	2 35	2 70	3 10	3 50	3 90	4 50	4 95	"
Double, Roller Bushed,	2 90	3 65	4 35	5 10	5 80	6 55	7 25	8 55	9 30	"
Triple, Roller Bushed,	4 10	5 10	6 10	7 10	8 10	9 10	10 10	12 00	13 00	"

☞ Orders should always designate the length of Blocks required, and state whether Rope Strapped, or Iron Strapped, are wanted ; also, whether they shall be Iron Bushed, or Roller Bushed. All Single Blocks have a Becket at the lower end, as shown in the Engravings. Double and Triple Blocks are made with Beckets for special orders.

See next page for List of Thick Mortise Blocks, Awning Blocks, Sheaves, Etc.

Tackle Blocks.
(CONTINUED.)
Rope Strapped Blocks, Hook and Thimble.
THICK MORTISE.

Length of Blocks,	9	10	11	12	Inch.
Diameter of Wheels,	5¾	6½	7¼	8	"
Diameter of Rope,	1¼	1⅜	1½	1½	"
Single, Iron Bushed,	$3.65	4.00	4.50	5.00	Each.
Double, Iron Bushed,	7.30	8.00	9.00	10.00	"
Single, Roller Bushed,	5.65	6.30	7.00	7.50	"
Double, Roller Bushed,	11.30	12.60	14.00	15.00	"

Plain Blocks, not Strapped.

IRON BUSHED.   ROLLER BUSHED.

4 to 10 inch........$0.10 per Inch.   4 to 10 inch........$0.25 per Inch.
10 to 12 inch......... 0.13   "   10 to 12 inch......... 0.28   "

Awning Blocks, Rope Strapped.

	1¼	1½	2	2½	3	3½	4 Inch.
Single,	.12	.12	.13	.14	.15	.17	.22 Each.
Double,	.24	.24	.26	.28	.30	.34	.44 "

Awning Blocks, Not Strapped.

	1¼	1½	2	2½	3	3½	4	Inch.
Single,	.10	.10	.11	.12	.13	.15	.18	Each.
Double,	.20	.20	.22	.24	.26	.30	.36	"

Sheaves for Blocks.

Length of Blocks,	4	5	6	7	8	9	10	11	12	Inch.
Diameter of Wheels,	2½	3	3½	4¼	5	5¾	6½	7¼	8	"
Iron Bushed,	.16	.20	.24	.28	.38	.42	.55	.65	.80	Each
Roller Bushed,	.60	.70	.80	.90	1.30	1.40	1.60	1.85	2.50	"

Pinking Irons.

	⅜	½	⅝	¾	⅞	1	1⅛	1¼	1½	Inch.
Diamond Pattern,	$3.00	3.00	3.00	3.00	3.00	3.00	3.50	3.50	4.00	Per Dozen.
Scolloped "		3.00	3.00	3.00	3.00	3.00	3.50	3.50	4.00	"

Assorted, ⅜ to 1 inch, 1 dozen in a box, $3.00 Per Dozen.

☞ **Please attach this to our Catalogue for 1874,**
And cancel entirely, or change the Prices, on pages 38, 39, 40, 41, 44, and 45.

SUPPLEMENT

TO

CATALOGUE

OF

TOOLS AND HARDWARE,

MANUFACTURED BY THE

STANLEY RULE & LEVEL CO.,

NEW BRITAIN, CONN.

WAREROOMS:

35 CHAMBERS-ST., NEW YORK.

ORDERS FILLED AT THE WAREROOMS, OR AT NEW BRITAIN.

JANUARY, 1876.

Bailey's Patent Planes.

[Manufactured only by the Stanley Rule & Level Co., under the original Patents.]

The sale of over **100,000** of these Planes already, and a constantly increasing demand for them, is the best evidence that all dealers in Mechanics' Tools can have of the estimation in which Bailey's Patent Adjustable Planes are held by the men who use them. A few of the numerous testimonials to the superiority of these Tools, from practical men in all branches of wood-working, will be found on page 68 of this Catalogue.

☞ The Planes are in perfect working order when sent into market; and a printed description of the method of adjustment accompanies each Plane.

Iron Planes.

No.						Each.
1.	Smooth Plane,	5½ inches in Length,	1¼ inch Cutter			$3 00
2.	" "	7 "	"	1½ "	"	3 50
3.	" "	8 "	"	1¾ "	"	4 00
4.	" "	9 "	"	2 "	"	4 50

5.	Jack Plane,	14 inches in Length,	2 inch Cutter			5 00
6.	Fore "	18 "	"	2⅜ "	"	6 00
7.	Jointer "	22 "	"	2⅜ "	"	7 00
8.	" "	24 "	"	2½ "	"	8 00

9. Block Plane, 10 inches in Length, 2 inch Cutter.......... 7 50
10. Carriage Makers' Rabbet Plane, 14 inches in Length,
    2⅛ inch Cutter.. 5 50
11. Belt Makers' Plane, 2⅜ inch Cutter........................ 3 50

IRON PLANES.—(CONTINUED.)
Bailey's Patent Adjustable Block Planes.

Each.
No. 9½ Excelsior Block Plane, 6 inches in length, 1¾ inch Cutter....... $2 00
15 Excelsior Block Plane, 7 inches in length, 1¾ inch Cutter....... 2 25

No. 9¾ Excelsior Block Plane, with Rosewood handle, 6 inches in length,
    1¾ inch Cutter... 2 25
  15½ Excelsior Block Plane, with Rosewood handle, 7 inches in length,
    1¾ inch Cutter... 2 50

These Block Planes are adjusted by a Screw and Lever movement, and the mouth can be opened wide, or made close, as the nature of the work to be done may require. The handles to Nos. 9¾ and 15½ are secured to the stock by the use of an iron nut, and can be easily attached to, or liberated from the Plane, as the convenience of the workman may require.

Bailey's Patent Adjustable Veneer Scraper.

No. 12. 3 inch Cutter...$4 50
Cast Steel, Hand, Veneer Scrapers, 3×5 inch, per dozen........... 4 00

Bailey's Patent Adjustable Circular Plane.

No. 13. 1¾ inch Cutter...$4 50

This Plane has a *Flexible Steel Face*, and by means of the thumb screws at each end of the Stock, can be easily adapted to plane circular work—either concave or convex.

Wood Planes.

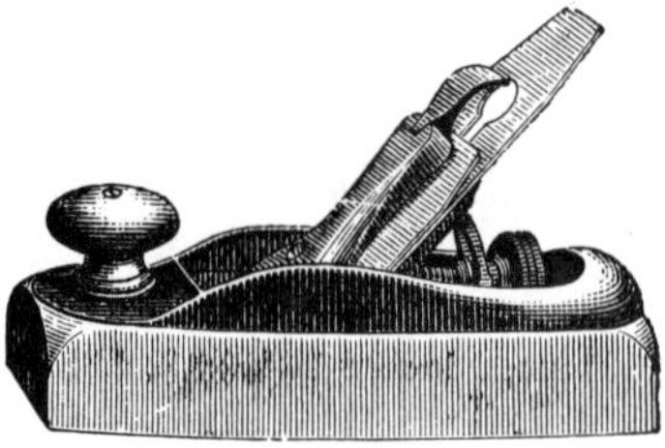

No.							Each.
21.	Smooth	Plane,	7	inches in Length,	$1\frac{3}{4}$	inch Cutter.......	$2 50
22.	"	"	8	" "	$1\frac{3}{4}$	" "	2 50
23.	"	"	9	" "	$1\frac{3}{4}$	" "	2 50
24.	"	"	8	" "	2	" "	2 75
25.	Block	"	$9\frac{1}{2}$	" "	$1\frac{3}{4}$	" "	2 75

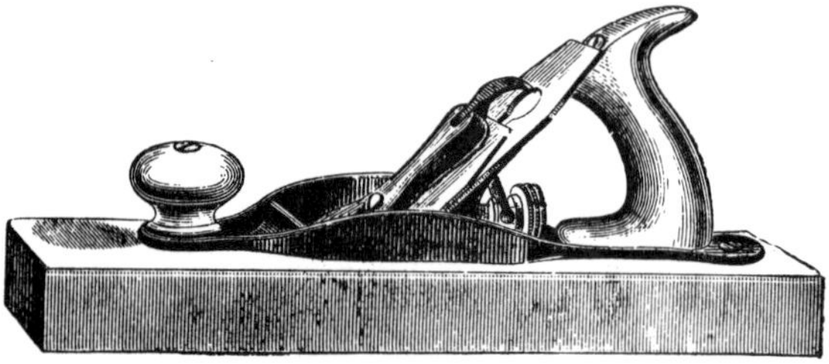

26.	Jack	Plane,	15	inches in Length,	2	inch Cutter........	3 00
27.	"	"	15	" "	$2\frac{1}{8}$	" "	3 25
28.	Fore	"	18	" "	$2\frac{3}{8}$	" "	3 50
29.	"	"	20	" "	$2\frac{3}{8}$	" "	3 50
30.	Jointer	"	22	" "	$2\frac{3}{8}$	" "	3 75
31.	"	"	24	" "	$2\frac{3}{8}$	" "	3 75
32.	"	"	26	" "	$2\frac{5}{8}$	" "	4 25
33.	"	"	28	" "	$2\frac{5}{8}$	" "	4 25
34.	"	"	30	" "	$2\frac{5}{8}$	" "	4 50

35.	Handle	Smooth,	9	inches in Length,	2	inch Cutter.....	3 50
36.	"	"	10	" "	$2\frac{3}{8}$	" "	3 75
37.	Jenny	"	13	" "	$2\frac{5}{8}$	" "	4 00

☞ Extra Plane-woods, of every style, can be supplied cheaply.

Bailey's Patent Plane Irons.

☞ These Plane-Irons have stood the severest tests applied to them by practical workmen, and are by the manufacturers fully WARRANTED.

PRICES:

SINGLE IRONS.

Inches,	1¼	1⅝	1¾	2	2⅛	2⅜	2⅝
Each,	30	34	38	42	44	48	50 cts.

DOUBLE IRONS.

Each,	50	56	62	68	74	80	84 cts.

☞ Orders for DOUBLE IRONS should state whether they are wanted for Iron Planes, or for Wood Planes.

TESTIMONIALS.

"Works splendidly on our Yellow Pine."—H. C. HALL, *Jacksonville, Florida.*

"Far superior to any I have seen used."—JOHN ENGLISH, *North Milwaukee, Wis., Car Shops.*

"The BEST PLANE now in use."—CHICKERING & SONS, *Piano Forte Manufacturers, Boston, Mass.*

"Superior to any others used in this Shop."—W. S. TOWN, *Master of Car Repairs, Hudson River R. R. Co.*

"All that is wanted to sell your planes here is to show them."—E. WATERS, *Manufacturer of Sporting Boats, Troy, N. Y.*

"We know of no other plane that would be a successful subtitute for them."—R. BURDETT & Co., *Organ Manufacturers, Erie, Pa.*

"The best plane ever introduced in carriage-making."—JAS. L. MORGAN, *at J. B. Brewster & Co's., Carriage Manufactory, New York.*

"From the satisfaction they give would like to introduce them here."—J. ALEXANDER, *at Bell's Melodeon Factory, Guelph, Canada.*

"I have used your planes and have never found their equal."—B. F. FURRY, *Foreman McLear & Kendall's Coach Works, Wilmington, Del.*

"Bailey's Planes are used in our Factory. We can get only one verdict, and that is, the men would not be without them."—STEINWAY & SON, *Piano Forte Manufacturers, New York.*

"I have owned a set of Bailey's Planes for six months, and can cheerfully say that I would not use any others, for I believe they are the best in use."—C. C. HARRIS, *Stair Builder, St. Louis, Mo.*

"We have Bailey's Planes in our Factory, and there is but one opinion about them. They are the best and cheapest tool we have ever used. Good tools are always the cheapest."—JOS. BECKHAUS, *Carriage Builder, Philadelphia.*

"I have used a set of Bailey's Planes about six months, and would not use any other now. There is inquiry about them here every day. Send me your descriptive Catalogue."—EDWARD CARVILLE, *Sacramento City, California.*

A Boston Mechanic says: "I always tell my shopmates, when they wish to use my Plane, not to borrow a BAILEY PLANE unless they intend to buy one, as they will never be satisfied with any other Plane after using this."

MILLER'S PATENT COMBINED
Plow, Filletster and Matching Plane.

This Tool embraces, in a most ingenious and successful combination, the common Carpenter's Plow, an adjustable Filletster, and a perfect Matching Plane. The entire assortment can be kept in smaller space, or made more portable, than an ordinary Carpenter's Plow.

☞ Each Plane is accompanied by a Tonguing tool (¼ inch) a Filletster Cutter, and eight Plow-bits (1-8, 3-16, 1-4, 5-16, 3-8, 7-16, 1-2 and 5-8 inch.)

PRICES, Including Plow Bits, Tonguing and Grooving Tools.
No. 41. Iron Stock and Fence...... $10 00 | No. 42. Gun Metal Stock and Fence...... $12 50

COMBINED
Plow and Matching Plane.

The above engraving represents the Tool, adjusted for use as a Plow. With each Plow eight Bits (1-8, 3-16, 1-4, 5-16, 3-8, 7-16, 1-2 and 5-8 inch) are furnished; also a Tonguing Tool (¼ inch), and by use of the latter, together with the ¼ inch Plow Bit for grooving, a perfect Matching Plane is made.

PRICES, Including Plow Bits, Tonguing and Grooving Tools.
No. 43. Iron Stock and Fence...... $7 50 | No. 44. Gun Metal Stock and Fence...... $10 00
☞ The Tool is packed in a box, and a printed description accompanies each one.

TRAUT'S PATENT ADJUSTABLE

Dado, Filletster, Plow, Etc.

The Tool here represented, consists of two sections:—A main stock, with two bars, or arms; and a sliding section, having its bottom, or face, level with that of the main stock.

It can be used as a Dado of any required width, by inserting the bit into the main stock, and bringing the sliding section snugly up to the edge of the bit. The two spurs, one on each section of the plane, will thus be brought exactly in front of the edges of the bit. The gauge on the sliding section will regulate the depth to which the tool will cut.

By attaching the Guard-plate, shown above, to the sliding section, the Tool may be readily converted into a Plow, a Filletster, or a Matching Plane—as explained in the printed instructions which go in every box.

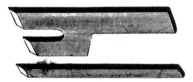

The Tool is accompanied by eight Plow Bits, ($\frac{3}{16}$, $\frac{1}{4}$, $\frac{5}{16}$, $\frac{3}{8}$, $\frac{1}{2}$, $\frac{5}{8}$, $\frac{7}{8}$, and $1\frac{1}{4}$ inch), a Filletster Cutter, and a Tonguing Tool. All these tools are secured in the main stock on a *skew*.

PRICE, Including Plow Bits, Tonguing and Grooving Tools.
No. 46. IRON Stock and Fence..$10 00

Adjustable Dado.

It can be used as a Dado of any required width, by inserting the bit into the main stock, and bringing the sliding section snugly up to the edge of the bit. The two spurs, one on each section of the plane, will thus be brought exactly in front of the edges of the bit. The gauge on the sliding section will regulate the depth to which the tool will cut.

PRICE, Including Bits (3-8, 1-2, 5-8, 7-8 and 1 1-4 Inch.)
No. 47. IRON Stock and Fence,................. ....$5 00

Patent Tonguing and Grooving Plane.

Two separate Tools, which are always used in connection with each other, are here combined in one; thus affording to the owner two superior Tools, in a cheap form, and occupying no more space, in a chest or on a bench, than one ordinary Tonguing or Grooving Tool.

The stock of this Tool is made of metal, and it has two cutters fastened into the stock by thumb-screws. The guide, or fence, when set as shown in the above engraving, allows both of the cutters to act; and the cutters being placed at a suitable distance apart, a perfect Tonguing Plane is made. The guide, or fence, which is hung on a pivot at its center, may be easily swung around, end for end; thus one of the cutters will be covered, and the guide held in a new position, thereby converting the Tool into a Grooving Plane. A groove will thus be cut to exactly match the tongue which is made by the other adjustment of the Tool.

The guide, or fence, is hung for grooving boards planed from 1 inch stuff; and on these the tongue and groove will both come in the center of the board. Boards varying from ¾ to 1¼ inch in thickness can be matched equally well, by working the Plane so that the tongue and groove shall both come at the regular distance from one edge of the boards to be matched, leaving the distance to the other edge to vary as it may. One extra width cutter accompanies the Tool, to be used on the outer side of the tongue, in tonguing boards thicker than those planed from 1 inch stuff.

The ingenuity and simplicity of this Tool, together with its compact form and durability, will commend it to the favorable regard of all wood-working mechanics.

PRICE, Including Tonguing and Grooving Tools.

No. 48. IRON Stock and Fence.................................... $2 50

Patent Improved Mitre Box.

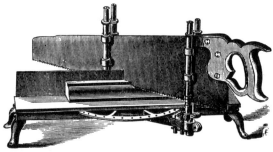

The peculiar features which distinguish this Mitre Box above all others in market, are referred to below:

The frame is made of a single casting, and is subject to no change of position; being finished accurately at first, it must always remain true. The slot in the back of the frame, through which the saw passes, is only one-eighth of an inch wide, thereby obviating any liability to push short pieces of work through the slot, when the saw is in motion.

This Mitre Box can be used with a Back Saw, or a Panel Saw, equally well. If a Back Saw is used, both links which connect the rollers, or guides, are left in the upper grooves, and the back of the saw is passed through under the links. If a Panel Saw is used, the link which connects the rollers on the back spindle, is changed to the lower groove; and then the blade of the saw will be stiffly supported by both sets of rollers, and be made to serve as well as a Back Saw.

By slightly raising or lowering the spindles, when necessary, the leaden rolls at the bottom may be adjusted to stop the saw at the proper depth; and, by the use of a set-screw, the spindles on which the guides revolve, may be turned sufficiently to make the rollers bear firmly on the sides of a saw blade of any thickness.

If a narrow saw blade is used, or if the saw blade becomes narrower from use, the rollers may be lowered on the spindles by removing some of the brass rings from under them.

PRICES.

Mitre Box, 20 inches..$7 00
Mitre Box, 20 inches, with 20 inch Disston's Back Saw...........................10 00

INDEX.

	Page.
FRONTISPIECE, View of Factory at New Britain,	
Auger Handles,	48
Awl Hafts,	53
Awls, Patent Pegging,	53
Brad Awl Handles,	47
Brad Awls, Handled,	47
Bevels, Sliding T,	32
Bevels, Patent Flush, Bailey's,	33
Bevels, Patent Flush, Eureka,	34
Cabinet Makers' Clamps,	55
Chalk-line Reels and Awls,	47
Carpenters' Tool Handles,	53
Cattle Ties,	59
Chisel Handles,	48
Cross Cut Saw Handles,	48
Countersinks, Wheeler's Patent,	43
Dado, Filletster, Plow, Etc., Combined,	45
Door Stops,	57
Door Fenders, Floor,	57
File Handles,	48
Gauges,	35 to 37
Hand Screws,	54
Level Glasses,	27
Mallets, Hickory and Lignumvitæ,	50 and 51
Mitre Squares, Improved,	30
Mitre Try Squares, Improved,	30
Miscellaneous Articles, Stanley's,	14 and 15
Miscellaneous Articles, Stearns',	21

INDEX.

	Page.
Planes, Bailey's Patent,	38 to 40
Plane Irons, Bailey's Patent,	41
Plane Handles,	49
Pinking Irons,	64
Plumbs and Levels,	24
Plumbs and Levels, Patent Adjustable,	25
Plumbs and Levels, Nicholson's Patent,	26
Plumbs and Levels, Iron Frame,	26
Pocket Levels,	27
Plow, Filletster, and Matching Plane, Combined,	44
Plow and Matching Plane, Combined,	44
Plumb Bobs, Adjustable,	56
Rules, Stanley's,	5 to 13
Rules, Stearns',	16 to 21
Sash Frame Pulleys,	58
Sash Cord Irons,	58
Scratch Awls, Handled,	47
Saw Handles,	49
Screw Drivers,	46
Steak Hammers,	61
Spoke Shaves, Bailey's Patent,	42 and 43
Spoke Shave Cutters, Bailey's Patent,	43
Tackle Blocks,	62 to 64
Tack Hammers,	60
Trammel Points,	56
Tool Handles and Tools, Excelsior,	52
Try Squares, No. 2,	28
Try Squares, Improved,	29
Try Squares, Plumb and Level,	28
Try Square and Bevel, Patent Combination,	31
Upholsterers' Hammers,	60
Veneer Scrapers, Bailey's Patent,	39

THE STANLEY
Adjustable Planes.

PATENTED.

These Planes are adjusted by the use of a Compound Lever, and are equally well adapted to coarse or fine work. The Planes are thoroughly tested at our Factory, and are in perfect working order when sent into market. The Plane Irons are made of the best English Cast Steel, and are fully Warranted.

Planes Nos. 104 and 105, have a Wrought Steel Stock, and are commended for their lightness of weight, and the ease with which they can be worked.

Steel Planes.

No.		Each.
104.	Smooth Plane, 9 inches in Length, $2\frac{1}{8}$ inch Cutter	$3 00

| 105. | Jack Plane, 14 inches in Length, $2\frac{1}{8}$ inch Cutter | 3 50 |

Iron Block Planes.

| 110. | Block Plane, $7\frac{1}{2}$ inches in Length, $1\frac{3}{4}$ inch Cutter | 70 |

| 120. | Block Plane, Adjustable, $7\frac{1}{2}$ inches in Length, $1\frac{3}{4}$ inch Cutter | 1 00 |
| | CUTTERS, for above Block Planes, warranted Cast Steel....per dozen, | 2 00 |

(Reproduced from the 1877 Catalogue, see Note inside front cover of this Reprint)

THE STANLEY ADJUSTABLE PLANES. 41

Wood Planes.

No.   Each.
122.   Smooth Plane, 8 inches in Length, 1¾ in. Cutter.........$1 50

135.   Handle Smooth, 10 inches in Length, 2⅛ in. Cutter....... 2 00

127.   Jack Plane,   15 inches in Length, 2⅛ in. Cutter......... 2 00
129.   Fore Plane,   20 inches in Length, 2⅜ in. Cutter......... 2 25
132.   Jointer Plane, 26 inches in Length, 2⅝ in. Cutter......... 2 75

☞ The STANLEY PLANE IRONS are the same prices, for corresponding widths, as per list of Plane Irons, page 39.

Patent Improved Rabbet Planes.
WITH STEEL CASE.

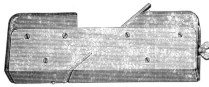

No.   Each.
80.   Rabbet Plane, Skew, 9 inches in Length, 1½ in. Cutter.....$1 10
90.   Rabbet Plane, Skew, 9 inches in Length, with Spur, 1½ in. Cutter.. 1 25

(Reproduced from the 1877 Catalogue, see Note inside front cover of this Reprint)

1877 STANLEY RULE & LEVEL Co. CATALOG and PRICE LIST

While there is not a great deal of difference in content between the 1877 Catalog and the previous 1874 issue, five pages contain entirely new illustrations of products.

These include the N° 104, N° 105, N° 122, N° 135, N° 127, N° 129 and N° 132 Liberty Bell Patented Planes; N° 80 and N° 90 Patent Improved Rabbet Planes; N° 101, N° 102, and N° 103 Block Planes; N° 62 Patent Reversible Spoke Shave and the Patent Adjustable Box Scraper, and a new design of Patent Excelsior Tool Handles. Twenty four pages of the same content reveal price reductions from the 1874 issue.

Ken Roberts
1980

PRICE LIST

OF

U. S. STANDARD

BOXWOOD AND IVORY

RULES,

PLUMBS AND LEVELS, TRY SQUARES, BEVELS,
GAUGES, MALLETS, HANDLES,
AWL HAFTS, SCREW DRIVERS, HAND
SCREWS, IRON AND WOOD PLANES,
SPOKE SHAVES, ETC.

MANUFACTURED BY THE

STANLEY
RULE AND LEVEL CO.

NEW BRITAIN, CONN., U. S. A.

WAREROOMS:

No. 35 CHAMBERS STREET, NEW YORK.

ORDERS FILLED AT THE WAREROOMS, OR AT NEW BRITAIN.

JANUARY, 1877.

Numbers of Rules manufactured by

STANLEY RULE & LEVEL CO.,

Corresponding with Numbers of other Manufacturers.

Stanley.	Stearns.	Stephens & Co.	Chapin.	Hubbard.	Belcher.
1	....	....	....	....	....
2	7	13	41	2	85
4	6½	15	44	4	87
5	....	17	45	5	88
6	....	....	50	....	....
8	....	....	....	....	....
12	....	....	48	....	....
14	....	....	....	....	....
15	....	27	49	....	....
16	....	....	51	....	....
18	11	2	39	18	81
22	....	5	40	22	....
26	9	9	46	26	P91
27	....	....	47	....	....
29	....	1	38	29	80
32	....	....	72	....	....
33	....	....	....	....	....
34	....	....	....	....	....
35	....	....	....	....	....
36	13½	95	70	36	P99
37	....	....	....	....	....
38	55	95½	74	38	P299
39	....	98½	76	39	C221
40	....	99½	78	40	C224
41	....	....	....	....	....
42	....	31	85	42	901
43	....	30	85½	43	900
43½	....	....	....	....	....
44	....	....	....	....	....
45	....	....	....	....	....
45½	....	....	....	....	....
46	....	....	....	....	....
46½	....	....	....	....	....
47	....	....	....	....	....
47½	....	....	....	....	....
48	....	....	....	....	....
48½	....	....	....	....	....
49	....	....	....	....	....
49½	....	....	....	....	....
50	....	....	....	....	....
51	46	46	16	51	65
52	....	49¼	18	52	67
53	45	48	17	53	66
54	....	49	19	54	68
55	....	73	6	55	5
56	....	74	7	56	6
57	....	75	9	57	8
58	....	....	....	....	....
59	....	50	20	59	69
60	....	52	21	60	B69
61	38	42	11	61	61
61½	36	44	11½	61½	41
62	35	42½	15	62	64
63	37	45	12	63	62
63½	33	44½	13	63½	42
64	....	72	3	64	2
65	75	71	2	65	1
65½	74	72½	5	65½	4
66	....	32	84	66	....
66½	....	....	....	....	....
67	....	53	22	67	70
68	41	41	10	68	60
69	43	70	1	69	0
70	32	54	23	70	71
71	....	....	....	....	....
72	....	56	24	72	72
72½	....	54½	26	72½	74
73	21	57	27	73	75
75	20	59	28	75	76
76	15	60	30	76	78
77	17	61	31	77	79
78	....	....	....	....	....
78½	....	63	32	78½	B79
79	....	64	34	79	571
81	....	66	36	81	576
82	14	67	37	82	....
83	....	....	....	....	....
84	....	42¼	14	84	63
85	50	77	59	85	242
86	48	83	60	86	266
87	48B	84	61	87	268
88	57B	94	57	88	208
88½	....	93	56	88½	206
89	....	86	62	89	B269
90	59	89	52	90	100
91	....	....	58	....	....
92	57	91	55	92	202
92½	....	90½	54	92½	....
93	....	....	....	....	....
95	....	....	64	....	....
97	....	....	....	....	....
98	....	....	....	....	....
99	....	....	....	....	....

OFFICE OF THE

STANLEY RULE AND LEVEL CO.,

New Britain, Conn., Jan. 1, 1877.

Gentlemen:

Your attention is invited to this revised Catalogue and Price List of Improved Carpenters' Tools, etc., manufactured by us.

Many of the lines represented are already familiar to the entire Hardware Trade, and stand approved in the estimation of Mechanics everywhere. Such changes have been made in prices, as improvements in our facilities for producing goods warrant; and new Tools have been added, to complete the assortment necessary to meet the fullest requirements of dealers in our line of goods.

We aim to secure for our goods the best practical qualities, and the highest standard of style and finish; and we ask for them the closest scrutiny and comparison. The goods themselves shall be their own best recommendation.

Respectfully,

STANLEY RULE & LEVEL CO.

PRICE LIST

☞ All Rules embraced in the following Lists, bear the Invoice Number by which they are sold, and are graduated to correspond with the description given of each.

Stanley's Boxwood Rules.
One Foot, Four Fold, Narrow.

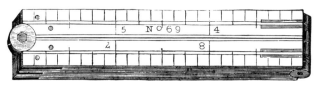

No.   Per Dozen.
69. Round Joint, Middle Plates, 8ths and 16ths of inches.............................$\frac{5}{8}$ in. wide, $3 00

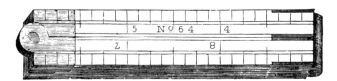

65. Square Joint, Middle Plates, 8ths and 16ths of inches.............................$\frac{5}{8}$ in. wide,   3 50
64. Square Joint, Edge Plates, 8ths and 16ths of inches.............................$\frac{5}{8}$ "   5 00
65½. Square Joint, Bound, 8ths and 16ths of inches, $\frac{5}{8}$ "   11 00

55. Arch Joint, Middle Plates, 8ths and 16ths of inches.............................$\frac{5}{8}$ in. wide,   4 00
56. Arch Joint, Edge Plates, 8ths and 16ths of inches.............................$\frac{5}{8}$ "   6 00
57. Arch Joint, Bound, 8ths and 16ths of inches..$\frac{5}{8}$ "   12 00

Two Feet, Four Fold, Narrow.

No.			Per Dozen.
68.	Round Joint, Middle Plates, 8ths and 16ths of inches ....1 in. wide,		$4 00
8.	Round Joint, Middle Plates, Extra Thick, 8ths and 16ths of inches 1	"	4 50
61.	Square Joint, Middle Plates, 8ths and 16ths of inches 1	"	5 00
63.	Square Joint, Edge Plates, 8ths, 10ths, and 16ths of inches, Drafting Scales..........1	"	7 00
84.	Square Joint, Half Bound, 8ths, 10ths, and 16ths of inches, Drafting Scales..........1	"	12 00
62.	Square Joint, Bound, 8ths, 10ths, and 16ths of inches, Drafting Scales................1	"	15 00
51.	Arch Joint, Middle Plates, 8ths, 10ths, and 16ths of inches, Drafting Scales..........1	"	6 00
53.	Arch Joint, Edge Plates, 8ths, 10ths, and 16ths of inches, Drafting Scales............1	"	8 00
52.	Arch Joint, Half Bound, 8ths, 10ths, and 16ths of inches, Drafting Scales............1	"	13 00
54.	Arch Joint, Bound, 8ths, 10ths, and 16ths of inches, Drafting Scales 1	"	16 00
59.	Double Arch Joint, Bitted, 8ths, 10ths, and 16ths of inches, Drafting Scales..........1	"	9 00
60.	Double Arch Joint, Bound, 8ths, 10ths, and 16ths of inches, Drafting Scales..........1	"	21 00

Two Feet, Four Fold, Extra Narrow.

61½.	Square Joint, Middle Plates, 8ths and 16ths of inches ....$\frac{3}{4}$ in. wide,		5 50
63½.	Square Joint, Edge Plates, 8ths, 10ths, and 16ths of inches..... ....$\frac{3}{4}$	"	8 00
62½.	Square Joint, Bound, 8ths, 10ths, 12ths, and 16ths of inches........................$\frac{3}{4}$	"	15 00

STANLEY'S BOXWOOD RULES.

Stanley's Two Feet, Four Fold, Narrow Rules.

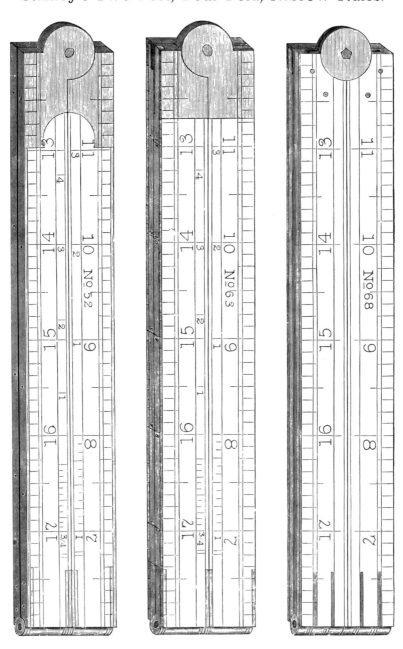

Two Feet, Four Fold, Broad.

No.			Per Dozen.
67.	Round Joint, Middle Plates, 8ths and 16ths of inches	$1\frac{3}{8}$ in. wide,	$5 00
70.	Square Joint, Middle Plates, 8ths and 16ths of inches, Drafting Scales	$1\frac{3}{8}$ "	7 00
72.	Square Joint, Edge Plates, 8ths, 10ths, and 16ths of inches, Drafting Scales	$1\frac{3}{8}$ "	9 00
$72\frac{1}{2}$.	Square Joint, Bound, 8ths, 10ths, and 16ths of inches, Drafting Scales	$1\frac{3}{8}$ "	18 00
73.	Arch Joint, Middle Plates, 8ths, 10ths, and 16ths of inches, Drafting Scales	$1\frac{3}{8}$ "	9 00
75.	Arch Joint, Edge Plates, 8ths, 10ths, and 16ths of inches, Drafting Scales	$1\frac{3}{8}$ "	11 00
76.	Arch Joint, Bound, 8ths, 10ths, and 16ths of inches, Drafting Scales	$1\frac{3}{8}$ "	20 00
77.	Double Arch Joint, Bitted, 8ths, 10ths, and 16ths of inches, Drafting Scales	$1\frac{3}{8}$ "	12 00
78.	Double Arch Joint, Half Bound, 8ths, 10ths, and 16ths of inches, Drafting Scales	$1\frac{3}{8}$ "	20 00
$78\frac{1}{2}$.	Double Arch Joint, Bound, 8ths, 10ths, and 16ths of inches, Drafting Scales	$1\frac{3}{8}$ "	24 00
83.	Arch Joint, Edge Plates, Slide, 8ths, 12ths, and 16ths of inches, 100ths of a foot, and Octagonal Scales	$1\frac{3}{8}$ "	14 00

Board Measure, Two Feet, Four Fold.

79.	Square Joint, Edge Plates, 12ths and 16ths of inches, Drafting Scales	$1\frac{3}{8}$ in. wide,	11 00
81.	Arch Joint, Edge Plates, 12ths and 16ths of inches, Drafting Scales	$1\frac{3}{8}$ "	13 00
82.	Arch Joint, Bound, 12ths and 16ths of inches, Drafting Scales	$1\frac{3}{8}$ "	22 00

Stanley's Two Feet, Four Fold, Broad Rules.

Two Feet, Two Fold.

No.			Per Dozen.
29. Round Joint, 8ths and 16ths of inches	1¾ in. wide,		$3 50
18. Square Joint, 8ths and 16ths of inches	1½	"	5 00
22. Square Joint, Bitted, Board Measure, 10ths and 16ths of inches, and Octagonal Scales	1½	"	8 00
1. Arch Joint, 8ths and 16ths of inches, Octagonal Scales	1½	"	7 00
2. Arch Joint, Bitted, 8ths, 10ths, and 16ths of inches, Octagonal Scales	1½	"	8 00
4. Arch Joint, Bitted, Extra Thin, 8ths and 16ths of inches, Drafting and Octagonal Scales	1½	"	10 00
5. Arch Joint, Bound, 8ths, 10ths, and 16ths of inches, Drafting and Octagonal Scales	1½	"	16 00

Two Feet, Two Fold, Slide.

26. Square Joint, Slide, 8ths, 10ths, and 16ths of inches, Octagonal Scales	1¼ in. wide,		9 00
27. Square Joint, Bitted, Gunter's Slide, 8ths, 10ths, and 16ths of inches, 100ths of a foot, Drafting and Octagonal Scales	1¼	"	12 00
12. Arch Joint, Bitted, Gunter's Slide, 8ths, 10ths, and 16ths of inches, 100ths of a foot, Drafting and Octagonal Scales	1¼	"	14 00
15. Arch Joint, Bound, Gunter's Slide, 8ths, 10ths, and 16ths of inches, Drafting and Octagonal Scales	1½	"	24 00
6. Arch Joint, Bitted, Gunter's Slide, Engineering, 8ths, 10ths, and 16ths of inches, 100ths of a foot, Octagonal Scales	1½	"	18 00
16. Arch Joint, Bound, Gunter's Slide, Engineering, 8ths, 10ths, and 16ths of inches, Octagonal Scales	1½	"	28 00

☞ With recently constructed machinery, we can furnish, to order, Rules marked with Spanish graduations, or with Metric graduations. See opposite page, for List of Meters, and of Rules graduated English and Metric.

Meters, and Rules, Graduated English and Metric.

Boxwood.

No.			Per Dozen.
10. Arch Joint, Four Fold, Graduated Mm. Cm. and Dm. ½ Meter in Length		20 Mm. wide,	$12 00
20. Arch Joint, Four Fold, Graduated Mm. Cm. and Dm. 1 Meter in Length		25 Mm. "	17 00
30. Arch Joint, Four Fold, Graduated Mm. Cm. and Dm. 1 Meter in Length		34 Mm. "	20 00
165. Square Joint, Four Fold, English on one side, and Metric on the other, 12 inch		⅝ in. wide,	5 00
161. Square Joint, Four Fold, English on one side, and Metric on the other, 24 "		1 "	7 00
151. Arch Joint, Four Fold, English on one side, and Metric on the other, 24 "		1 "	8 00
173. Arch Joint, Four Fold, English on one side, and Metric on the other, 24 "		1⅜ "	11 00
101. Arch Joint, Two Fold, English on one side, and Metric on the other, 24 "		1¼ "	9 00

Stanley's Two Feet, Two Fold, Slide Rules.

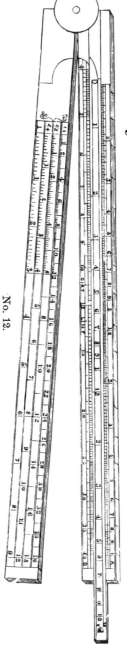

No. 12.

☞ We have an improved Treatise on the Gunter's Slide and Engineers' Rules, showing their utility, and containing full and complete instructions, enabling Mechanics to make their own calculations. It is also particularly adapted to the use of persons having charge of cotton or woolen machinery, surveyors, and others. 200 pages bound in cloth. Price $1.00, net. Sent by mail, postpaid, on receipt of the price.

Boxwood Caliper Rules.

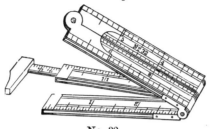

No. 32.

No.			Per Dozen.
36.	Square Joint, Two Fold, 6 inch, 8ths, 10ths, and 16ths of inches.................... ⅞ in. wide,		$7 00
36½.	Square Joint, Two Fold, 12 inch, 8ths, 10ths, 12ths, and 16ths of inches............1¾ "		12 00
32.	Arch Joint, Edge Plates, Four Fold, 12 inch, 8ths, 10ths, 12ths, and 16ths of inches.....1 "		12 00
32½.	Arch Joint, Bound, Four Fold, 12 inch, 8ths, 10ths, 12ths, and 16ths of inches..........1 "		20 00

Two Feet, Six Fold Rules.

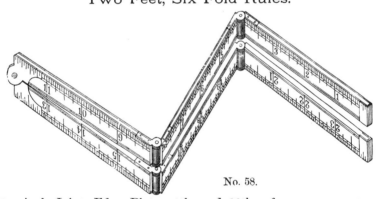

No. 58.

58.	Arch Joint, Edge Plates, 8ths and 16ths of inches................................... ¾ in. wide,	13 00

Three Feet, Four Fold Rules.

66.	Arch Joint, Middle Plates, Four Fold, 16ths of inches outside, and Yard Stick graduations on inside........................1 in. wide,	8 00
66½.	Arch Joint, Middle Plates, Four Fold, 8ths and 16ths of inches ....1 "	8 00

Ship Carpenters' Bevels.

42.	Boxwood, Double Tongue, 8ths and 16ths of inches,	6 00
43.	Boxwood, Single Tongue, 8ths and 16ths of inches,	6 00

Stanley's Ivory Rules.

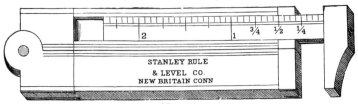

No. 38.

Ivory Caliper.

No.		Per Dozen.
38.	Square Joint, German Silver, Two Fold, Caliper, 6 inch, 8ths, 10ths, and 16ths of inches..⅞ in. wide,	$15 00
39.	Square Joint, Edge Plates, German Silver, Four Fold, Caliper, 12 inch, 8ths, 10ths, 12ths, and 16ths of inches.............⅞ "	38 00
40.	Square Joint, German Silver, Bound, Four Fold, Caliper, 12 inch, 8ths and 16ths of inches..⅝ "	44 00

Ivory, One Foot, Four Fold.

90.	Round Joint, Brass, Middle Plates, 8ths and 16ths of inches......................	10 00
92½.	Square Joint, German Silver, Middle Plates, 8ths and 16ths of inches...............⅝ in. wide,	14 00
92.	Square Joint, German Silver, Edge Plates, 8ths and 16ths of inches...............⅝ "	17 00
88½.	Arch Joint, German Silver, Edge Plates, 8ths and 16ths of inches...................⅝ "	21 00
88.	Arch Joint, German Silver, Bound, 8ths and 16ths of inches.....................⅝ "	32 00
91.	Square Joint, German Silver, Edge Plates, 8ths, 10ths, 12ths, and 16ths of inches....¾ "	23 00

Ivory, Two Feet, Four Fold.

85.	Square Joint, German Silver, Edge Plates, 8ths, 10ths, 12ths, and 16ths of inches.....⅞ in. wide,	54 00
86.	Arch Joint, German Silver, Edge Plates, 8ths, 10ths, and 16ths of inches, 100ths of a foot, Drafting Scales ....1 "	64 00
87.	Arch Joint, German Silver, Bound, 8ths, 10ths, and 16ths of inches, Drafting Scales......1 "	80 00
89.	Double Arch Joint, German Silver, Bound, 8ths, 10ths, and 16ths of inches, Drafting Scales 1 "	92 00
95.	Arch Joint, German Silver, Bound, 8ths, 10ths, and 16ths of inches, Drafting Scales......1⅜ "	102 00
97.	Double Arch Joint, German Silver, Bound, 8ths, 10ths, and 16ths of inches, Drafting Scales 1⅜ "	116 00

Miscellaneous Articles.

Bench Rules.

No.		Per Dozen.
34.	Bench Rule, Maple, Brass Tips....................2 feet,	$3 00
35.	" Board Measure, Brass Tips,............2 "	6 00
31.	" Satin Wood, Brass Capped Ends......2 "	12 00

Board, Log, and Wood Measures.

46.	Board Stick, Octagon, Brass Caps, 8 to 23 feet......2 feet,	8 00
46½.	" " Square, " 8 to 23 " ......2 "	8 00
47.	" " Octagon, " 8 to 23 " ......3 "	12 00
47½.	" " Square, " 8 to 23 " ......3 "	12 00
43½.	" " Flat, Hickory, Cast Brass Head and Tip, 6 Lines, 12 to 22 feet.....................3 "	12 00
49.	Board Stick, Flat, Hickory, Steel Head, Brazed, Extra Strong, 6 Lines, 12 to 22 feet............3 "	26 00
48.	Walking Cane, Board Measure, Octagon, Hickory, Solid Cast Brass Head and Tip, 8 Lines, 9 to 16 feet ....3 "	12 00
48¼.	Walking Cane, Scribner's Log Measure, Octagon, Hickory, Solid Cast Brass Head and Tip. ....3 "	15 00
71.	Wood Measure, Brass Caps, 8ths of inches and 10ths of a foot.............. ....4 "	8 00

Yard Sticks.

33.	Yard Stick, Polished...........................	2 00
41.	" " Brass Tips, Polished..................	3 50
50.	" " Hickory, Brass Cap'd Ends, Polished...	4 50

Wantage and Gauging Rods.

44.	Wantage Rod, 8 lines..........................	5 00
37.	" " 12 lines.........................	7 00
45.	Gauging, " 120 gallons..................3 feet,	7 00
45½.	" " Wantage Tables.................4 "	18 00
49½.	Forwarding Stick ....5 "	24 00

Scholars' Rules.

23.	Maple, 12 inch, ¾ inch wide, Beveled Edge, 16ths of inches.................	1 00
98.	Boxwood, 12 inch, ¾ inch wide, Beveled Edge, 8ths and 16ths of inches.........................	1 25
99.	Boxwood, 12 inch, ¾ inch wide, Beveled Edge, 10ths and 16ths of inches.........................	1 50

Board and Log Measures.

NOTE.—Nos. 43½, 46, 46½, 47, 47½, 48, and 49 give the contents in Board Measure of 1 inch Boards.

DIRECTIONS.—Place the stick across the flat surface of the Board, bringing the inside of the Cap snugly to the edge of the same; then follow with the eye the Column of figures in which the length of the Board is given as the first figure under the Cap, and at the mark nearest the opposite edge of the Board will be found the contents of the Board in feet.

No. 47

No. 47½

No. 43½

No. 49

No. 48

NOTE.—The Log Measure (No. 48½) has Scribner's Tables, and gives the number of feet of one inch square edged boards, which can be sawed from a log of any size, from 12 to 36 inches in diameter, and of any length. The figures immediately under the head of the Cane, are for the length of Logs in feet. Under these figures, on the same line, at the mark nearest the diameter of the Log, will be found the number of feet the Log will make. If the Log to be measured is not over 15 feet long, the diameter should be taken at the small end; if over 15 feet, at the middle of the Log.

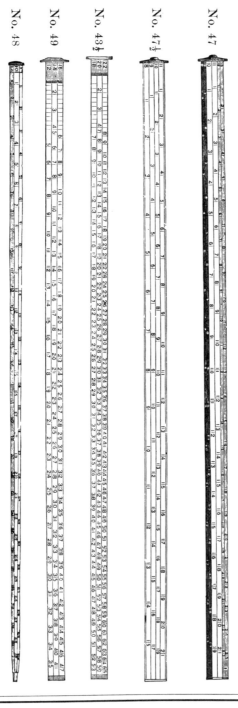

Stearns' Boxwood Rules.

Two Feet, Two Fold.

No.   Per Dozen.

1. Arch Joint, Bound, Slide, Engineering, 10ths and 16ths of inches, 100ths of a foot, Drafting Scales.................................$1\frac{1}{2}$ in. wide, $28 00
2. Arch Joint, Bitted, Slide, Engineering, 10ths, 12ths, and 16ths of inches, 100ths of a foot, Drafting Scales.......................$1\frac{1}{2}$ " 18 00
3. Arch Joint, Bound, Gunter's Slide, 10ths, 12ths, and 16ths of inches, 100ths of a foot, Drafting and Octagonal Scales...........$1\frac{1}{2}$ " 24 00
5. Arch Joint, Bitted, Gunter's Slide, 10ths, 12ths, and 16ths of inches, 100ths of a foot, Drafting and Octagonal Scales...........$1\frac{1}{2}$ " 14 00
7. Arch Joint, Bitted, 8ths and 16ths of inches, Drafting and Octagonal Scales...........$1\frac{1}{2}$ " 8 00
9. Square Joint, Slide, 8ths and 16ths of inches, Drafting and Octagonal Scales...........$1\frac{1}{2}$ " 9 00
11. Square Joint, 8ths and 16ths of inches.......$1\frac{1}{2}$ " 5 00

Two Feet, Four Fold, Broad.

14. Arch Joint, Bound, Board Measure, 10ths, 12ths, and 16ths of inches, Drafting and Octagonal Scales [see Engraving, p. 15]...$1\frac{1}{2}$ in. wide, 24 00
15. Arch Joint, Bound, 8ths, 10ths, 12ths, and 16ths of inches, Drafting and Octagonal Scales..$1\frac{1}{2}$ " 22 00
17. Arch Joint, Arch Back, Edge Plates, 8ths, 10ths, 12ths, and 16ths of inches, 100ths of a foot, Drafting and Octagonal Scales.....$1\frac{1}{2}$ " 15 00
18. Arch Joint, Triple Plated Edge Plates, Board Measure, 10ths, 12ths, and 16ths of inches, 100ths of a foot, Drafting and Octagonal Scales....................................$1\frac{1}{2}$ " 14 00
19. Arch Joint, Triple Plated Edge Plates, Slide, 8ths, 10ths, 12ths, and 16ths of inches, 100ths of a foot, Drafting and Octagonal Scales..$1\frac{1}{2}$ " 15 00
20. Arch Joint, Triple Plated Edge Plates, 8ths, 10ths, 12ths, and 16ths of inches, 100ths of a foot, Drafting and Octagonal Scales.....$1\frac{1}{2}$ " 13 00
21. Arch Joint, Middle Plates, Bitted, 8ths and 16ths of inches, Drafting and Octagonal Scales.....................................$1\frac{1}{2}$ " 11 00
32. Square Joint, Middle Plates, Bitted, 8ths and 16ths of inches, Drafting and Octagonal Scales.....................................$1\frac{1}{2}$ " 9 00

STEARNS' BOXWOOD RULES. 15

Stearns' Two Feet, Four Fold Rules.

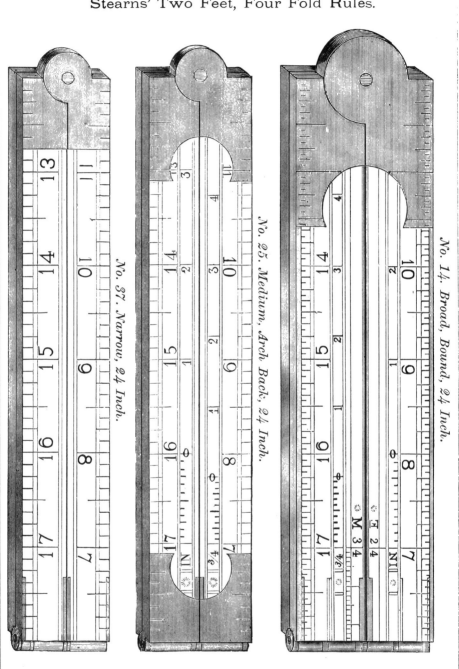

No. 37. Narrow, 24 Inch.

No. 25. Medium, Arch Back, 24 Inch.

No. 14. Broad, Bound, 24 Inch.

Two Feet, Four Fold, Medium.

No.				Per Dozen.
22.	Arch Joint, Bound, Board Measure, 16ths of inches, Drafting Scales	$1\frac{1}{8}$	in. wide,	$20 00
23.	Arch Joint, Bound, 8ths, 10ths, 12ths, and 16ths of inches, Drafting Scales	$1\frac{1}{8}$	"	18 00
25.	Arch Joint, Arch Back, Edge Plates, 8ths, 10ths, and 16ths of inches, 100ths of a foot, Drafting and Octagonal Scales [see Engraving, p. 15]	$1\frac{1}{8}$	"	13 00
26.	Arch Joint, Edge Plates, Board Measure, 16ths of inches, Drafting Scales	$1\frac{1}{8}$	"	11 00
27.	Arch Joint. Middle Plates, Bitted, 8ths and 16ths of inches, Drafting and Octagonal Scales	$1\frac{1}{8}$	"	9 00

Two Feet, Four Fold, Narrow.

45.	Arch Joint, Edge Plates, 8ths and 16ths of inches	1	in. wide,	8 00
46.	Arch Joint, Middle Plates, 8ths and 16ths of inches	1	"	6 00
34.	Square Joint, Square Back, Edge Plates, 8ths and 16ths of inches	1	"	10 00
35.	Square Joint, Bound, 8ths and 16ths of inches, Drafting and Octagonal Scales	1	"	15 00
37.	Square Joint, Edge Plates, 8ths and 16ths of inches [see Engraving, p. 15]	1	"	7 00
38.	Square Joint, Middle Plates, 8ths and 16ths of inches	1	"	5 00
41.	Round Joint, Middle Plates, 8ths and 16ths of inches	1	"	4 00

Two Feet, Four Fold, Extra Narrow.

31.	Arch Joint, Edge Plates, 8ths, 10ths, and 16ths of inches	$\frac{3}{4}$	in. wide,	9 00
72.	Arch Joint, Bound, 8ths, 10ths, 12ths, and 16ths of inches	$\frac{3}{4}$	"	17 00
33.	Square Joint, Edge Plates, 8ths, 10ths, and 16ths of inches	$\frac{3}{4}$	"	8 00
36.	Square Joint, Middle Plates, 8ths and 16ths of inches	$\frac{3}{4}$	"	5 50

Two Feet, Six Fold, Extra Narrow.

$28\frac{1}{2}$.	Arch Joint, Edge Plates, 8ths, 10ths, 12ths, and 16ths of inches, 100ths of a foot	$\frac{3}{4}$	in. wide,	13 50

STEARNS' BOXWOOD RULES.

One Foot, Four Fold.

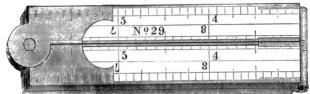

No.		Per Dozen.
29.	Arch Joint, Bound, 8ths, 10ths, 12ths, and 16ths of inches................13-16ths in. wide,	$13 00
30.	Arch Joint, Edge Plates, 8ths, 10ths, 12ths, and 16ths of inches, 100ths of a foot....13-16ths "	8 00
39.	Square Joint, Edge Plates, 8ths and 16ths of inches..........................13-16ths "	6 00
40.	Round Joint, Middle Plates, 8ths and 16ths of inches.........................13-16ths "	3 25
74.	Square Joint, Bound, 8ths and 16ths of inches.............................5-8ths "	12 00
75.	Square Joint, Middle Plates, 8ths and 16ths of inches.............................5-8ths "	4 00
43.	Round Joint, Middle Plates, 8ths and 16ths of inches.............................5-8ths "	3 00

One Foot, Four Fold, Caliper.

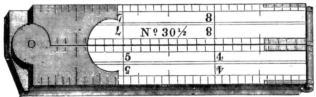

29½.	Arch Joint, Bound, 8ths, 10ths, 12ths, and 16ths of inches...............13-16ths in. wide,	18 00
30½.	Arch Joint, Edge Plates, 8ths, 10ths, 12ths, and 16ths of inches, 100ths of a foot...13-16ths "	12 50

Six Inch, Two Fold, Caliper.

12.	Square Joint, Brass Case, Spring Caliper, 8ths and 16ths of inches....................1⅛ in. wide,	12 00
13.	Square Joint, 8ths and 16ths of inches.......1⅛ "	8 00
13½.	Square Joint, 8ths and 16ths of inches...13-16ths "	7 00

Stearns' Ivory Rules.
Ivory, One Foot, Four Fold.

		Per Dozen.	
No.		Unbound.	Bound.
51.	Arch Joint, Edge Plates, German Silver, 8ths, 10ths, 12ths, and 16ths of inches (100ths of a foot on edges of unbound) 13-16ths in. wide,	$30 00	$40 00
52.	Square Joint, Edge Plates, German Silver, 8ths, 10ths, 12ths, and 16ths of inches (100ths of a foot on edges of unbound) 13-16ths in. wide,	27 00	37 00

Ivory, One Foot, Four Fold, Caliper.

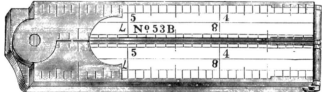

53.	Arch Joint, Edge Plates, German Silver, 8ths, 10ths, 12ths, and 16ths of inches (100ths of a foot on edges of unbound) 13-16ths in. wide,	35 00	45 00
54.	Square Joint, Edge Plates, German Silver, 8ths, 10ths, 12ths, and 16ths of inches (100ths of a foot on edges of unbound) 13-16ths in. wide,	32 00	42 00

Ivory, Six Inch, Two Fold, Caliper.

55.	Square Joint, German Silver, 8ths and 16ths of inches..............13-16ths in. wide,	15 00
55½.	Square Joint, German Silver Case, Spring Caliper, 8ths and 16ths of inches, 13-16ths in. wide,	18 00

Ivory, One Foot, Four Fold.

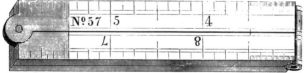

57.	Square Joint, Edge Plates, German Silver, 8ths and 16ths of inches...5-8ths in. wide,	18 00	28 00
58.	Square Joint, Edge Plates, Brass, 8ths and 16ths of inches............5-8ths in. wide,	15 00	
59.	Round Joint, Middle Plates, Brass, 8ths and 16ths of inches........................	12 00	

STEARNS' IVORY RULES, ETC. 19

Ivory, Two Feet, Four Fold, Broad.

		Per Dozen.	
No.		Unbound.	Bound.

47. Arch Joint, Triple Plated Edge Plates, German Silver, 8ths, 10ths, 12ths, and 16ths of inches, 100ths of a foot, Drafting and Octagonal Scales.......................1½ in. wide, $95 00 $110 00

Ivory, Two Feet, Four Fold, Medium.

48. Arch Joint, Edge Plates, German Silver, 8ths, 10ths, 12ths, and 16ths of inches (100ths of a foot on edges of unbound), Drafting Scales.......................1⅛ in. wide,   75 00   85 00

Ivory, Two Feet, Four Fold, Narrow.

50. Square Joint, Edge Plates, German Silver, 8ths, 10ths, 12ths, and 16ths of inches (100ths of a foot on edges of unbound), Drafting Scales.......................1 in. wide,   65 00   75 00

Ivory, Two Feet, Four Fold, Extra Narrow.

56. Arch Joint, Edge Plates, German Silver, 8ths, 10ths, 12ths, and 16ths of inches (100ths of a foot on edges of unbound)........¾ in. wide,   55 00   65 00

Ivory, Two Feet, Six Fold, Extra Narrow.

60. Arch Joint, Edge Plates, German Silver, 8ths, 10ths, 12ths, and 16ths of inches (100ths of a foot on edges of unbound)........¾ in. wide,   80 00   90 00

Miscellaneous Articles.

No.   Per Dozen.

63½. Bench Rule, Boxwood, Bound, 8ths and 16ths of inches..............................24 in. long, $18 00

64. Bench Rule, Maple, Capped Ends, 8ths and 16ths of inches........................24 "   4 00

71. Yardstick, Maple, Capped Ends..............36 "   4 00

80. Saddlers' Rule, Maple, Capped Ends, 8ths and 16ths of inches...............1½ in. wide, 36 "   9 00

81. Pattern Makers' Shrinkage Rule, Boxwood, 8ths and 16ths of inches...................24¼ "   15 00

82. Pattern Makers' Shrinkage Rule, Two Fold, Boxwood, Triple Plated Edge Plates, 8ths and 16ths of inches........................24¼ "   18 00

DESCRIPTION OF THE

Patent Improved Adjustable Plumbs and Levels,

MANUFACTURED BY THE

STANLEY RULE AND LEVEL CO.

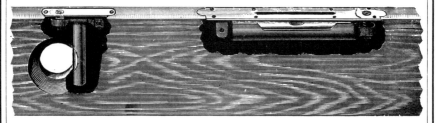

[SECTIONAL DRAWING.]

The Spirit-glass (or bubble tube), in the Level, is set in a Metallic Case, which is attached to the Brass Top-plate above it—at one end by a substantial hinge, and at the opposite end by an Adjusting Screw, which passes down through a flange on the Metallic Case. Between this flange and the Top-plate above, is inserted a stiff spiral spring; and by driving or slackening the Adjusting Screw, should occasion require, the Spirit-glass can be instantly adjusted to a position parallel with the base of the Level.

The Spirit-glass in the Plumb, is likewise set in a Metallic Case attached to the Brass Top-plate at its outer end. By the use of the Adjusting Screw at the lower end of the Top-plate, the Plumb-glass can be as readily adjusted to a right angle with the base of the Level, if occasion requires, and by the same method as adopted for the Level-glass.

The simplicity of this improved method of adjusting the Spirit-glasses will commend itself to every Mechanic. But one screw is used in the operation, and the action of the Brass Spring is perfectly reliable under all circumstances.

☞ Our Tipped Plumbs and Levels are now made with solid Brass Tips, of a handsome design, which entirely cover the ends of the Level Stock; and the finish on all our Levels has been materially improved, by the introduction of new methods in their manufacture.

PLUMBS AND LEVELS.

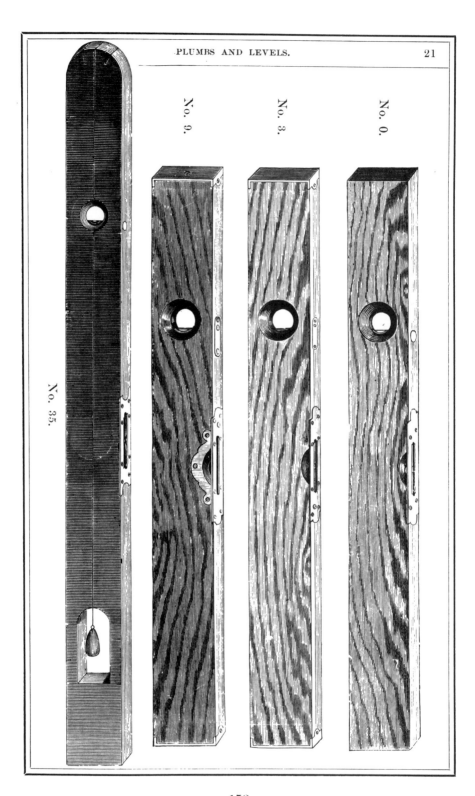

Plumbs and Levels.

No.		Per Dozen.
102.	Levels, Arch Top Plate, Two Side Views, Polished, Assorted .... 10 to 16 in.,	$9 00
103.	Levels, Arch Top Plate, Two Side Views, Polished, Assorted.................. 18 to 24 "	12 00
104.	Plumb and Level, Arch Top Plate, Two Side Views, Polished, Assorted, ... 12 to 18 "	14 00
1½.	Mahogany Plumb and Level, Arch Top Plate, Two Side Views, Polished, Assorted. ..... 18 to 24 "	16 50
1¾.	Mahogany Plumb and Level, Arch Top Plate, Two Brass Lipped Side Views, Polished and Tipped, Assorted .... 12 to 18 "	27 00
00.	Plumb and Level, Arch Top Plate, Two Side Views, Polished, Assorted.............. 18 to 24 "	16 00
0.	Plumb and Level, Arch Top Plate, Two Side Views, Polished, Assorted [see Engraving, p. 21]................................. 24 to 30 "	18 00
01.	Mahogany Plumb and Level, Arch Top Plate, Two Side Views, Polished, Assorted...... 24 to 30 "	22 50
02.	Plumb and Level, Arch Top Plate, Two Brass Lipped Side Views, Polished, Assorted.... 24 to 30 "	24 00
03.	Plumb and Level, Arch Top Plate, Two Side Views, Polished and Tipped, Assorted.... 24 to 30 "	28 00
04.	Plumb and Level, Arch Top Plate, Two Brass Lipped Side Views, Polished and Tipped, Assorted................................. 24 to 30 "	35 00
06.	Mahogany Plumb and Level, Arch Top Plate, Two Brass Lipped Side Views, Polished, Assorted............................. 24 to 30 "	30 00
7.	Masons' Plumb and Level, Arch Top Plate, Two Side Views, Polished and Tipped...... 36 "	36 00
12.	Machinists' Brass Bound Rosewood Plumb and Level, Two Brass Side Views, Polished.... 20 "	115 00

PATENT IMPROVED
Adjustable Plumbs and Levels.

No.   Per Dozen.

1. Patent Adjustable Mahogany Plumb and Level, Arch Top Plate, Two Side Views, Polished, Assorted.....................26 to 30 in., $27 00
2. Patent Adjustable Plumb and Level, Arch Top Plate, Two Brass Lipped Side Views, Polished, Assorted....................... 26 to 30 " 27 00
3. Patent Adjustable Plumb and Level, Arch Top Plate, Two Side Views, Polished and Tipped, Assorted [see Engraving, p. 21]..26 to 30 " 32 00
4. Patent Adjustable Plumb and Level, Arch Top Plate, Two Brass Lipped Side Views, Polished and Tipped, Assorted...........26 to 30 " 39 00
5. Patent Adjustable Plumb and Level, Triple Stock, Arch Top Plate, Two Ornamental Brass Lipped Side Views, Polished and Tipped, Assorted......................26 to 30 " 48 00
6. Patent Adjustable Mahogany Plumb and Level, Arch Top Plate, Two Brass Lipped Side Views, Polished, Assorted............26 to 30 " 33 00
9. Patent Adjustable Mahogany Plumb and Level, Arch Top Plate, Two Ornamental Brass Lipped Side Views, Polished and Tipped, Assorted [see Engraving, p. 21]..26 to 30 " 48 00
10. Patent Adjustable Mahogany Plumb and Level, Triple Stock, Two Ornamental Brass Lipped Side Views, Arch Top Plate, Polished and Tipped, Assorted................26 to 30 " 60 00
11. Patent Adjustable Rosewood Plumb and Level, Arch Top Plate, Two Ornamental Brass Lipped Side Views, Polished and Tipped, Assorted.............. ....26 to 30 " 90 00
25. Patent Adjustable Mahogany Plumb and Level, Arch Top Plate, Improved Double Adjusting Side Views, Polished and Tipped, 30 " 54 00
32. Patent Adjustable Mahogany Graduated Plumb and Level (easily adjusted to work at any angle or elevation required)........28 " 100 00
35. Patent Adjustable Masons' Plumb and Level [see Engraving, p. 21].......3¾ in. wide, 42 " 36 00

NICHOLSON'S PATENT
Metallic Plumbs and Levels.

The chief points of excellence in these Levels are the following : Once accurately constructed, they always remain so ; the sides and edges are perfectly true, thus combining with a Level a convenient and reliable straight edge ; shafting, or other work overhead, may be lined with accuracy by sighting the bubble from below, through an aperture in the base of the Level. Other points are their lightness, strength, and convenience of handling. All Levels, being thoroughly tested, are warranted reliable.

No.		Per Dozen.
13.	14 inches long	$18 00
14.	20 " "	21 00
15.	24 " "	27 00

IMPROVED
Iron Frame Plumbs and Levels.

These Levels have a Cast-Iron Frame, thoroughly braced, and combine the greatest strength with the least possible weight of metal. They are made absolutely correct in their form, and are subject to no change from warping or shrinking. The sides are inlaid with Black Walnut, secured by the screws which pass from one to the other, through the Iron braces, and the sides may be easily removed, if occasion requires.

No.			Per Dozen.
48.	Patent Adjustable Plumb and Level, Iron Frame, with Black Walnut (Inlaid) Sides	12 inch,	$24 00
49.	Patent Adjustable Plumb and Level, Iron Frame, with Black Walnut (Inlaid) Sides	18 "	30 00

Pocket Levels.

No. 41.

No.		Per Dozen.
40.	Cast Iron Top Plate, Japanned, 1 dozen in a box	$2 50
41.	Brass Top Plate, 1 dozen in a box	3 00
42.	All Brass, Pocket Level, 1 dozen in a box	8 00

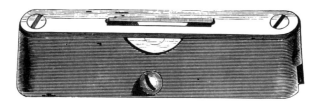

| 46. | Iron, Brass Top Plate, a superior article | 3 50 |

Machinists' Iron Levels.

38.	Iron Level, Brass Top Plate	4 inch,	3 00
39.	Iron Level, Brass Top Plate	6 "	4 50
43.	Iron Plumb and Level, Two Side Views, Brass Top Plate	9 "	10 00
45.	Iron Level, Brass Top Plate, Extra Finish	9 "	12 00

LEVEL GLASSES.

Per Gross.

Level Glasses, packed in ¼ gross boxes	1¾ inch,	$9 50
Level Glasses, packed in ¼ gross boxes	2 "	10 00
Level Glasses, packed in ¼ gross boxes	2½ "	10 50
Level Glasses, packed in ¼ gross boxes	3 "	11 50
Level Glasses, packed in ¼ gross boxes	3½ "	13 00
Level Glasses, packed in ¼ gross boxes	4 "	14 50
Level Glasses, packed in ¼ gross boxes	4½ "	16 00
Assorted, 1¾, 3, and 3½ inch, in ¼ gross boxes		12 00

Try Squares, No. 2.

GRADUATED STEEL BLADES.

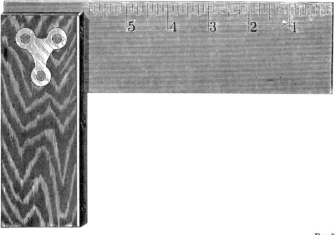

			Per Dozen.
Rosewood, 1 dozen in a box	3	inch,	$3 00
Rosewood, ½ "	4½	"	3 75
Rosewood, ½ "	6	"	5 00
Rosewood, ½ "	7½	"	5 75
Rosewood, ½ "	9	"	6 50
Rosewood, ½ "	12	"	8 50
Rosewood, with Rest, ½ dozen in a box	15	"	12 50
Rosewood, " ½ "	18	"	15 50

Plumb and Level Try Squares.

A very convenient Tool for Mechanics. A spirit-glass being set in the inner edge of the Try Square Handle, constitutes the Handle a Level; and when the Handle is brought to an exact level, the blade of the square will be upright, and become a perfect Plumb.

No.		Per Dozen.
1.	12 inch, Rosewood, with extra heavy Brass Mountings	$15 00
2.	7½ inch, Rosewood, with extra heavy Brass Mountings	7 50

½ dozen in a box.

Improved Try Squares, No. 1.

Iron Frame Handle, with Rosewood (Inlaid) Sides, Steel Blades, Square inside and out.

Inches,	3	4	6	8	9	10	12
Per Dozen,	$6 00	7 00	9 00	11 50	13 00	16 00	20 00

½ dozen in a box.

PATENT

Improved Try Squares, No. 2.

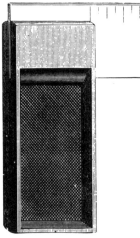

Iron Handle, Graduated Steel Blade, Square inside and out.

Inches,	4	6	8	10	12
Per Dozen,	$3 00	4 00	5 00	6 00	7 50

½ dozen in a box.

Improved Mitre Try Squares.

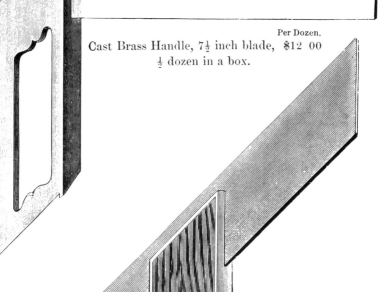

Per Dozen.
Cast Brass Handle, 7½ inch blade, $12 00
½ dozen in a box.

IMPROVED
Mitre Squares.

Iron Frame Handle, with Black Walnut (Inlaid) Sides.

Inches,	8	10	12
Per Dozen,	$7 00	8 00	9 00

½ dozen in a box.

WINTERBOTTOM'S PATENT
Combined Try and Mitre Square.

This Tool can be used with equal convenience and accuracy, as a Try Square, or a Mitre Square. By simply changing the position of the handle, and bringing the mitred face at the top of the handle against one edge of the work in hand, a perfect mitre, or angle of forty-five degrees, can be struck from either edge of the blade.

No 1. Iron Frame Handle, with Black Walnut (Inlaid) Sides, Graduated Steel Blades.

Inches,	4	6	8
Per Dozen,	$6 00	7 50	9 00

½ dozen in a box.

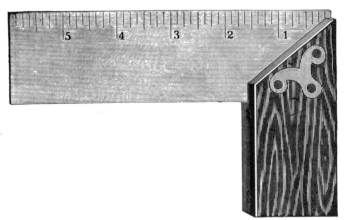

No. 2. Rosewood Handle, Graduated Steel Blades.

Inches,	4½	6	7½	9	12
Per Dozen,	$4 00	5 00	6 00	7 00	9 00

½ dozen in a box.

PATENT COMBINATION

Try Square and Bevel.

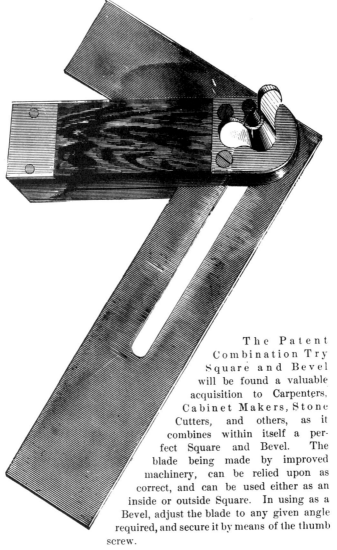

The Patent Combination Try Square and Bevel will be found a valuable acquisition to Carpenters, Cabinet Makers, Stone Cutters, and others, as it combines within itself a perfect Square and Bevel. The blade being made by improved machinery, can be relied upon as correct, and can be used either as an inside or outside Square. In using as a Bevel, adjust the blade to any given angle required, and secure it by means of the thumb screw.

Inches,	4	6	8	10	12
Per Dozen,	$5 00	6 00	7 50	9 00	10 50

½ dozen in a box.

Sliding T Bevels.

ROSEWOOD, WITH BRASS THUMB SCREW.

6 inch	per doz.,	$5	50
8 "	"	6	00
10 "	"	6	50
12 "	"	7	00
14 "	"	7	50

½ dozen in a box.

Patent Eureka Flush T Bevel.

IRON HANDLE, STEEL BLADE, WITH PARALLEL EDGES.

The Blade is easily secured at any angle, by turning the Thumb Screw at the lower end of the Handle.

Inches,	6	8	10
Per Dozen,	$6 00	6 50	7 50

½ dozen in a box.

Gauges.

☞ All Marking Gauges, excepting Nos. 0, 61, and 61½ have an Adjusting Point of finely-tempered steel, which may be readily removed and replaced if it needs sharpening. The point can be thrust down as it wears away, or if by any means it be broken off, it can be easily repaired.

All Gauges with Brass Thumb Screws, excepting No. 68, have also a Brass Shoe inserted in the head, under the end of the Thumb Screw. This shoe protects the gauge-bar from being dented by the action of the screw; and the broad surface of the shoe being in contact with the bar, the head is held more firmly in position than by any other method, and with less wear of the screw threads.

The engraving below (No. 60) shows an Iron Marking and Cutting Gauge. No. 60½ is the same in form, but has a Reversible Brass Slide slotted into the face of the bar. When a Mortise Gauge is required, the brass slide may be turned over in the bar. The point in the brass slide may be moved to any position, and the slide will be secured by a single turn of the screw which fastens the head of the Gauge.

No. 60.

No.		Per Dozen.
60.	Patent Iron Marking and Cutting Gauge, Oval Bar, Marked, Adjusting Steel Point, equally well adapted for use on Metals, or Wood, ¼ dozen in a box.......	$5 00
60½.	Patent Iron Reversible Gauge, Mortise, Marking and Cutting combined, Brass Slide, Oval Bar, Marked, Adjusting Steel Point, equally well adapted for use on Metals, or Wood, ½ dozen in a box............	8 00

No. 61.

0.	Marking Gauge, Beechwood, Boxwood Thumb Screw, Marked, Steel Point, 1 dozen in a box.............	75
61.	Marking Gauge, Beechwood, Boxwood Thumb Screw, Oval Bar, Marked, Steel Point, 1 dozen in a box....	1 00
61½.	Marking Gauge, Beechwood, Boxwood Thumb Screw, Oval Head and Bar, Marked, Steel Point, 1 dozen in a box,	1 25

GAUGES.

No. 65.

No.		Per Dozen.
62.	Patent Marking Gauge, Beechwood, Polished, Boxwood Thumb Screw, Oval Bar, Marked, Adjusting Steel Point, 1 dozen in a box	$2 00
64.	Patent Marking Gauge, Polished, Plated Head, Boxwood Thumb Screw, Oval Bar, Marked, Adjusting Steel Point, 1 dozen in a box.............................	2 75
64½.	Patent Marking Gauge, Polished, Oval Plated Head, Brass Thumb Screw and Shoe, Oval Bar, Marked, Adjusting Steel Point, ½ dozen in a box...................	4 50
65.	Patent Marking Gauge, Boxwood, Polished, Plated Head, Brass Thumb Screw and Shoe, Oval Bar, Marked, Adjusting Steel Point, ½ dozen in a box................	5 00
66.	Patent Marking Gauge, Rosewood, Oval Plated Head and Bar, Brass Thumb Screw and Shoe, Oval Bar, Marked, Adjusting Steel Point, ½ dozen in a box............	6 00

No. 71.

71.	Patent Double Gauge (Marking and Mortise Gauge combined), Beechwood, Polished, Plated Head and Bars, Brass Thumb Screws and Shoes, Oval Bars, Marked, Steel Points, ½ dozen in a box.......................	8 00
72.	Patent Double Gauge (Marking and Mortise Gauge combined), Beechwood, Polished, Boxwood Thumb Screws, Oval Bars, Marked, Steel Points, ½ doz. in a box,	4 00
74.	Patent Double Gauge (Marking and Mortise Gauge combined), Boxwood, Polished, Full Plated Head and Bars, Brass Thumb Screws and Shoes, Oval Bars, Marked, Steel Points, ½ dozen in a box.......................	14 00
70.	Cutting Gauge, Mahogany, Polished, Plated Head, Boxwood Thumb Screw, Oval Bar, Marked, Steel Cutter, 1 dozen in a box................................	4 00
83.	Handled Slitting Gauge, 17-inch Bar, Marked, ½ dozen in a box...	9 00
84.	Handled Slitting Gauge, with Roller, 17-inch Bar, Marked, ½ dozen in a box......................................	10 00

GAUGES.

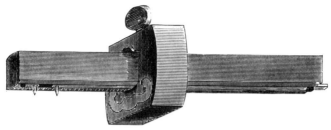

No. 73.

No.		Per Dozen.
73.	Patent Mortise Gauge, Boxwood, Polished, Plated Head, Brass Slide, Brass Thumb Screw and Shoe, Oval Bar, Marked, Steel Points, ½ dozen in a box.............	$8 00
76.	Patent Mortise Gauge, Boxwood, Polished, Plated Head, Screw Slide, Brass Thumb Screw and Shoe, Oval Bar, Marked, Steel Points, ½ dozen in a box.............	11 00
80.	Patent Mortise Gauge, Boxwood, Full Plated Head, Plated Bar, Screw Slide, Brass Thumb Screw and Shoe, Marked, Steel Points, ½ dozen in a box.........	18 00
67.	Mortise Gauge, Adjustable Wood Slide, Boxwood Thumb Screw, Oval Bar, Marked, Steel Points, ½ dozen in a box	4 00
68.	Mortise Gauge, Plated Head, Adjustable Wood Slide, Brass Thumb Screw, Oval Bar, Marked, Steel Points, ½ dozen in a box...................................	6 00

No. 77.

77.	Patent Mortise and Marking Gauge, Rosewood, Plated Head, Improved Screw Slide, Brass Thumb Screw and Shoe, Oval Bar, Marked, Steel Points, ½ dozen in a box	10 00
78.	Patent Mortise Gauge, Rosewood, Plated Head, Screw Slide, Brass Thumb Screw and Shoe, Oval Bar, Marked, Steel Points, ½ dozen in a box	11 00
79.	Patent Mortise Gauge, Rosewood, Plated Head and Bar, Screw Slide, Brass Thumb Screw and Shoe, Marked, Steel Points, ½ dozen in a box......................	13 00
85.	Panel Gauge, Beechwood, Boxwood Thumb Screw, Oval Bar, Steel Point, 1 dozen in a box.......	3 20
85½.	Panel Gauge, Rosewood, Plated Head and Bar, Brass Thumb Screw, Steel Point, ½ dozen in a box.........	18 00
90.	Williams' Patent Combination Gauge, ½ dozen in a box,	18 00

PATENT IMPROVED

Solid Cast Steel Screw Drivers.

By the Aid of improved machinery, we are producing Screw Drivers which are superior to any others in the market—and at the prices paid for the ordinary Tools.

The Blades are made from the best quality of Cast Steel, and are tempered with great care. They are ground down to a correct taper, and pointed at the end, by special machinery; thus procuring perfect uniformity in size, form, and strength, while the peculiar shape of the point gives it unequalled firmness in the screw-head when in use.

The shanks of the Blades are properly slotted to receive a patent metallic fastening, which secures them permanently in the Handles. The Handles are of the most approved pattern, the Brass Ferrules, of the thimble form, extra heavy, and closely fitted.

No. 1 Screw Drivers, have Black Enameled Handles. No. 2 Screw Drivers, have the same Blade, and are sold from the same list as No. 1. The difference in discount will be noted in our Discount Sheets.

PRICES.

Sizes	$1\frac{1}{2}$	2	3	4	5	6	7	8	10	Inches.
	$1.00	1.50	2.00	2.50	3.00	3.50	4.00	4.75	6.00	Per Dozen.

Sizes $1\frac{1}{2}$, 2, 3, and 4 inch, are packed 1 dozen in a box; all other sizes $\frac{1}{2}$ dozen in a box.

Sewing-Machine Screw Drivers.

Special Prices will be quoted for these in bulk.

Screw Driver Handles.

Assorted Sizes, 1 dozen in a box.................per dozen, $1 00

Bailey's Patent Adjustable Planes

Manufactured only by the Stanley Rule & Level Co., under the original Patents.

Since we commenced the manufacture of these Planes, they have earned for themselves an enviable reputation, which has already become world-wide. Our facilities for the production of the Planes, were never so good as now; and with an extended demand for them in all markets (our sales already exceed 125,000 Planes), we are enabled to offer them at reduced prices.

☞ Each Plane is thoroughly tested at our Factory, and put in perfect working order, before being sent into market. A printed description of the method of adjustment accompanies each Plane.

Iron Planes.

No.		Each.
1.	Smooth Plane, $5\frac{1}{2}$ inches in Length, $1\frac{1}{4}$ inch Cutter	$2 50
2.	Smooth Plane, 7 inches in Length, $1\frac{5}{8}$ inch Cutter	3 00
3.	Smooth Plane, 8 inches in Length, $1\frac{3}{4}$ inch Cutter	3 50
4.	Smooth Plane, 9 inches in Length, 2 inch Cutter	4 00
5.	Jack Plane, 14 inches in Length, 2 inch Cutter	4 50
6.	Fore Plane, 18 inches in Length, $2\frac{3}{8}$ inch Cutter	5 50
7.	Jointer Plane, 22 inches in Length, $2\frac{3}{8}$ inch Cutter	6 50
8.	Jointer Plane, 24 inches in Length, $2\frac{5}{8}$ inch Cutter	7 50
9.	Block Plane, 10 inches in Length, 2 inch Cutter	7 00
10.	Carriage Makers' Rabbet Plane, 14 inches in Length, $2\frac{1}{8}$ inch Cutter	5 00
11.	Belt Makers' Plane, $2\frac{3}{8}$ inch Cutter	3 00

Iron Planes.
Bailey's Patent Adjustable Block Planes.

These Block Planes are adjusted by a Screw and Lever movement, and the mouth can be opened wide, or made close, as the nature of the work to be done may require. The handles to Nos. 9¾ and 15½ are secured to the stock by the use of an iron nut, and can be easily attached to or liberated from the Plane, as the convenience of the workman may require.

No.		Each.
9½.	Excelsior Block Plane, 6 inches in Length, 1¾ inch Cutter.........	$1 50
15.	Excelsior Block Plane, 7 inches in Length, 1¾ inch Cutter.........	1 75

9¾. Excelsior Block Plane, with Rosewood Handle, 6 inches in Length, 1¾ inch Cutter... 1 75
15½. Excelsior Block Plane, with Rosewood Handle, 7 inches in Length, 1¾ inch Cutter... 2 00
CUTTERS, for above Block Planes, warranted Cast Steel, ...per dozen, 3 00

Bailey's Adjustable Veneer Scraper.

12. 3 inch Cutter... 4 00
CAST STEEL, HAND, VENEER SCRAPERS, 3 x 5 inch, ...per dozen. 3 50

Bailey's Patent Adjustable Circular Plane.

13. 1¾ inch Cutter 4 00

This Plane has a Flexible Steel Face, and by means of the thumb screws at each end of the Stock, can be easily adapted to plane circular work—either concave or convex.

Wood Planes.

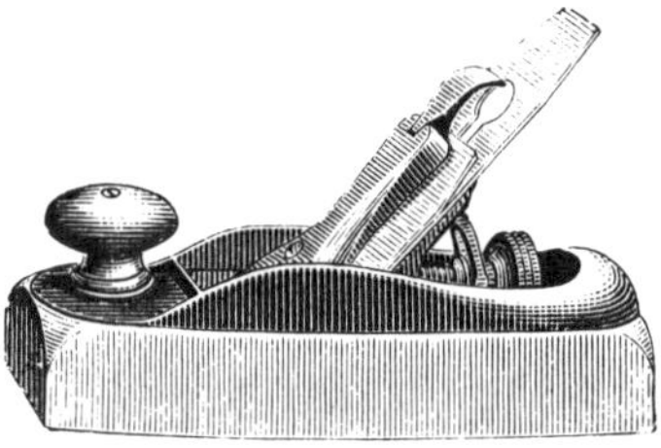

No.				Each.
21.	Smooth Plane,	7 inches in Length,	$1\frac{3}{4}$ inch Cutter......	$2 00
22.	Smooth Plane,	8 inches in Length,	$1\frac{3}{4}$ inch Cutter......	2 00
23.	Smooth Plane,	9 inches in Length,	$1\frac{3}{4}$ inch Cutter......	2 00
24.	Smooth Plane,	8 inches in Length,	2 inch Cutter......	2 25
25.	Block Plane,	$9\frac{1}{2}$ inches in Length,	$1\frac{3}{4}$ inch Cutter......	2 25

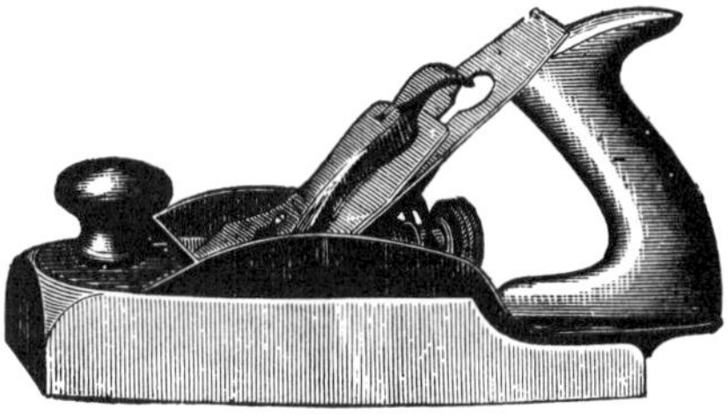

35.	Handle Smooth,	9 inches in Length,	2 inch Cutter.....	3 00
36.	Handle Smooth,	10 inches in Length,	$2\frac{3}{8}$ inch Cutter.....	3 25
37.	Jenny Smooth,	13 inches in Length,	$2\frac{5}{8}$ inch Cutter.....	3 50

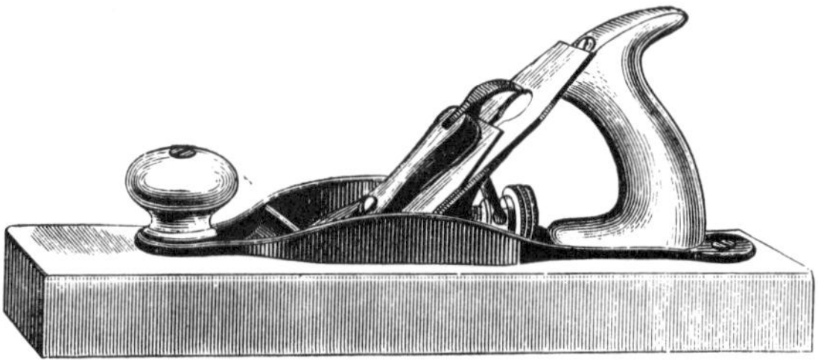

26.	Jack Plane,	15 inches in Length,	2 inch Cutter.......	2 50
27.	Jack Plane,	15 inches in Length,	$2\frac{1}{8}$ inch Cutter.......	2 75
28.	Fore Plane,	18 inches in Length,	$2\frac{3}{8}$ inch Cutter.......	3 00
29.	Fore Plane,	20 inches in Length,	$2\frac{3}{8}$ inch Cutter.......	3 00
30.	Jointer Plane,	22 inches in Length,	$2\frac{3}{8}$ inch Cutter.......	3 25
31.	Jointer Plane,	24 inches in Length,	$2\frac{3}{8}$ inch Cutter.......	3 25
32.	Jointer Plane,	26 inches in Length,	$2\frac{5}{8}$ inch Cutter.......	3 75
33.	Jointer Plane,	28 inches in Length,	$2\frac{5}{8}$ inch Cutter.......	3 75
34.	Jointer Plane,	30 inches in Length,	$2\frac{5}{8}$ inch Cutter.......	4 00

☞ Extra Plane-woods of every style, can be supplied cheaply.

Bailey's Plane Irons.

☞ These Plane Irons have stood the severest tests applied to them by practical workmen, and are by the manufacturers fully Warranted.

PRICES:

SINGLE IRONS.

Inches,	1¼	1⅝	1¾	2	2⅛	2⅜	2⅝
Each,	25	30	33	37	40	44	48 cts.

DOUBLE IRONS.

Each,	45	50	55	60	65	70	75 cts.

☞ Orders for Double Irons should state whether they are wanted for Iron Planes, or Wood Planes.

TESTIMONIALS.

" Works splendidly on our Yellow Pine."—H. C. HALL, Jacksonville, Florida.

" Far superior to any I have seen used."—JOHN ENGLISH, North Milwaukee, Wis., Car Shops.

" The Best Plane now in use."—CHICKERING & SONS, Piano Forte Manufacturers, Boston, Mass.

" Superior to any others used in this Shop."—W. S. TOWN, Master of Car Repairs, Hudson River R. R. Co.

" All that is wanted to sell your planes here is to show them."—E. WATERS. Manufacturer of Sporting Boats, Troy, N. Y.

" We know of no other plane that would be a successful substitute for them." R. BURDETT & Co., Organ Manufacturers, Erie, Pa.

" The best plane ever introduced in carriage-making."—JAS. L. MORGAN, at J. B. Brewster & Co.'s, Carriage Manufactory, New York.

" From the satisfaction they give would like to introduce them here."—J. ALEXANDER, at Bell's Melodeon Factory, Guelph, Canada.

" I have used your planes and have never found their equal."—B. F. FURRY. Foreman McLear & Kendall's Coach Works, Wilmington, Del.

" Bailey's Planes are used in our Factory. We can get only one verdict, and that is, the men would not be without them."—STEINWAY & SONS, Piano Forte Manufacturers, New York.

" I have owned a set of Bailey's Planes for six months, and can cheerfully say that I would not use any others, for I believe they are the best in use."—C. C. HARRIS, Stair Builder, St. Louis, Mo.

" We have Bailey's Planes in our Factory, and there is but one opinion about them. They are the best and cheapest tool we have ever used. Good tools are always the cheapest."—JOS. BECKHAUS, Carriage Builder, Philadelphia.

" I have used a set of Bailey's Planes about six months, and would not use any other now. There is inquiry about them here every day. Send me your descriptive Catalogue."—EDWARD CARVILLE, Sacramento City, California.

A Boston Mechanic says: "I always tell my shopmates, when they wish to use my Plane, not to borrow a Bailey Plane unless they intend to buy one, as they will never be satisfied with any other Plane after using this."

THE STANLEY
Adjustable Planes.

PATENTED.

These Planes are adjusted by the use of a Compound Lever, and are equally well adapted to coarse or fine work. The Planes are thoroughly tested at our Factory, and are in perfect working order when sent into market. The Plane Irons are made of the best English Cast Steel, and are fully Warranted.

Planes Nos. 104 and 105, have a Wrought Steel Stock, and are commended for their lightness of weight, and the ease with which they can be worked.

Steel Planes.

No.		Each.
104.	Smooth Plane, 9 inches in Length, $2\frac{1}{8}$ inch Cutter.......	$3 00

| 105. | Jack Plane, 14 inches in Length, $2\frac{1}{8}$ inch Cutter........ | 3 50 |

Iron Block Planes.

| 110. | Block Plane, $7\frac{1}{2}$ inches in Length, $1\frac{3}{4}$ inch Cutter....... | 70 |

| 120. | Block Plane, Adjustable, $7\frac{1}{2}$ inches in Length, $1\frac{3}{4}$ inch Cutter .... | 1 00 |
| | CUTTERS, for above Block Planes, warranted Cast Steel....per dozen, | 2 00 |

THE STANLEY ADJUSTABLE PLANES. 40½

Iron Block Planes.

No.		Each.
101.	Block Plane, 3½ inches in Length, 1 inch Cutter	$0 25

This Plane was first introduced as a convenient tool for amateurs, and others, who work with scroll saws, picture framing, etc. It has proved so valuable to mechanics in all the lighter kinds of wood working, and so useful about offices, stores, and dwellings, for making slight repairs of windows, doors, furniture, etc., that it seems likely to be wanted in every household.

CUTTERS for above Block Plane, warranted Cast Steel, per doz., $1 00

No.		Each.
102.	Block Plane, 5½ inches in Length, 1¼ inch Cutter	$0 50
103.	Adjustable Block Plane, 5½ in. in Length, 1¼ in. Cutter	75

CUTTERS for above Block Planes, warranted Cast Steel, per doz., 1 50

Tonguing and Grooving Planes.
(PATENTED.)

☞ Same pattern as No. 48 (see page 44), but adapted to matching boards varying from ⅜ to ⅝ inch in thickness.

PRICE, including Tonguing and Grooving Tools.

No. 49. Iron Stock and Fence..............................$2 50

EXPLANATORY.

☞ Our Patent Improved Solid Cast Steel Screw Drivers (see page 35) have all been brought up to a uniform quality, and the Blades to all are fastened securely in the Handles. They are no longer designated as Nos. 1 and 2 (the quality of all being now No. 1), but they are called "Black Handles," and "Varnished Handles." Both styles are sold from the former List, but at different discounts, as will be seen by reference to our Discount Sheets.

Wood Planes.

No.		Each.
122.	Smooth Plane, 8 inches in Length, 1¾ in. Cutter	$1 50

| 135. | Handle Smooth, 10 inches in Length, 2⅛ in. Cutter | 2 00 |

127.	Jack Plane, 15 inches in Length, 2⅛ in. Cutter	2 00
129.	Fore Plane, 20 inches in Length, 2⅜ in. Cutter	2 25
132.	Jointer Plane, 26 inches in Length, 2⅝ in. Cutter	2 75

☞ The STANLEY PLANE IRONS are the same prices, for corresponding widths, as per list of Plane Irons, page 39.

Patent Improved Rabbet Planes.

WITH STEEL CASE.

No.		Each.
80.	Rabbet Plane, Skew, 9 inches in Length, 1½ in. Cutter	$1 10
90.	Rabbet Plane, Skew, 9 inches in Length, with Spur, 1½ in. Cutter	1 25

MILLER'S PATENT COMBINED
Plow, Filletster & Matching Plane.

This tool embraces, in a most ingenious and successful combination, the common Carpenters' Plow, an adjustable Filletster, and a perfect Matching Plane. The entire assortment can be kept in smaller space, or made more portable, than an ordinary Carpenters' Plow.

☞ Each Plane is accompanied by a Tonguing tool (¼ inch), a Filletster Cutter, and eight Plow-bits (1-8, 3-16, 1-4, 5-16, 3-8, 7-16, 1-2 and 5-8 inch).

PRICES, including Plow Bits, Tonguing and Grooving Tools.
No. 41.   Iron Stock and Fence..$ 9 00
No. 42.   Gun Metal Stock and Fence................................. 12 00

COMBINED
Plow and Matching Plane.

The above engraving represents the Tool, adjusted for use as a Plow. With each Plow eight Bits (1 8, 3-16, 1-4, 5-16, 3-8, 7-16, 1-2, and 5-8 inch) are furnished; also a Tonguing Tool (¼ inch), and by use of the latter, together with the ¼ inch Plow Bit for grooving, a perfect Matching Plane is made.

PRICES, including Plow Bits, Tonguing and Grooving Tools.
No. 43.   Iron Stock and Fence..$ 7 00
No. 44.   Gun Metal Stock and Fence................................. 10 00

The Tool is packed in a box, and a printed description accompanies each one.

TRAUT'S PATENT ADJUSTABLE

Dado, Filletster, Plow, Etc.

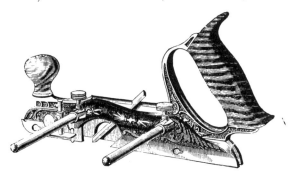

The Tool here represented, consists of two sections: A main stock, with two bars, or arms; and a sliding section, having its bottom, or face, level with that of the main stock.

It can be used as a Dado of any required width, by inserting the bit into the main stock, and bringing the sliding section snugly up to the edge of the bit. The two spurs, one on each section of the plane, will thus be brought exactly in front of the edges of the bit. The gauge on the sliding section will regulate the depth to which the tool will cut.

By attaching the Guard-plate, shown above, to the sliding section, the Tool may be readily converted into a Plow, a Filletster, or a Matching Plane—as explained in the printed instructions which go in every box.

The Tool is accompanied by eight Plow Bits, (3-16, 1-4, 5-16, 3-8, 1-2, 5-8, 7-8, and 1 1-4 inch), a Filletster Cutter, and a Tonguing Tool. All these tools are secured in the main stock on a skew.

PRICE, including Plow Bits, Tonguing and Grooving Tools.
No. 46. Iron Stock and Fence..$8 00

Adjustable Dado.

It can be used as a Dado of any required width, by inserting the bit into the main stock, and bringing the sliding section snugly up to the edge of the bit. The two spurs, one on each section of the plane, will thus be brought exactly in front of the edges of the bit. The gauge on the sliding section will regulate the depth to which the tool will cut.

PRICE, including Bits (3-8, 1-2, 5-8, 7-8 and 1 1-4 inch).
No. 47. Iron Stock and Fence..$5 00

Tonguing and Grooving Plane.

PATENTED.

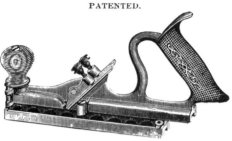

The stock of this tool is made of metal, and it has two cutters fastened into the stock by thumb screws. The guide, or fence, when set as shown in the above engraving, allows both of the cutters to act; and the cutters being placed at a suitable distance apart, a perfect Tonguing Plane is made. The guide, or fence, which is hung on a pivot at its centre, may be easily swung around, end for end; thus one of the cutters will be covered, and the guide held in a new position, thereby converting the Tool into a Grooving Plane. A groove will be cut to exactly match the tongue which is made by the other adjustment of the Tool.

The guide, or fence, is hung for grooving boards planed from 1 inch stuff; and on these the tongue and groove will both come in the centre of the board. Boards varying from ¾ to 1¼ inch in thickness can be matched equally well, by working the Plane so that the tongue and groove shall both come at the regular distance from one edge of the boards to be matched, leaving the distance to the other edge to vary as it may. One extra width cutter accompanies the Tool, to be used on the outer side of the tongue, in tonguing boards thicker than those planed from 1 inch stuff.

The ingenuity and simplicity of this Tool, together with its compact form and durability, will commend it to the favorable regard of all wood-working mechanics.

PRICE, including Tonguing and Grooving Tools.

No. 48. Iron Stock and Fence.. $2 50

Improved Mitre Box.

The peculiar features which distinguish this Mitre Box above all others in market, are referred to below:

The frame is made of a single casting, and is subject to no change of position; being finished accurately at first, it must always remain true. The slot in the back of the frame, through which the saw passes, is only three-eighths of an inch wide, thereby obviating any liability to push short pieces of work through the slot, when the saw is in motion.

This Mitre Box can be used with a Back Saw, or a Panel Saw, equally well. If a Back Saw is used, both links which connect the rollers, or guides, are left in the upper grooves, and the back of the saw is passed through under the links. If a Panel Saw is used, the link which connects the rollers on the back spindle, is changed to the lower groove; and then the blade of the saw will be stiffly supported by both sets of rollers, and be made to serve as well as a Back Saw.

By slightly raising or lowering the spindles, when necessary, the leaden rolls at the bottom may be adjusted to stop the saw at the proper depth; and, by the use of a set screw, the spindles on which the guides revolve, may be turned sufficiently to make the rollers bear firmly on the sides of a saw blade of any thickness.

If a narrow saw blade is used, or if the saw blade becomes narrower from use, the rollers may be lowered on the spindles by removing some of the brass rings from under them.

PRICES.

Mitre Box, 20 inches $7 00
Mitre Box, 20 inches, with 20 inch Disston's Back Saw................... 10 00

Improved Trammel Points.

These Tools are used by Millwrights, Machinists, Carpenters, and all Mechanics having occasion to strike arcs, or circles, larger than can be conveniently done with ordinary Compass dividers. They may be used on a straight wooden bar of any length, and when secured in position by the thumb screws, all circular work can be readily laid out by their use. They are made of Bronze Metal, and have Steel Points, either of which can be removed, and replaced by the pencil socket which accompanies each pair, should a pencil mark be preferred in laying out work.

PRICES.

No.			Per Pair.
1.	(Small) Br'ze Metal, Steel Points,	$1	25
2.	(Medium) "	"	1 50
3.	(Large) "	"	2 25

Adjustable Plumb Bobs.

These Plumb Bobs are constructed with a reel at the upper end, upon which the line may be kept; and by dropping the bob with a slight jerk, while the ring is held in the hand, any desired length of line may be reeled off. A spring, which has its bearing on the reel, will check and hold the bob firmly at any point on the line. The pressure of the spring may be increased, or decreased, by means of the screw which passes through the reel. A suitable length of line comes already reeled on each Plumb Bob.

PRICES.

No.					Each.
1.	(Small) Bronze Metal, with Steel Point				$1 50
2.	(Large) "	"	"	"	2 00
5.	(Large) Iron	"	"	"	1 00

Bailey's Iron Spoke Shaves.

☞ The Spoke Shaves in the following List, are superior in style, quality, and finish, to any in market. The Cutters are made of the best English Cast Steel, tempered and ground by an improved method, and are in perfect working order when sent from the Factory.

No.		Per Dozen.
51.	Double Iron, Raised Handle, 10 inch, $2\frac{1}{8}$ in. Cutter,	$4 00
52.	Double Iron, Straight Handle, 10 inch, $2\frac{1}{8}$ in. Cutter	4 00
53.	Adjustable, Raised Handle, 10 inch, $2\frac{1}{8}$ in. Cutter	5 00
54.	Adjustable, Straight Handle, 10 inch, $2\frac{1}{8}$ in. Cutter,	5 00
55.	Model Double Iron, Hollow Face, 10 inch, $2\frac{1}{8}$ in. Cutter.	4 00
56.	Coopers' Spoke Shave, 18 inch, $2\frac{5}{8}$ inch Cutter	8 00
$56\frac{1}{2}$.	Coopers' Spoke Shave (Heavy), 19 inch, 4 inch Cutter	10 00
57.	Coopers' Spoke Shave (Light), 18 inch, $2\frac{1}{8}$ inch Cutter	5 00
58.	Model Double Iron, 10 inch, $2\frac{1}{8}$ inch Cutter	3 50
59.	Single Iron (Pattern of No. 56), 10 inch, $2\frac{1}{8}$ in. Cutter	4 00
60.	Double Cutter, Hollow and Straight, 10 in. $1\frac{1}{2}$ in. Cutters,	5 00

Price List of Spoke Shave Cutters.

Nos. 51, 52, 53, 54, 55, 57, 58, 59.................... 1 00
    No. 60 (in pairs), $1 50—No. 56, $1 50—No. $56\frac{1}{2}$, $2 25.

Patent Reversible Spoke Shave.

This Spoke Shave can be worked to and from the person using it, without changing position.

No.		Per Dozen.
62.	Raised Handle, (Heavy), Double Cutter, 10 inch, 2¼ inch Cutter	$6 50

Patent Adjustable Box Scraper.

70. Malleable Iron, 2 inch Steel Cutter..................... 7 00

WHEELER'S
Patent Countersink.
FOR WOOD.
OVER 35,000 HAVE BEEN SOLD.

The Bit of this Countersink is in the shape of a hollow eccentric cone, thus securing a cutting edge of uniform draft from the point to the base of the tool, and obviating the tendency of such a tool to lead off into the wood at its cutting edge, and to leave an angular line where it ceases to cut. The form of the tool between the cutter and the shank is that of a hollow half-cone, inverted, thus leaving ample space, just back of and above the cutter, for the free escape of shavings. This Countersink works equally well for every variety of screw, the pitch of the cone being the same as the taper given to the heads of all sizes of screws, thereby rendering only a single tool necessary for every variety of work. The Countersink cuts rapidly, and is easily sharpened by drawing a thin file lengthwise inside of the cutter. The ingenious method of attaching a gauge to a Countersink will be observed by reference to the engraving. By fastening the gauge at a given point, any number of screws may be driven so as to leave the heads flush with the surface, or at a uniform depth below it. The gauge can be easily moved, or detached entirely, by means of the set-screw; and the Countersinks are sold with or without the gauge.

PRICES.

Per Dozen.
Countersinks, 1 dozen in a box............ $3 50
Countersinks, 1 dozen in a box, with Gauge, 5 00

Chalk-Line Reels, Etc.

	Per Dozen.
Chalk-Line Reels, 3 dozen in a box	$0 45
Chalk-Line Reels, with 60 feet best quality Chalk-Line, 1 dozen in a box	2 00
Chalk-Line Reels, with Steel Scratch Awls, 1 dozen in a box	1 10
Chalk-Line Reels, with Steel Scratch Awls, and 60 feet best quality Chalk-Line, 1 dozen in a box	2 75

Handled Scratch Awls.

No.		Per Gross.
1.	Handled, Steel Scratch Awls, 1 dozen in a box	$7 50
2.	Handled, Steel Scratch Awls, Large, 1 dozen in a box	9 00

Handled Brad Awls, Etc.

	Per Gross.
Handled Brad Awls, assorted, 1 dozen in a box	$8 00
Handled Brad Awls, assorted, Large, 1 dozen in a box	9 50

Brad Awl Handles, Brass Ferrules, assorted, 3 dozen in a box,	3 75

Chisel Handles, Etc.

Per Gross.
Polished Hickory Firmer Chisel Handles, ass'd, 1 doz. in a box, $5 25
Polished Hickory Firmer Chisel Handles, " Large, "   "   6 25
Polished Apple   Firmer Chisel Handles,   "   "   "   6 00
Polished Apple   Firmer Chisel Handles, " Large, "   "   7 00

Polished Socket Firmer Chisel Handles, asst'd, 3 doz. in a box,   3 50
Polished Apple Socket Firmer Chisel Handles, assorted, 3 doz.
    in a box... 4 50

Polished Hickory Socket Framing Chisel Handles, Iron Ferrules,
    assorted, 1 dozen in a box.......................... 6 00

File Handles, Brass Ferrules, assorted, 3 dozen in a box....... 3 50
File Handles, Brass Ferrules, assorted, Large, 3 dozen in a box, 4 00

Polished Auger Handles, assorted.............................. 6 00
Polished Auger Handles, assorted, Large....................... 7 00
      Packed 2 gross in a box.

Hickory Cross Cut Saw Handles, 15 inches long, 70 cents per dozen.

Patent Excelsior Tool Handles.

DIRECTIONS.—Unscrew the Cap of the Handle, and select the Tool needed; with the thumb the centre bolt may be thrust down sufficiently to open the clamp at the small end of the Handle, into which the Tool can be inserted; then in replacing the cover and screwing it down to its place, the clamp will be closed, and the Tool firmly secured for use.

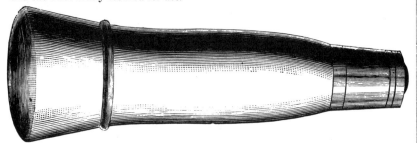

No.		Per Dozen.
1.	Turkey Boxwood Handle, with Twenty Tools	$11 00

½ dozen in a box.

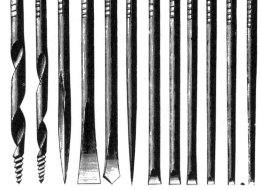

2.	Iron Handle, with Twelve Tools	5 50
3.	Iron Handle, with Twenty Tools (same as in No. 1)	9 00

½ dozen in a box.

HANDLES, AWL HAFTS, ETC. 51

Carpenters' Tool Handles.

No.   Per Dozen.
8. Carpenters' Tool Handle, Steel Screw, and Nut, with Iron Wrench, 1 dozen in a box ....$ 1 00
8½. Carpenters' Tool Handle, Steel Screw, and Nut, with Iron Wrench, and 10 Brad Awls, assorted sizes. One handle with brad awls, packed in box, and 12 boxes in package,   4 00

Awl Hafts.

No.   Per Gross.
5. Hickory, Pegging Awl Haft, Plain Top, Steel Screw, and Nut, with Iron Wrench, 1 dozen in a box $10 00
6. Hickory, Pegging Awl Haft, Leather Top, Steel Screw, and Nut, with Iron Wrench, 1 dozen in a box 12 00
7. Hickory, Pegging Awl Haft, Leather Top, Extra Large, Steel Screw, and Nut, with Iron Wrench, 1 doz. in a box, 14 00

6½. Appletree Sewing Awl Haft, to hold any size Awl, Steel Screw, and Nut, with Iron Wrench, 1 dozen in a box. 12 00

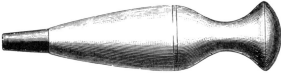

10. Common Sewing Awl Haft, Brass Ferrule, 3 dozen in a box,   3 50
11. Common Pegging Awl Haft, Brass Ferrule, 3 dozen in a box,   3 50
12. Saddlers' Sewing Awl Haft, Oval, Black Enameled, 3 dozen in a box 4 50

PATENT PEGGING AWLS.

PATENT PEGGING AWLS (short start, for Handles Nos. 5 and 6), assorted Nos. 00, 0, 1, 2, 3, 4, 5, and 6, 1 gross in a box,   75

Mallets.—Mortised Handles.

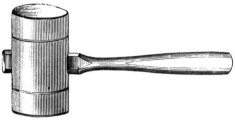

No.			Per Dozen.
1.	Round Hickory,	Mortised, 5 in. long, 3 in. diam....	$1 50
2.	Round Hickory,	Mortised, 5½ in. long, 3½ in. diam....	2 00
3.	Round Hickory,	Mortised, 6 in. long, 4 in. diam....	2 50
5.	Round Lignumvitæ,	Mortised, 5 in. long, 3 in. diam....	3 00
6.	Round Lignumvitæ,	Mortised, 5½ in. long, 3½ in. diam....	4 00
7.	Round Lignumvitæ,	Mortised, 6 in. long, 4 in. diam....	5 00

8.	Square Hickory,	Mortised, 6 in. long, 2½ by 3½ in....	2 00
9.	Square Hickory,	Mortised, 6½ in. long, 2¾ by 3¾ in....	2 50
10.	Square Hickory,	Mortised, 7 in. long, 3 by 4 in....	3 00
11.	Square Lignumvitæ,	Mortised, 6 in. long, 2½ by 3½ in....	3 75
12.	Square Lignumvitæ,	Mortised, 6½ in. long, 2¾ by 3¾ in....	4 75
13.	Square Lignumvitæ,	Mortised, 7 in. long, 3 by 4 in....	5 75

14.	Round Mallet, Mortised, Iron Rings, 6 in. long, 4 in. diam.	6 00
14½.	Round Mallet, Mortised, Iron Rings, 5½ in. long, 3½ in. diam.	4 00

Mallets.—Mortised Handles.

No.		Per Dozen.
15.	Round Iron Mallet, Mortised, Hickory Ends, 2½ in. diam.,	$4 00

No.		Per Dozen.
16.	Round Mallet, Heavy Malleable Iron Socket, Mortised, Hickory Ends, 3 in. diameter........................	8 00

Tinners' Mallets.

No.		Per Dozen.
4.	Round Hickory, 5½ in. long, assorted, 2¼ and 2½ in. diameter,	$1 00

Hand Screws.

BEADED JAWS.

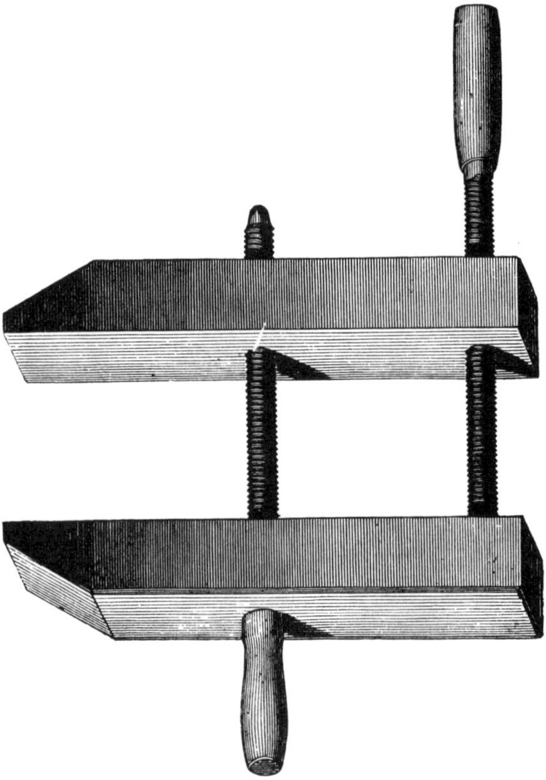

Diameter of Screws.	Length of Screws.	Length of Jaws.	Size of Jaws.	Per Dozen.
$1\frac{1}{4}$ inch.	24 inch.	20 inch.	$2\frac{7}{8}$ by $2\frac{7}{8}$ inch.	$10 50
$1\frac{1}{8}$ inch.	20 inch.	18 inch.	$2\frac{5}{8}$ by $2\frac{5}{8}$ inch.	8 50
1 inch.	18 inch.	16 inch.	$2\frac{3}{8}$ by $2\frac{3}{8}$ inch.	6 50
$\frac{7}{8}$ inch.	16 inch.	14 inch.	2 by 2 inch.	4 75
$\frac{3}{4}$ inch.	12 inch.	10 inch.	$1\frac{5}{8}$ by $1\frac{5}{8}$ inch.	3 25
$\frac{5}{8}$ inch.	10 inch.	$8\frac{1}{2}$ inch.	$1\frac{3}{8}$ by $1\frac{3}{8}$ inch.	2 50
$\frac{1}{2}$ inch.	10 inch.	8 inch.	$1\frac{1}{4}$ by $1\frac{1}{4}$ inch.	2 25

Hand Screws, $\frac{1}{2}$, $\frac{5}{8}$, and $\frac{3}{4}$ inch, packed 2 dozen in a case ; all other sizes, 1 dozen in a case.

Molders' Flask Screws, Rammers, and Mallets made to order.

Cabinet Makers' Clamps.

Per Dozen.
2 feet, inside of Jaws ....$ 9 00
3 feet, inside of Jaws 10 00
4 feet, inside of Jaws 11 00
5 feet, inside of Jaws 12 00

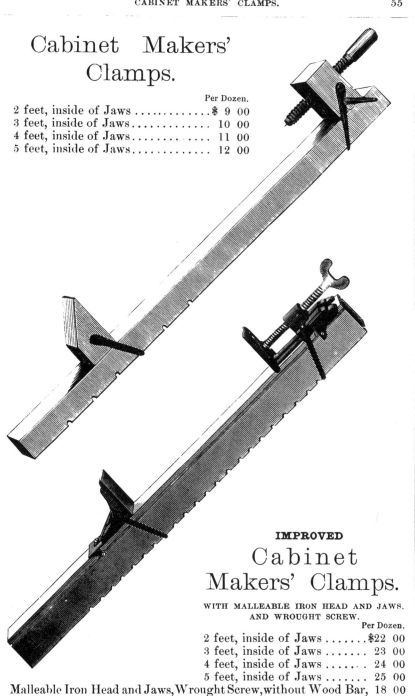

IMPROVED

Cabinet Makers' Clamps.

WITH MALLEABLE IRON HEAD AND JAWS, AND WROUGHT SCREW.

Per Dozen.
2 feet, inside of Jaws $22 00
3 feet, inside of Jaws 23 00
4 feet, inside of Jaws 24 00
5 feet, inside of Jaws 25 00
Malleable Iron Head and Jaws, Wrought Screw, without Wood Bar, 18 00

Saw Handles.

All full sizes, SPEAR & JACKSON's pattern, of perfect timber, well seasoned, and every way superior and reliable goods.

No.		Per Dozen.
1.	Full size, Cherry, Varnished Edges	$1 65
2.	Full size, Beech, Varnished Edges	1 40
3.	Full size, Beech, Plain Edges	1 20
4.	Small Panel, Beech, Varnished Edges, for 16 to 20 in. Saws,	1 35
5.	Meat Saw, Beech, Varnished Edges	1 35
6.	Compass Saw, Beech, Varnished Edges	1 25
7.	Back Saw, Beech, Varnished Edges	1 35

Packed 2 gross in a case.

Plane Handles.

	Per Dozen.
Jack Plane Handles, 5 gross in a case	$0 42
Fore, or Jointer Handles, 3¾ gross in a case	75

Door Stops.
Plain Door Stops, with Iron Screws.

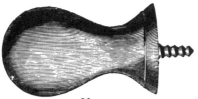

No. 6.

	Birch.	B. Walnut.	
2½ inch,	$3 75	$4 75	Per Gross.
3 inch,	4 25	5 25	Per Gross.

Packed 3 dozen in a box.

Improved Rubber Tip Door Stops.

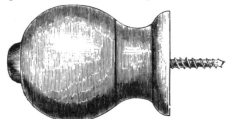

No. 8.

	Birch.	Chestnut.	B. Walnut.	Brown Enam.	White Enam.	
2½ inch,	$5 50	$6 50	$6 50	$6 50	$10 00	Per Gross.
3 inch,	6 00	7 00	7 00	7 00	11 00	Per Gross.

Packed 3 dozen in a box.

Floor Door Fenders

No. 10.

WITH RUBBER CUSHION.

2¼ Inch.

Chestnut.	B. Walnut.
$8 00	$8 00

No. 11.

WITH RUBBER RING.

	Chestnut.	B. Walnut.	
2¼ inch,	$10 00	$10 00	Per Gross.

Packed 3 dozen in a box.

Improved Sash Frame Pulleys.

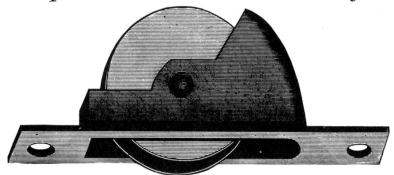

Superiority is claimed for these Pulleys over any others in use, in these essential particulars:

First. In respect to wear and friction. Turkey Boxwood is used for the Wheels, and large smooth Iron Axles, instead of two grinding surfaces of rough cast-iron; thus producing little friction, and no perceptible wear.

Second. The wheels being bored and turned true, the irregular motion and noise of the iron wheel is avoided.

Third. The groove of the wheel is turned out smooth, and of proper shape to receive the cord, which is thus prevented from wearing, and made to last much longer than in other pulleys.

	Per Dozen.
Sash Frame Pulley, 1¾ inch, 1 dozen in a box	$0 50
Sash Frame Pulley, 2 inch, 1 dozen in a box	65
Sash Frame Pulley, 2¼ inch, 1 dozen in a box	90

SMITH'S
Sash Cord Iron.

These Irons can be more easily put into the edge of a Sash than any other, requiring only the boring of a round hole; and no screws are needed to keep them in position. When the windows are taken out for cleaning, or any other purpose, the Irons can be removed or replaced without the use of any Tool.

In boxes of 1 gross .... 75 cts.

CATTLE TIES.

Patent Improved Cattle Tie.

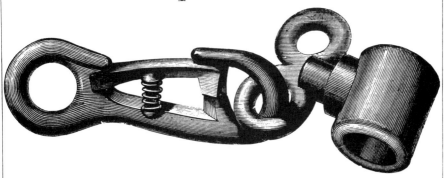

It is believed that this Cattle Tie, in both its parts, embraces more truly valuable features than any similar article heretofore sold. The snap is properly proportioned for the greatest strength required, and is made of Malleable iron. The form of the hooked end prevents the snap becoming unfastened by any accidental means; while the Improved Spiral Spring, which is made of brass, and not liable to rust from being wet, or to break in cold weather, acts with certainty, and is protected in its position under an independent lug, or bar, upon which all shocks or pressure must first come.

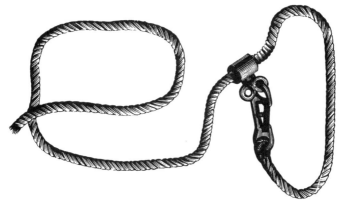

The other part of the Cattle Tie consists of an iron socket, which may be secured at any desired position on a rope, by means of a Malleable thumb screw in one side of the socket; the thumb screw having a perforated head, through which the snap is readily hooked. The socket may at once be changed from one position on the rope to another, and secured perfectly without the use of a screw-driver, or other Tool, thus adapting the length of the rope to the size of the neck or horns of any animal. The extreme simplicity of both parts insures great durability.

Per Dozen.
Cattle Tie, with Rope, Japanned..$4 00
Cattle Tie, without Rope, Japanned, 1 dozen in a box..................... 1 35
Improved Spiral Spring Snaps (only), Japanned................. 70

Steak Hammer and Ice Pick.

No.		Per Dozen
7.	Iron Japanned, ½ dozen in a box	$2 25
8.	X Plated, Polished Handle, ½ dozen in a box	3 25

Improved Steak Hammer and Ice Pick.

No.		Per Dozen.
9.	Malleable Iron, Japanned, Inlaid Handle, ½ dozen in a box	$4 25
10.	Malleable Iron, X Plated, Inlaid Handle, ¼ dozen in a box	5 25

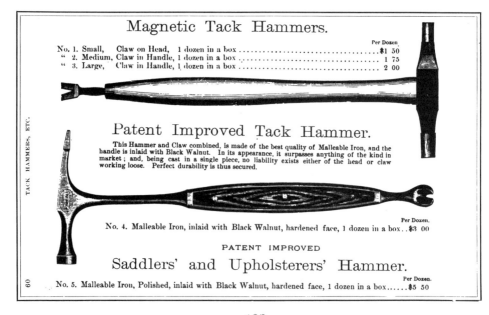

Magnetic Tack Hammers.

			Per Dozen
No. 1.	Small,	Claw on Head, 1 dozen in a box	$1 50
" 2.	Medium,	Claw in Handle, 1 dozen in a box	1 75
" 3.	Large,	Claw in Handle, 1 dozen in a box	2 00

Patent Improved Tack Hammer.

This Hammer and Claw combined, is made of the best quality of Malleable Iron, and the handle is inlaid with Black Walnut. In its appearance, it surpasses anything of the kind in market; and, being cast in a single piece, no liability exists either of the head or claw working loose. Perfect durability is thus secured.

Per Dozen.
No. 4. Malleable Iron, inlaid with Black Walnut, hardened face, 1 dozen in a box..$3 00

PATENT IMPROVED
Saddlers' and Upholsterers' Hammer.

Per Dozen.
No. 5. Malleable Iron, Polished, inlaid with Black Walnut, hardened face, 1 dozen in a box......$5 50

Tackle Blocks.

Rope Strapped Blocks.

HOOK AND THIMBLE.

Length of Blocks,	4	5	6	7	8	9	10	11	12	Inch.
Diameter of Wheels,	2½	3	3½	4¼	5	5¾	6½	7¼	8	"
Diameter of Rope,	½	⅝	¾	⅞	1	1	1⅛	1⅛	1¼	"
Single, Iron Bushed,	$0 75	0 95	1 15	1 35	1 85	2 15	2 40	2 80	3 25	Each.
Double, Iron Bushed,	1 30	1 60	2 00	2 30	3 00	3 70	4 50	5 40	6 00	"
Triple, Iron Bushed,	1 65	2 05	2 45	2 85	3 95	4 95	5 80	7 00	8 00	"
Single, Roller Bushed,	1 25	1 50	2 00	2 35	2 95	3 25	3 75	4 35	4 85	"
Double, Roller Bushed,	2 50	3 00	3 70	4 35	5 50	6 00	6 75	8 00	9 00	"
Triple, Roller Bushed,	3 50	4 25	5 00	5 85	7 55	8 50	9 25	11 20	12 50	"

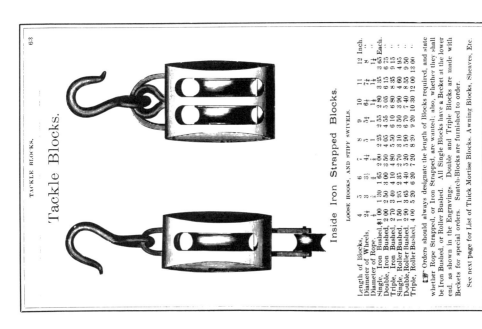

Tackle Blocks.

Inside Iron Strapped Blocks.

LOOSE HOOKS, AND STIFF SWIVELS.

Length of Blocks,	4	5	6	7	8	9	10	11	12	Inch.
Diameter of Wheels,	2½	3	3½	4¼	5	5¾	6½	7¼	8	"
Diameter of Rope.	½	⅝	¾	⅞	1	1	1⅛	1⅛	1¼	"
Single, Iron Bushed,	$1 00	1 30	1 65	2 00	2 25	2 55	2 80	3 35	3 65	Each.
Double, Iron Bushed,	2 00	2 50	3 00	3 50	4 05	4 55	5 05	6 15	6 75	"
Triple, Iron Bushed,	2 70	3 40	4 10	4 80	5 50	6 10	6 80	8 35	9 15	"
Single, Roller Bushed,	1 50	1 95	2 35	2 70	3 10	3 50	3 90	4 60	4 95	"
Double, Roller Bushed,	2 90	3 65	4 40	5 20	5 90	6 70	7 40	8 55	9 50	"
Triple, Roller Bushed,	4 00	5 20	6 20	7 20	8 20	9 20	10 30	12 00	13 00	"

☞ Orders should always designate the length of Blocks required, and state whether Rope Strapped, or Iron Strapped, are wanted; also, whether they shall be Iron Bushed, or Roller Bushed. All Single Blocks have a Becket at the lower end, as shown in the Engravings. Double and Triple Blocks are made with Beckets for special orders. Snatch-Blocks are furnished to order.

See next page for List of Thick Mortise Blocks, Awning Blocks, Sheaves, Etc.

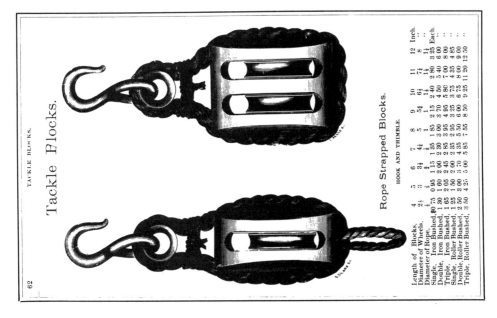

Tackle Blocks.

Rope Strapped Blocks, Hook and Thimble.
THICK MORTISE.

Length of Blocks,	9	10	11	12	Inch.
Diameter of Wheels,	5¾	6½	7¼	8	"
Diameter of Rope,	1¼	1⅜	1½	1½	"
Single, Iron Bushed,	$3 30	$3 65	$4 00	$4 45	Each.
Double, Iron Bushed,	6 60	7 30	8 00	8 90	"
Single, Roller Bushed,	5 15	5 70	6 25	6 85	"
Double Roller Bushed,	10 30	11 40	12 50	13 70	"

Plain Blocks, Not Strapped.
IRON BUSHED.

4 to 10 Inch..................................10 cts. per Inch.
11 and 12 Inch..................................14 " "

ROLLER BUSHED.

4 to 10 Inch..................................24 cts. per Inch.
11 and 12 Inch ....30 " "

Awning Blocks.

Awning Blocks, Rope Strapped.

	1¼	1½	2	2½	3	3½	4	Inch.
Single,	.10	.10	.11	.13	.15	.17	.20	Each.
Double,	.20	.20	.22	.26	.30	.34	.40	"

Awning Blocks, Not Strapped.

	1¼	1½	2	2½	3	3½	4	Inch.
Single,	.07	.07	.08	.10	.11	.13	.16	Each.
Double,	.14	.14	.16	.20	.22	.26	.32	"

Sheaves for Blocks.

Length of Blocks,	4	5	6	7	8	9	10	11	12 Inch.
Diameter of Wheels,	2½	3	3½	4¼	5	5¾	6½	7¼	8 "
Iron Bushed,	.16	.20	.24	.28	.38	.42	.55	.65	.80 Each.
Roller Bushed,	.55	.65	.80	.90	1.10	1.25	1.50	1.75	2.10 "

INDEX.

	Page.
Auger Handles,	49
Awl Hafts,	51
Awls, Patent Pegging,	51
Awning Blocks,	64
Brad Awl Handles,	48
Brad Awls, Handled,	48
Bevels, Sliding T,	31
Bevels, Patent Flush, Eureka,	31
Box Scraper, Adjustable,	47
Cabinet Makers' Clamps,	55
Chalk-Line Reels and Awls,	48
Carpenters' Tool Handles,	51
Cattle Ties,	59
Chisel Handles,	49
Cross Cut Saw Handles,	49
Countersinks, Wheeler's Patent,	47
Dado, Filletster, Plow, Etc., Combined,	43
Dado, Adjustable,	43
Door Stops,	57
Door Fenders, Floor,	57
File Handles,	49
Gauges,	32 to 34
Hand Screws,	54
Level Glasses,	25
Mallets, Hickory and Lignumvitæ,	52 and 53
Mitre Box, Improved,	44
Mitre Squares, Improved,	28
Mitre Try Squares, Improved,	28
Miscellaneous Articles, Stanley's,	12 and 13
Miscellaneous Articles, Stearns',	19

INDEX.

	Page.
Planes, Bailey's Patent Adjustable,	36 to 38
Plane Irons, Bailey's,	39
Planes, The Stanley Adjustable,	40 and 41
Planes, Rabbet,	41
Planes, Tonguing and Grooving,	44
Plane Handles,	56
Plumbs and Levels, Non-Adjustable,	22
Plumbs and Levels, Patent Adjustable,	23
Plumbs and Levels, Nicholson's Patent,	24
Plumbs and Levels, Iron Frame,	24
Plumbs and Levels, Machinists',	25
Pocket Levels,	25
Plow, Filletster and Matching Plane, Combined,	42
Plow and Matching Plane, Combined,	42
Plumb Bobs, Adjustable,	45
Rules, Stanley's,	3 to 11
Rules, Stearns',	14 to 19
Rules, with Metric Graduations,	9
Sash Frame Pulleys,	58
Sash Cord Irons,	58
Scratch Awls, Handled,	48
Saw Handles,	56
Screw Drivers,	35
Screw Driver Handles,	35
Steak Hammers,	61
Spoke Shaves, Bailey's,	46
Spoke Shave, Patent Reversible,	47
Spoke Shave Cutters,	46
Tackle Blocks,	62 to 64
Tack Hammers,	60
Trammel Points,	45
Tool Handles and Tools, Excelsior,	50
Try Squares, No. 2,	26
Try Squares, Improved,	27
Try Squares, Plumb and Level,	26
Try Square and Bevel, Patent Combination,	30
Try and Mitre Squares, Winterbottom's Patent,	29
Upholsterers' Hammers,	60
Veneer Scrapers,	37

1884 STANLEY RULE & LEVEL Co. CATALOG and PRICE LIST

In January 1884, when this Catalog and Price List was first published, only six new products appeared from the previous 1879 Revised Catalog. These were: Traut's N° 45 Adjustable Beading, Rabbet and Slitting Plane; his N° 50 Beading Plane; the N° 4½ Bailey Plane; the N° 130 Block Plane; the N° 65 Chamfer Spoke Shave; and the Adjustable Try Square.

Soon after the 1884 Catalog and Price List was issued, and through 1886, several important planes and other tools were added to the Catalog on insert sheets. The most important of these new additions were: N° 66 Hand Beader; N° 78 Duplex Rabbet Plane; N°ˢ 180-182, 190-192 Improved Rabbet Planes; N° 71 Woodworker's Hand Scraper; Lateral Adjustment of Bailey Planes; N° 10½ Carriage Maker's Plane; N° 112 Adjustable Scraper; N° 74 Floor Plane; N° 72 and 72½ Improved Chamfer Planes; N°ˢ 141 and 143 Bull Nose Plow, Filletster & Matching Plane; Hollow & Round Attachments for the N° 45 Combination Plane; and the N° 30½ Marking Gauge. All of the above are included as insert sheets in this 1884 Price List.

Ken Roberts
1980

PRICE LIST

OF

U. S. STANDARD

BOXWOOD AND IVORY

RULES,

PLUMBS AND LEVELS, TRY SQUARES,
BEVELS, GAUGES, MALLETS,
IRON AND WOOD ADJUSTABLE PLANES,
SPOKE SHAVES, SCREW DRIVERS,
AWL HAFTS, HANDLES, ETC.

MANUFACTURED BY THE

STANLEY
RULE AND LEVEL CO.,

NEW BRITAIN, CONN., U. S. A.

WAREROOMS:

No. 29 CHAMBERS STREET, NEW YORK.

ORDERS FILLED AT THE WAREROOMS, OR AT NEW BRITAIN.

JANUARY, 1884.

Office of the Stanley Rule and Level Co.,

NEW BRITAIN, CONN.,

January 1st, 1884.

We invite attention to the within Catalogue and Price List of Improved Carpenters' Tools manufactured by us.

Buyers will notice the omission of Stearns' Boxwood Rules from this Catalogue. The improvement of our regular line of Boxwood Rules up to its present high standard, has rendered the carrying of the two lines by us quite unnecessary.

We retain the full line of Stearns' Ivory Rules, so long known and favorably regarded by dealers in fine tools. These goods are mostly distinctive in style and finish from any other similar line. The prices of Stearns' Ivory Rules have now been changed to admit of a uniform discount with the line of Stanley's Ivory Rules.

A few valuable additions to our line of Special Tools have been made; and many improvements in the manufacture of all have been introduced, until we believe that our Company's name on a Carpenter's Tool is a Trade-Mark, and a pledge to the buyer of the excellence of the tool.

With our manufacturing facilities greatly enlarged, we may now confidently expect to meet promptly the demands of our many friends.

PRICE LIST.

☞ All Rules embraced in the following Lists bear the Invoice Number, by which they are sold, and are graduated to correspond with the description given of each.

STANLEY'S BOXWOOD RULES.
One Foot Four Fold, Narrow.

No.		Per Doz.
69.	Round Joint, Middle Plates, 8ths and 16ths of inches................................$\frac{5}{8}$ in. wide,	$3 00

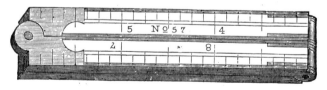

65.	Square Joint, Middle Plates, 8ths and 16ths of inches ....$\frac{5}{8}$ in. wide,		3 50
64.	Square Joint, Edge Plates, 8ths and 16ths of inches................................$\frac{5}{8}$ "		5 00
65½.	Square Joint, Bound, 8ths and 16ths of inches..$\frac{5}{8}$ "		11 00

55.	Arch Joint, Middle Plates, 8ths and 16ths of inches................................$\frac{5}{8}$ in. wide,		4 00
56.	Arch Joint, Edge Plates, 8ths and 16ths of inches................................$\frac{5}{8}$ "		6 00
57.	Arch Joint, Bound, 8ths and 16ths of inches..$\frac{5}{8}$ "		12 00

STANLEY'S BOXWOOD RULES.

Two Feet, Four Fold, Narrow.

No.		Per Dozen.
68.	Round Joint, Middle Plates, 8ths and 16ths of inches..............................1 in. wide,	$4 00
8.	Round Joint, Middle Plates, Extra Thick, 8ths and 16ths of inches......................1 "	4 50
61.	Square Joint, Middle Plates, 8ths and 16ths of inches..................................1 "	5 00
63.	Square Joint, Edge Plates, 8ths, 10ths, 12ths and 16ths of inches, Drafting Scales........1 "	7 00
84.	Square Joint, Half Bound, 8ths, 10ths, 12ths and 16ths of inches, Drafting Scales........1 "	12 00
62.	Square Joint, Bound, 8ths, 10ths, 12ths and 16ths of inches, Drafting Scales............1 "	15 00
51.	Arch Joint, Middle Plates, 8ths, 10ths, 12ths and 16ths of inches, Drafting Scales........1 "	6 00
53.	Arch Joint, Edge Plates, 8ths, 10ths 12ths and 16ths of inches, Drafting Scales........1 "	8 00
53½.	Arch Joint, Edge Plates, 8ths, 10ths, 12ths and 16ths of inches, with inside Beveled Edges, and Architect's Drafting Scales.....1 " (See cut on page 9.)	20 00
52.	Arch Joint, Half Bound, 8ths, 10ths, 12ths and 16ths of inches, Drafting Scales............1 "	13 00
54.	Arch Joint, Bound, 8ths, 10ths, 12ths and 16ths of inches, Drafting Scales.................1 "	16 00
59.	Double Arch Joint, Bitted, 8ths, 10ths, 12ths and 16ths of inches, Drafting Scales........1 "	9 00
60.	Double Arch Joint, Bound, 8ths, 10ths 12ths and 16ths of inches, Drafting Scales........1 "	21 00

Two Feet, Four Fold, Extra Narrow.

61½.	Square Joint, Middle Plates, 8ths and 16ths of inches ....¾ in. wide,	5 50
63½.	Square Joint, Edge Plates, 8ths, 10ths and 16ths of inches..........................¾ "	8 00
62½.	Square Joint, Bound, 8ths, 10ths, 12ths and 16ths of inches..........................¾ "	15 00

Stanley's Two Feet, Four Fold, Narrow Rules.

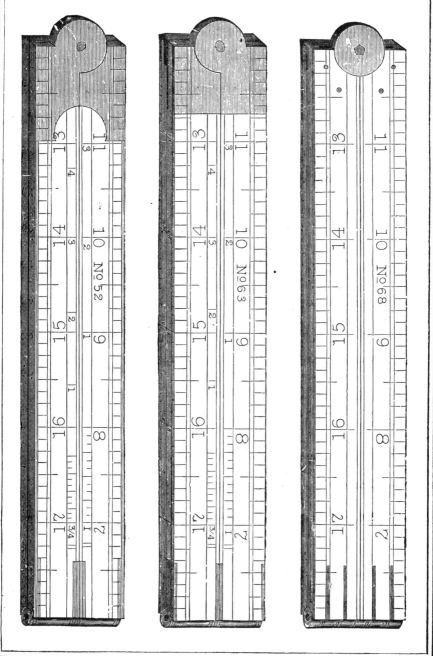

Two Feet, Four Fold, Broad.

No.   Per Dozen.

67. Round Joint, Middle Plates, 8ths and 16ths of inches............................$1\frac{3}{8}$ in. wide, $5 00

70. Square Joint, Middle Plates, 8ths and 16ths of inches, Drafting Scales.....................$1\frac{3}{8}$ " 7 00

72. Square Joint, Edge Plates, 8ths, 10ths and 16ths of inches, Drafting Scales............$1\frac{3}{8}$ " 9 00

72½. Square Joint, Bound, 8ths, 10ths and 16ths of inches, Drafting Scales.....................$1\frac{3}{8}$ " 18 00

73. Arch Joint, Middle Plates, 8ths, 10ths and 16ths of inches, Drafting Scales...........$1\frac{3}{8}$ " 9 00

75. Arch Joint, Edge Plates, 8ths, 10ths and 16ths of inches, Drafting Scales.................$1\frac{3}{8}$ " 11 00

76. Arch Joint, Bound, 8ths, 10ths and 16ths of inches, Drafting Scales.....................$1\frac{3}{8}$ " 20 00

77. Double Arch Joint, Bitted, 8ths, 10ths and 16ths of inches, Drafting Scales............$1\frac{3}{8}$ " 12 00

78. Double Arch Joint, Half Bound, 8ths, 10ths and 16ths of inches, Drafting Scales........$1\frac{3}{8}$ " 20 00

78½. Double Arch Joint, Bound, 8ths, 10ths and 16ths of inches, Drafting Scales............$1\frac{3}{8}$ " 24 00

83. Arch Joint, Edge Plates, Slide, 8ths, 12ths and 16ths of inches, 100ths of a foot, and Octagonal Scales........................$1\frac{3}{8}$ " 14 00

Board Measure, Two Feet, Four Fold.

79. Square Joint, Edge Plates, 12ths and 16ths of inches, Drafting Scales...................$1\frac{3}{8}$ in wide, 11 00

81. Arch Joint, Edge Plates, 12ths and 16ths of inches, Drafting Scales....................$1\frac{3}{8}$ " 13 00

82. Arch Joint, Bound, 12ths and 16ths of inches, Drafting Scales........................$1\frac{3}{8}$ " 22 00

Carriage Makers', Four Feet, Four Fold.

94. Arch Joint, Bound, 8ths and 16ths of incnes..$1\frac{1}{2}$ in. wide, 48 00

Stanley's Two Feet, Four Fold, Broad Rules.

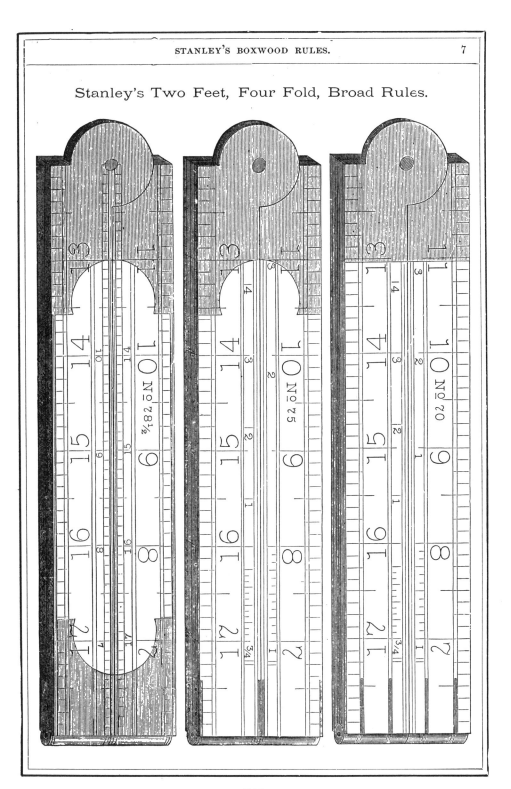

Two Feet, Two Fold.

No.   Per Dozen.

29. Round Joint, 8ths and 16ths of inches.........1⅜ in. wide, $3 50
18. Square Joint, 8ths and 16ths of inches.........1½   "   5 00
22. Square Joint, Bitted, Board Measure, 10ths, 12ths and 16ths of inches, Octagonal Scales..1½   "   8 00
1. Arch Joint, 8ths and 16ths of inches, Octagonal Scales.................................1½   "   7 00
2. Arch Joint, Bitted, 8ths, 10ths and 16ths of inches, Octagonal Scales...................1½   "   8 00
4. Arch Joint, (plates on outside of wood) Bitted, Extra Thin, 8ths and 16ths of inches, Drafting and Octagonal Scales....................1½   "   10 00
5. Arch Joint, Bound, 8ths, 10ths and 16ths of inches, Drafting and Octagonal Scales........1½   "   16 00

Two Feet, Two Fold, Slide.

26. Square Joint, Slide, 8ths, 10ths and 16ths of inches, Octagonal Scales..................1½ in. wide, 9 00
27. Square Joint, Bitted, Gunter's Slide, 8ths, 10ths and 16ths of inches, 100ths of a foot, Drafting and Octagonal Scales.................1½   "   12 00
12. Arch Joint, Bitted, Gunter's Slide, 8ths, 10ths and 16ths of inches, 100ths of a foot, Drafting and Octagonal Scales..................1½   "   14 00
15. Arch Joint, Bound, Gunter's Slide, 8ths, 10ths and 16ths of inches, Drafting and Octagonal Scales......................................1½   "   24 00
6. Arch Joint, Bitted, Gunter's Slide, Engineering, 8ths, 10ths and 16ths of inches, 100ths of a foot, Octagonal Scales..................1½   "   18 00
16. Arch Joint, Bound, Gunter's Slide, Engineering, 8ths, 10ths and 16ths of inches, Octagonal Scales...............................1½   "   28 00

☞ With recently constructed machinery, we can furnish, to order, Rules marked with Spanish graduations, or with Metric graduations.

STANLEY'S SLIDE RULES AND ARCHITECTS' RULES.

ARCHITECTS' RULES, WITH BEVELED EDGES.

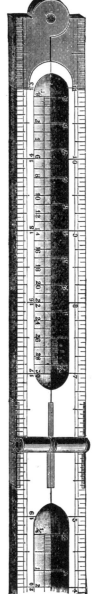

Nos. 53½ and 86½.

☞ For description of No. 53½ (Boxwood) see page 4, and for description of No. 86½ (Ivory) see page 11.

STANLEY'S TWO FEET, TWO FOLD, SLIDE RULES.

No. 12.

☞ We have an improved Treatise on the Gunter's Slide and Engineer's Rules, showing their utility, and containing full and complete instructions, enabling Mechanics to make their own calculations. It is also particularly adapted to the use of persons having charge of cotton or woolen machinery, surveyors, and others. 200 pages, bound in cloth. Price, $1.00, net. Sent by mail, postpaid, on receipt of the price.

Boxwood Caliper Rules

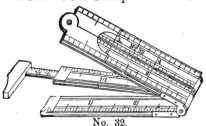

No. 32.

No.		Per Dozen.
36. Square Joint, Two Fold, 6 inch, 8ths, 10ths, 12ths and 16ths of inches............ ....⅞ in. wide,		$7 00
13. Square Joint, Two Fold, 6 inch, 8ths and 16ths of inches.... ....1⅛ "		10 00
36½. Square Joint, Two Fold, 12 inch, 8ths, 10ths, 12ths and 16ths of inches..................1⅜ "		12 00
32. Arch Joint, Edge Plates, Four Fold, 12 inch, 8ths, 10ths, 12ths and 16ths of inches.......1 "		12 00
32½. Arch Joint, Bound, Four Fold, 12 inch, 8ths, 10ths, 12ths and 16ths of inches...........1 "		20 00

Two Feet, Six Fold Rules.

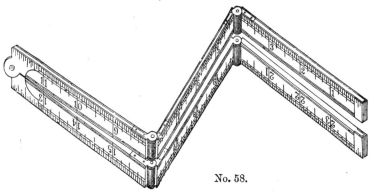

No. 58.

58. Arch Joint, Edge Plates, 8ths, 10ths, 12ths and 16ths of inches..........................¾ in. wide,		13 00
58½. Arch Joint, Bound, 8ths, 10ths, 12ths and 16ths of inches..................................¾ in. wide,		36 00

Three Feet, Four Fold Rules.

66. Arch Joint, Middle Plates, Four Fold, 16ths of inches outside, and Yard Stick graduations on inside.................................1 in. wide,		8 00
66½. Arch Joint, Middle Plates, Four Fold, 8ths and 16ths of inches ....1 "		8 00

Ship Carpenters' Bevels.

42. Boxwood, Double Tongue, 8ths and 16ths of inches,		6 00
43. Boxwood, Single Tongue, 8ths and 16ths of inches,		6 00

STANLEY'S IVORY RULES.

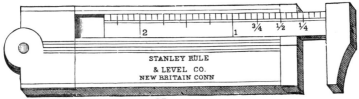

No. 38.

Ivory Caliper. *Per Dozen.*

38. Square Joint, German Silver, Two Fold Caliper, 6 inch, 8ths, 10ths, 12ths and 16ths of inches............................⅞ in. wide, $15 00
39. Square Joint, Edge Plates, German Silver, Four Fold, Caliper, 12 inch, 8ths, 10ths, 12ths, and 16ths of inches...............⅞ " 38 00
40. Square Joint, German Silver, Bound, Four Fold, Caliper, 12 inch, 8ths and 16ths of inches...⅝ " 44 00

Ivory, One Foot, Four Fold.

90. Round Joint, Brass, Middle Plates, 8ths and 16ths of inches......................... 10 00
92½. Square Joint, German Silver, Middle Plates, 8ths and 16ths of inches.................⅝ in. wide, 14 00
92. Square Joint, German Silver, Edge Plates, 8ths and 16ths of inches.................⅝ " 17 00
88½. Arch Joint, German Silver, Edge Plates, 8ths and 16ths of inches.....................⅝ " 21 00
88. Arch Joint, German Silver, Bound, 8ths and 16ths of inches.........................⅝ " 32 00
91. Square Joint, German Silver, Edge Plates, 8ths, 10ths, 12ths and 16ths of inches......¾ " 23 00

Ivory, Two Feet, Four Fold.

85. Square Joint, German Silver, Edge Plates, 8ths, 10ths, 12ths and 16ths of inches......⅞ in. wide, 54 00
86. Arch Joint, German Silver, Edge Plates, 8ths, 10ths, 12ths and 16ths of inches, 100ths of a foot, Drafting Scales....................1 " 64 00
86¼. Arch Joint, German Silver, Edge Plates, 8ths, 10ths, 12ths and 16ths of inches, with Inside Beveled Edges, and Architects' Drafting Scales (See cut on page 9.) ....1 " 96 00
87. Arch Joint, German Silver, Bound, 8ths, 10ths, 12ths and 16ths of inches, Drafting Scales..1 " 80 00
89. Double Arch Joint, German Silver, Bound, 8ths, 10ths, 12ths and 16ths of ins., Drafting Scales.1 " 92 00
95. Arch Joint, German Silver, Bound, 8ths, 10ths, 12ths and 16ths of inches, Drafting Scales.1⅜ " 102 00
97. Double Arch Joint, German Silver, Bound, 8ths, 10ths, 12ths and 16ths of ins, Drafting Scales.1⅜ " 116 00

MISCELLANEOUS ARTICLES.

Bench Rules.

No.		Per Dozen.
34.	Bench Rule, Maple, Brass Tips..................2 feet,	$3 00
35.	" " Board Measure, Brass Tips.........2 "	6 00

Board, Log and Wood Measures.

(For explanations, see Notes on opposite page.)

46.	Board Stick, Octagon, Brass Caps, 8 to 23 feet....2 feet,	8 00
46½.	" " Square, " 8 to 23 feet....2 "	8 00
47.	" " Octagon, " 8 to 23 feet....3 "	12 00
47½.	" " Square, " 8 to 23 feet....3 "	12 00
43½.	" " Flat, Hickory, Cast Brass Head and Tip, 6 Lines, 12 to 22 feet...................3 "	12 00
49.	Board Stick, Flat, Hickory, Steel Head, Brazed, Extra Strong, 6 Lines, 12 to 22 feet..........3 "	26 00
48.	Walking Cane, Board Measure, Octagon, Hickory, Solid Cast Brass Head and Tip, 8 Lines, 9 to 16 feet......................................3 "	12 00
48½.	Walking Cane, Log Measure (Doyle's Revised); Octagon, Hickory, Solid Cast Brass Head and Tip..3 "	15 00
71.	Wood Measure, Brass Caps, 8ths of inches and 10ths of a foot.............................4 "	8 00
49½.	Forwarding Stick, Cast Brass (T) Head..........5 "	24 00

Yard Sticks.

33.	Yard Stick, Polished...........................	2 00
41.	" " Brass Tips, Polished...............	3 50
50.	" " Hickory, Brass Cap'd Ends, Polished.	4 50

Wantage and Gauging Rods.

44.	Wantage Rod, 8 Lines........................	5 00
37.	" " 12 "	7 00
45.	Gauging " 120 gallons.....................3 feet,	7 00
45½.	" " 180 " and Wantage Tables....4 "	18 00

Saddlers' Rule.

80.	Saddlers' Rule, Maple, Capped Ends, 8ths and 16ths of inches.............1½ in. wide, 36 inches,	9 00

Pattern Makers' Shrinkage Rules.

30.	Pattern Makers' Shrinkage Rule, Boxwood, 8ths and 16ths of inches..................24¼ inches,	15 00
31.	Pattern Makers' Shrinkage Rule, Two Fold, Boxwood, Triple Plated Edge Plates, 8ths and 16ths of inches......................24¼ "	18 00

BOARD AND LOG MEASURES.

NOTE—Nos. 43½, 46, 46½, 47, 47½, 48, and 49 give the contents in Board Measure of 1 inch Boards.

DIRECTIONS.—Place the stick across the flat surface of the Board, bringing the inside of the Cap snugly to the edge of the same; then follow with the eye the Column of figures in which the length of the Board is given as the first figure under the Cap, and at the mark nearest the opposite edge of the Board will be found the contents of the Board in feet.

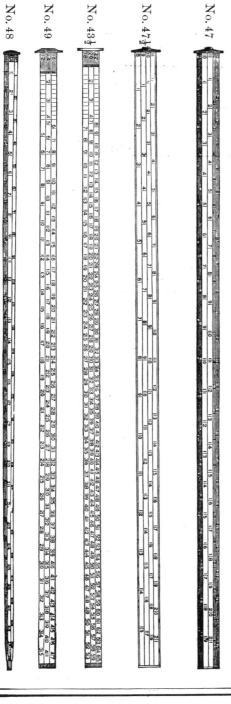

No. 47

No. 47½

No. 43½

No. 49

No. 48

NOTE.—The Log Measure (No 48½) has Doyle's Revised Tables, and gives the number of feet of one inch square edged boards, which can be sawed from a log of any size, from 12 to 36 inches in diameter, and of any length. The figures immediately under the head of the Cane, are for the lengths of Logs in feet. Under these figures, on the same line at the mark nearest the diameter of the Log, will be found the number of feet the Log will make.

STEARNS' IVORY RULES.

Formerly Manufactured in Brattleboro' Vt., by C. L. Mead, successor to E. A Stearns & Co.

Ivory, One Foot, Four Fold.

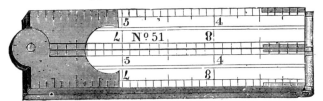

No.		Per Dozen. Unbound.	Bound.
51.	Arch Joint, Edge Plates, German Silver, 8ths, 10ths, 12ths, and 16ths of inches (100ths of a foot on edges of unbound) 13-16ths in. wide,	$33.00	$44.00
52.	Square Joint, Edge Plates, German Silver, 8ths, 10ths, 12ths and 16ths of inches (100ths of a foot on edges of unbound) 13-16ths in. wide,	30.00	40.00

Ivory, One Foot, Four Fold, Caliper.

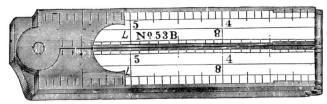

53.	Arch Joint, Edge Plates, German Silver, 8ths, 10ths, 12ths and 16ths of inches (100ths of a foot on edges of unbound) 13-16ths in. wide,	38.00	48.00
54.	Square Joint, Edge Plates, German Silver, 8ths, 10ths, 12ths and 16ths of inches (100ths of a foot on edges of unbound) 13-16ths in. wide,	35.00	45.00

Ivory, Six Inch, Two Fold, Caliper.

55.	Square Joint, German Silver, 8ths and 16ths of inches................13-16ths in. wide,	17.00	
55½.	Square Joint, German Silver Case, Spring Caliper, 8ths, and 16ths of inches, 13-16 in. wide,	21.00	

Ivory, One Foot, Four Fold.

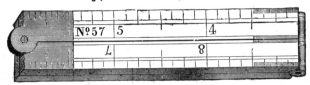

57.	Square Joint, Edge Plates, German Silver, 8ths and 16ths of inches....5-8ths in. wide,	20.00	30.00

STEARNS' IVORY RULES.

Ivory, Two Feet, Four Fold, Broad.

		Per Dozen.	
No.		Unbound.	Bound.
47.	Arch Joint, Triple Plated Edge Plates, German Silver, 8ths, 10ths, 12ths and 16ths of inches, 100ths of a foot, Drafting and Octagonal Scales...................1½ in. wide,	$100 00	$120 00

Ivory, Two Feet, Four Fold, Medium.

| 48. | Arch Joint, Edge Plates, German Silver, 8ths, 10ths, 12ths, and 16ths of inches (100ths of a foot on edges of unbound), Drafting Scales.......................1⅛ in wide, | 80 00 | 95 00 |

Ivory, Two Feet, Four Fold, Narrow.

| 50. | Square Joint, Edge Plates, German Silver, 8ths, 10ths, 12ths, and 16ths of inches (100ths of a foot on edges of unbound), Drafting Scales.......................1 in. wide, | 72 00 | 84 00 |

Ivory, Two Feet, Four Fold, Extra Narrow.

| 56. | Arch Joint, Edge Plates German Silver, 8ths, 10ths, 12ths, and 16ths of inches (100ths of a foot on edges of unbound).......¾ in wide, | 65 00 | 80 00 |

Ivory, Two Feet, Six Fold, Extra Narrow.

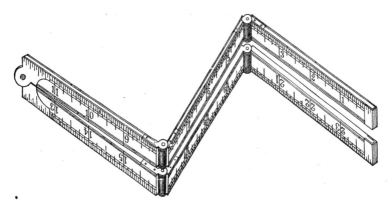

| 60. | Arch Joint, Edge Plates, German Silver, 8ths, 10ths, 12ths, and 16ths of inches (100ths of a foot on edges of unbound)......¾ in. wide, | 80 00 | 95 00 |

DESCRIPTION OF THE
PATENT IMPROVED ADJUSTABLE PLUMBS AND LEVELS

MANUFACTURED BY THE

STANLEY RULE AND LEVEL CO.

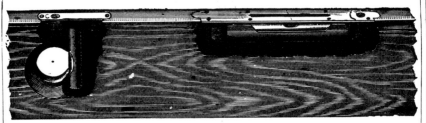

[SECTIONAL DRAWING.]

The Spirit-glass (or bubble tube), in the Level, is set in a Metallic Case, which is attached to the Brass Top-plate above it—at one end by a substantial hinge, and at the opposite end by an Adjusting Screw, which passes down through a flange on the Metallic Case. Between this flange and the Top-plate above, is inserted a stiff spiral spring; and by driving or slackening the Adjusting Screw, should occasion require, the Spirit-glass can be instantly adjusted to a position parallel with the base of the Level.

The Spirit-glass in the Plumb, is likewise set in a Metallic Case attached to the Brass Top-plate at its outer end. By the use of the Adjusting Screw at the lower end of the Top-plate, the Plumb-glass can be as readily adjusted to a right angle with the base of the Level, if occasion requires, and by the same method as adopted for the Level-glass.

The simplicity of this improved method of adjusting the Spirit-glasses will commend itself to every Mechanic. But one screw is used in the operation, and the action of the Brass Spring is perfectly reliable under all circumstances.

Dealers will do well to compare our method of adjustment with that of other so-called Adjustable Levels. The superiority of our Levels will thus be more apparent.

☞ Our Tipped Plumbs and Levels are now made with solid Brass Tips, of a handsome design, which entirely cover the ends of the Level Stock; and the finish on all our Levels has been materially improved by the introduction of new methods in their manufacture.

PLUMBS AND LEVELS. 17

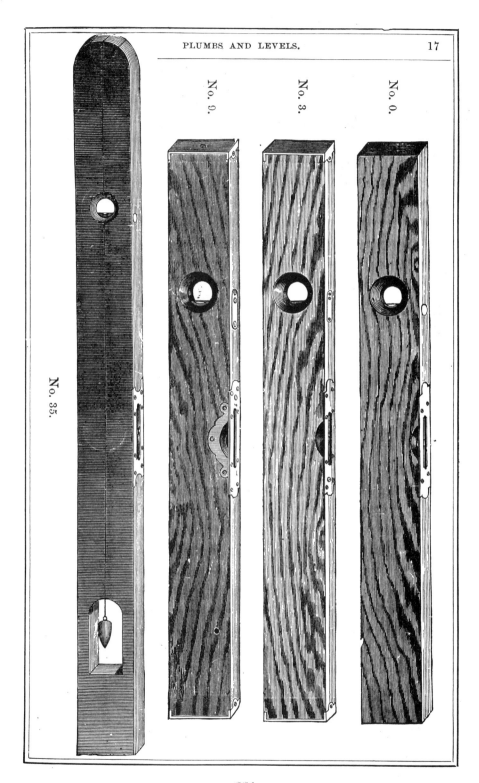

PLUMBS AND LEVELS.

No.		Per Dozen.
102.	Levels, Arch Top Plate, Two Side Views, Polished, Assorted...................10 to 16 in.,	$9 00
103.	Levels, Arch Top Plate, Two Side Views, Polished, Assorted...................18 to 24 "	12 00
104.	Plumb and Level, Arch Top Plate, Two Side Views, Polished, Assorted..............12 to 18 "	14 00
1½.	Mahogany Plumb and Level, Arch Top Plate, Two Side Views, Polished, Assorted.....18 to 24 "	16 50
1¾.	Mahogany Plumb and Level, Arch Top Plate, Two Brass Lipped Side Views, Polished and Tipped, Assorted ....12 to 18 "	27 00
00.	Plumb and Level, Arch Top Plate, Two Side Views, Polished, Assorted...............18 to 24 "	16 00
0.	Plumb and Level, Arch Top Plate, Two Side Views, Polished, Assorted [see Engraving p. 17]...................................24 to 30 "	18 00
01.	Mahogany Plumb and Level, Arch Top Plate, Two Side Views, Polished, Assorted.....24 to 30 "	22 50
02.	Plumb and Level, Arch Top Plate, Two Brass Lipped Side Views, Polished, Assorted...24 to 30 "	24 00
03.	Plumb and Level, Arch Top Plate, Two Side Views, Polished and Tipped, Assorted...24 to 30 "	28 00
04.	Plumb and Level, Arch Top Plate, Two Brass Lipped Side Views, Polished and Tipped, Assorted............................24 to 30 "	35 00
06.	Mahogany Plumb and Level, Arch Top Plate, Two Brass Lipped Side Views, Polished, Assorted............................24 to 30 "	30 00
7.	Masons' Plumb and Level, Arch Top Plate, Two Plumbs, Two Side Views, Polished and Tipped............................36 "	36 00
7½.	Masons' Plumb and Level, Arch Top Plate, Two Plumbs, Two Side Views, Polished..36 "	30 00

PATENT ADJUSTABLE PLUMBS AND LEVELS.

PATENT IMPROVED

ADJUSTABLE PLUMBS AND LEVELS.

No.			Per Dozen.
1.	Patent Adjustable Mahogany Plumb and Level, Arch Top Plate, Two Side Views, Polished, Assorted................26 to 30 in.,		$27 00
2.	Patent Adjustable Plumb and Level, Arch Top Plate, Two Brass Lipped Side Views, Polished, Assorted................26 to 30	"	27 00
3.	Patent Adjustable Plumb and Level, Arch Top Plate, Two Side Views, Polished and Tipped, Assorted [see Engraving, p. 17].26 to 30	"	30 00
4.	Patent Adjustable Plumb and Level, Arch Top Plate, Two Brass Lipped Side Views, Polished and Tipped, Assorted........26 to 30	"	39 00
5.	Patent Adjustable Plumb and Level, Triple Stock, Arch Top Plate, Two Ornamental Brass Lipped Side Views, Polished and Tipped, Assorted....................26 to 30	"	48 00
6.	Patent. Adjustable Mahogany Plumb and Level, Arch Top Plate, Two Brass Lipped Side Views, Polished, Assorted........26 to 30	"	33 00
9.	Patent Adjustable Mahogany Plumb and Level, Arch Top Plate, Two Ornamental Brass Lipped Side Views, Polished and Tipped, Assorted [see Engraving, p. 17].26 to 30	"	48 00
10.	Patent Adjustable Mahogany Plumb and Level, Triple Stock, Two Ornamental Brass Lipped Side Views, Arch Top Plate, Polished and Tipped, Assorted........26 to 30	"	60 00
11.	Patent Adjustable Rosewood Plumb and Level, Arch Top Plate, Two Ornamental Brass Lipped Side Views, Polished and Tipped, Assorted....................26 to 30	"	90 00
25.	Patent Adjustable Mahogany Plumb and Level, Arch Top Plate, Improved Double Adjusting Side Views, Polished and Tipped............................30	"	54 00
32.	Patent Adjustable Mahogany Graduated Plumb and Level (easily adjusted to work at any angle or elevation required).....28	"	100 00
35.	Patent Adjustable Masons' Plumb and Level, [see Engraving, p. 17]......3¾ in wide, 42	"	36 00

NICHOLSON'S PATENT

METALLIC PLUMBS AND LEVELS.

The chief points of excellence in these Levels are the following: Once accurately constructed, they always remain so; the sides and edges are perfectly true, thus combining with a Level a convenient and reliable straight edge; shafting, or other work overhead, may be lined with accuracy by sighting the bubble from below, through an aperture in the base of the Level. Other points are their lightness, strength, and convenience of handling. All Levels, being thoroughly tested, are warranted reliable.

No.		Per Dozen.
13.	14 inches long	$18 00
14.	20 " "	21 00
15.	24 " "	27 00

IMPROVED

IRON FRAME PLUMBS AND LEVELS.

These Levels have a Cast-Iron Frame, thoroughly braced, and combine the greatest strength with the least possible weight of metal. They are made absolutely correct in their form, and are subject to no change from warping or shrinking. The sides are inlaid with Black Walnut, secured by the screws which pass from one to the other, through the Iron braces, and the sides may be easily removed, if occasion requires.

No.		Per Dozen.
48.	Patent Adjustable Plumb and Level, Iron Frame, with Black Walnut (inlaid) Sides..12 inch,	$24 00
49.	Patent Adjustable Plumb and Level, Iron Frame, with Black Walnut (inlaid) Sides..18 "	30 00

POCKET LEVELS.

No. 41.

No.		Per Doz
40.	Iron, Iron Top Plate, Japanned, 1 dozen in a box	$2 50
41.	Iron, Brass Top Plate, 1 dozen in a box	3 00
42.	Brass, Brass Top Plate, 1 dozen in a box	8 00

46.	Iron, Brass Top Plate, a superior article	3 50

MACHINISTS' IRON LEVELS.

38.	Iron Level, Brass Top Plate	4 inch,	3 50
39.	Iron Level, Brass Top Plate	6 "	4 50

LEVEL GLASSES.

		Per Gross.
Level Glasses, packed 1 dozen in a box	1¾ inch,	$9 50
Level Glasses, packed 1 dozen in a box	2 "	10 00
Level Glasses, packed 1 dozen in a box	2½ "	10 50
Level Glasses, packed 1 dozen in a box	3 "	11 50
Level Glasses, packed 1 dozen in a box	3½ "	13 00
Level Glasses, packed 1 dozen in a box	4 "	14 50
Level Glasses, packed 1 dozen in a box	4½ "	16 00
Assorted, 1¾, 3, and 3½ inch, 1 dozen in a box		12 00

PATENT
IMPROVED TRY SQUARES.

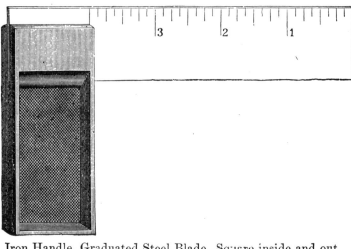

Iron Handle, Graduated Steel Blade, Square inside and out.

Inches	4	6	8	10	12
Per Dozen	$2 75	3 50	4 50	5 50	7 00

½ dozen in a box.

PATENT
ADJUSTABLE TRY SQUARES.

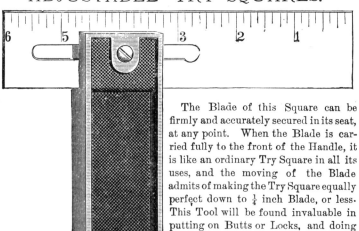

The Blade of this Square can be firmly and accurately secured in its seat, at any point. When the Blade is carried fully to the front of the Handle, it is like an ordinary Try Square in all its uses, and the moving of the Blade admits of making the Try Square equally perfect down to ¼ inch Blade, or less. This Tool will be found invaluable in putting on Butts or Locks, and doing short work about doors, windows etc.

Iron Handle, **Graduated Steel Blade,**

Inches	4	6
Per Dozen	$3 75	4 50

½ dozen in a box.

WINTERBOTTOM'S PATENT
COMBINED TRY AND MITRE SQUARE.

This Tool can be used with equal convenience and accuracy, as a Try Square, or a Mitre Square. By simply changing the position of the handle, and bringing the mitred face at the top of the handle against one edge of the work in hand, a perfect mitre, or angle of 45 degrees, can be struck from either edge of the blade.

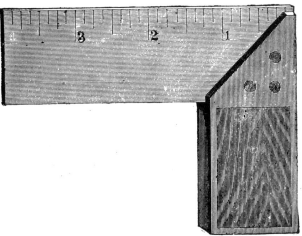

No. 1. Iron Frame Handle, with Black Walnut (Inlaid) Sides, Graduated Steel Blades, Square inside and out.

Inches,	4	6	8
Per Dozen,	$6 00	7 50	9 00

½ dozen in a box.

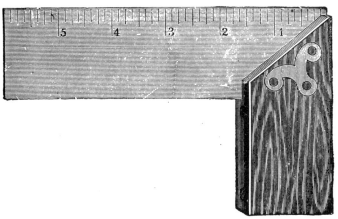

No. 2. Rosewood handle, Graduated Steel Blades, Square inside and out.

Inches	4½	6	7½	9	12
Per Dozen,	$4 50	5 00	6 00	7 00	9 00

½ dozen in a box

IMPROVED MITRE TRY SQUARE.

Per Dozen.
Cast Brass Handle, 7½ in. blade, $12 00
½ dozen in a box.

IMPROVED MITRE SQUARES.

Iron Frame Handle, with Black Walnut (Inlaid) Sides.

Inches	8	10	12
Per Dozen	$7 00	8 00	9 00

½ dozen in a box.

INLAID TRY SQUARES.

Iron Frame Handle, with Rosewood (Inlaid) Sides, Steel Blades, Square inside and out.

Inches,	3	4	6	8	9	10	12
Per Dozen,	$5 50	6 50	8 50	11 00	12 00	15 00	18 00

½ dozen in a box.

PATENT COMBINATION
TRY SQUARE AND BEVEL.

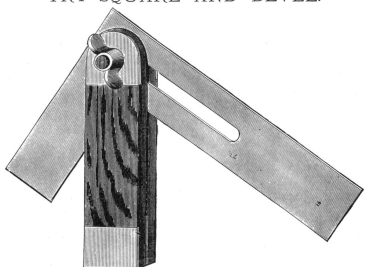

Inches,	4	6	8	10	12
Per Dozen,	$5 00	6 00	7 50	9 00	10 50

½ dozen in a box.

TRY SQUARES.

GRADUATED STEEL BLADES.

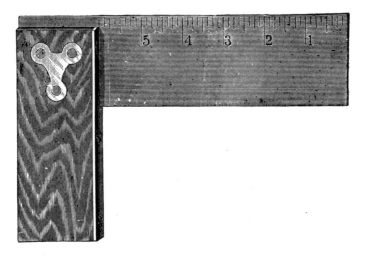

		Per Doz.
Rosewood, 1 dozen in a box .... 3 inch,		$3 00
Rosewood, ½ " .... 4½ "		3 75
Rosewood, ½ " .... 6 "		5 00
Rosewood, ½ " .... 7½ "		5 75
Rosewood, ½ " .... 9 "		6 50
Rosewood, ½ " ....10 "		7 25
Rosewood, ½ " ....12 "		8 50
Rosewood, with Rest, ½ dozen in a box ....15 "		12 50
Rosewood, " ½ " ....18 "		15 50

PLUMB AND LEVEL TRY SQUARES.

A very convenient tool for Mechanics. A spirit-glass being set in the inner edge of the Try Square Handle, constitutes the Handle a Level; and when the handle is brought to an exact level, the blade of the square will be upright, and become a perfect Plumb.

Per Doz.

7½ inch, Rosewood, ½ dozen in a box........................ $7 50

SLIDING T BEVELS.

ROSEWOOD HANDLE, WITH BRASS LEVER (FLUSH).

6 inch	per doz.,	$5 50
8 "	"	6 00
10 "	"	6 50
12 "	"	7 00
14 "	"	7 50

½ dozen in a box.

PATENT EUREKA FLUSH T BEVELS.

IRON HANDLE, STEEL BLADE, WITH PARALLEL EDGES.

The blade is easily secured at any angle by turning the Thumb Screw at the lower end of the Handle.

Inches,	6	8	10
Per Doz.,	$6 00	6 50	7 50

½ dozen in a box.

GAUGES.

☞ All Marking Gauges excepting No. 0, 61, and 61½, have an Adjusting Point of finely tempered steel, which may be readily removed and replaced if it needs sharpening. The Point can be thrust down as it wears away, or if by any means it be broken off, it can be easily repaired.

All Gauges with Brass Thumb Screws have also a Brass Shoe inserted in the head, under the end of the Thumb Screw. This shoe protects the gauge-bar from being dented by the action of the screw; and the broad surface of the shoe being in contact with the bar, the head is held more firmly in position than by any other method, and with less wear of the screw-threads.

The engraving below (No. 60) shows an Iron Marking and Cutting Gauge, No. 60½ is the same in form, but has a Reversible Brass Slide slotted into the face of the bar. When a Mortise Gauge is required, the brass slide may be turned over in the bar. The point in the brass slide may be moved to any position, and the slide will be secured by a single turn of the screw which fastens the head of the Gauge.

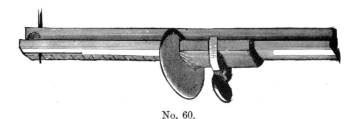

No. 60.

No.		Per Dozen.
60.	Patent Iron Marking and Cutting Gauge, Oval Bar, Marked, Adjusting Steel Point, equally well adapted for use on Metals, or Wood, ½ dozen in a box.......	$5 00
60½.	Patent Iron Reversible Gauge, Mortised, Marking and Cutting combined, Brass Slide, Oval Bar, Marked, Adjusting Steel Points, equally well adapted for use on Metals, or Wood, ½ dozen in a box.............	8 00

No. 61.

0.	Marking Gauge, Beechwood, Boxwood Thumb Screw, Marked, Steel Point, 1 dozen in a box.............	75
61.	Marking Gauge, Beechwood, Boxwood Thumb Screw, Oval Bar, Marked, Steel Point, 1 dozen in a box....	1 00
61¼.	Marking Guage, Beechwood, Boxwood Thumb Screw, Oval Head and Bar, Marked, Steel Point, 1 dozen in a box..	1 25

GAUGES.

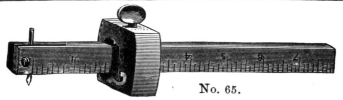

No. 65.

No.		Per Dozen.
62.	Patent Marking Gauge, Beechwood, Polished, Boxwood Thumb Screw, Oval Bar, Marked, Adjusting Steel Point, 1 dozen in a box.............	$2 00
64.	Patent Marking Gauge, Polished, Plated Head, Boxwood Thumb Screw, Oval Bar, Marked, Adjusting Steel Point, 1 dozen in a box...................	2 75
64½.	Patent Marking Gauge, Polished, Oval Plated Head, Brass Thumb Screw and Shoe, Oval Bar, Marked, Adjusting Steel Point, ½ dozen in a box...........	4 50
65.	Patent Marking Gauge, Boxwood, Polished, Plated Head, Brass Thumb Screw and Shoe, Oval Bar, Marked, Adjusting Steel Point, ½ dozen in a box...	5 00
66.	Patent Marking Gauge, Rosewood, Oval Plated Head and Bar, Brass Thumb Screw and Shoe, Oval Bar, Marked, Adjusting Steel Point, ½ dozen in a box....	6 00

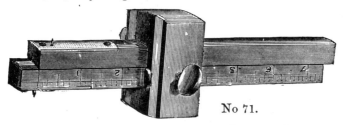

No 71.

71.	Patent Double Gauge (Marking and Mortise Gauge combined), Beechwood, Polished, Plated Head and Bars, Brass Thumb Screws and Shoes, Oval Bars, Marked, Steel Points, ½ dozen in a box............	8 00
72.	Patent Double Gauge (Marking and Mortise Gauge combined), Beechwood, Polished, Boxwood Thumb Screws, Oval Bars, Marked, Steel Points, ½ doz. in a box	4 00
74.	Patent Double Gauge (Marking and Mortise Gauge combined), Boxwood, Polished, Full Plated Head and Bars, Brass Thumb Screws and Shoes, Oval Bars Marked, Steel Points, ½ dozen in a box............	14 00
70.	Cutting Gauge, Mahogany, Polished, Plated Head, Boxwood Thumb Screw, Oval Bar, Marked, Steel Cutter, 1 dozen in a box...................	4 00
83.	Handled Slitting Gauge, 17-inch Bar, Marked, ½ dozen in a box.......................................	9 00
84.	Handled Slitting Gauge, with Roller, 17-inch Bar, Marked, ½ dozen in a box........................	10 00

GAUGES.

No. 73.

No.		Per Dozen.
73.	Patent Mortise Gauge, Boxwood, Polished, Plated Head, Brass Slide, Brass Thumb Screw and Shoe, Oval Bar, Marked, Steel Points, ½ dozen in a box....	$8 00
76.	Patent Mortise Gauge, Boxwood, Polished, Plated Head, Screw Slide, Brass Thumb Screw and Shoe, Oval Bar, Marked, Steel Points, ½ dozen in a box....	11 00
80.	Patent Mortised Gauge, Boxwood, Full Plated Head, Plated Bar, Screw Slide, Brass Thumb Screw and Shoe, Marked, Steel Points, ½ dozen in a box.......	18 00
67.	Mortise Gauge, Adjustable Wood Slide, Boxwood Thumb Screw, Oval Bar, Marked, Steel Points, ½ dozen in a box..................................	4 00
68.	Mortise Gauge, Plated Head, Adjustable Wood Slide, Brass Thumb Screw and Shoe, Oval Bar, Marked, Steel Points, ½ dozen in a box.....................	6 00

No. 77.

77.	Patent Mortise and Marking Gauge, Rosewood, Plated Head, Improved Screw Slide, Brass Thumb Screw and Shoe, Oval Bar, Marked, Steel Points, ½ dozen in a box	10 00
78.	Patent Mortise Gauge, Rosewood, Plated Head, Screw Slide, Brass Thumb Screw and Shoe, Oval Bar, Marked, Steel Points, ½ dozen in a box.............	11 00
79.	Patent Mortise Gauge, Rosewood, Plated Head and Bar, Screw Slide, Brass Thumb Screw and Shoe, Marked, Steel Points, ½ dozen in a box.............	13 00
85.	Panel Gauge, Beechwood, Boxwood Thumb Screw, Oval Bar, Steel Points, 1 dozen in a box............	3 20
85½.	Panel Gauge, Rosewood, Plated Head and Bar, Brass Thumb Screw and Shoe, Steel Point, ½ dozen in a box	18 00

STANLEY'S IMPROVED GAUGES. 30½

STANLEY'S PATENT
IMPROVED GAUGES.

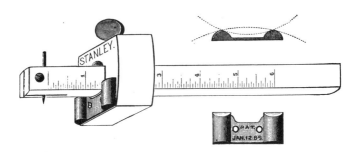

The Brass Face, with two ribs or projections, attached to one side of the Gauge-head (see cut), will enable the owner to run a gauge-line with perfect steadiness and accuracy around curves of any degree, and either concave or convex. The opposite side of the Gauge-head remains flat, and can be used for all ordinary work.

Any style of Gauge found in our Catalogue, pages 28–30—with this valuable improvement included—may be ordered as No. 161, 162, etc., instead of No. 61, 62, etc. The following Nos. are carried in stock, and have the most liberal sale.

MARKING GAUGES.

No. PER DOZ.
161. Beechwood, Boxwood Thumb Screw..$2 00
162. Beechwood, Boxwood Thumb Screw, Adjusting Steel Point............. 3 00
164. Beechwood, Boxwood Thumb Screw, Plated Head, Adjusting Steel Point.. 3 75
165. Boxwood, Brass Thumb Screw and Shoe, Plated Head, Adjusting Steel Point.. 6 00
166. Rosewood, Brass Thumb Screw and Shoe, Oval Plated Head, Adjusting Steel Point.. 7 00

MORTISE GAUGES.

172. Double Gauge (Marking and Mortise), Beechwood, Boxwood Thumb Screws..$5 00
173. Boxwood, Brass Slide, Plated Head, Brass Thumb Screw and Shoe...... 9 00
177. Rosewood, Screw Slide, Plated Head, Brass Thumb Screw and Shoe......11 00

PATENT IMPROVED.
SOLID CAST STEEL SCREW DRIVERS.

By the aid of improved machinery, we are producing Screw Drivers which are superior to any others in the market, both in style of finish, and working qualities, and at the prices paid for ordinary Tools.

The blades are made from the best quality of Cast Steel, and are tempered with great care. They are ground down to a correct taper, and shaped at the end by special machinery; thus procuring perfect uniformity in size and strength, while the peculiar form of the point gives it unequaled firmness in the screw-head when in use.

The shanks of the Blades are properly slotted to receive a patent metallic fastening, which secures them permanently in the Handles. The Handles are of the most approved pattern, the Brass Ferrules, of the thimble form, extra heavy, and closely fitted.

Our Screw Drivers have acquired an enviable reputation with Hardware Dealers; and they are sought by a class of Mechanics who require tools which will stand the severest tests, in practical use. The Screw Drivers are by us fully WARRANTED.

PRICES.

VARNISHED HANDLES.

Sizes,	$1\frac{1}{2}$	2	3	4	5	6	7	8	10	12	Inches.
	$1.00	1.50	2.00	2.50	3.00	3.50	4.00	4.75	6.00	8.00	Per Doz.

½ dozen in a box.

BLACK (ENAMELED) HANDLES.

Sizes,	$1\frac{1}{2}$	2	3	4	5	6	7	8	10	12	Inches.
	$1.00	1.50	2.00	2.50	3.00	3.50	4.00	4.75	6.00	8.00	Per Doz.

½ dozen in a box.

Sewing-Machine Screw Drivers.
SPECIAL PRICES WILL BE QUOTED FOR THESE IN BULK.

Screw Driver Handles.

Assorted Sizes, 1 dozen in a box.................per dozen, $1 00
Large Sizes, 1 dozen in a box.................per dozen, 1 25

BAILEY'S PATENT ADJUSTABLE PLANES.

Manufactured only by the Stanley Rule & Level Co., under the original Patents.
OVER 450,000 ALREADY SOLD.

These planes meet with universal approbation from the best Mechanics, as their extensive sale abundantly testifies. For beauty of style and finish, they are unequaled, and the superior methods for adjusting them readily in all their parts, render them economical to the owner.

☞ Each Plane is thoroughly tested at our Factory, and put in perfect working order, before being sent into the market. A printed description of the method of adjustment accompanies each plane. The Plane Irons are by us fully warranted.

Iron Planes.

No.		Each.
1.	Smooth Plane, 5½ inches in Length, 1¼ inch Cutter	$2 25
2.	Smooth Plane, 7 inches in Length, 1⅝ inch Cutter	2 75
3.	Smooth Plane, 8 inches in Length, 1¾ inch Cutter	3 00
4.	Smooth Plane, 9 inches in Length, 2 inch Cutter	3 25
4½.	Smooth Plane, 10 inches in Length, 2⅜ inch Cutter	3 75

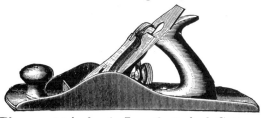

5.	Jack Plane, 14 inches in Length, 2 inch Cutter	3 75
6.	Fore Plane, 18 inches in Length, 2⅜ inch Cutter	4 75
7.	Jointer Plane, 22 inches in Length, 2⅜ inch Cutter	5 50
8.	Jointer Plane, 24 inches in Length, 2⅝ inch Cutter	6 50
9.	Block Plane, 10 inches in Length, 2 inch Cutter	6 50
10.	Carriage Makers' Rabbet Plane, 14 in. Length, 2⅛ in. Cutter	4 50
11.	Belt Makers' Plane, 2⅜ inch Cutter	3 00

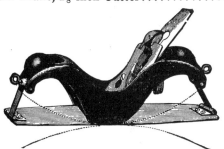

13.	Circular Plane, 1¾ inch Cutter	4 00

This Plane has a Flexible Steel Face, and by means of the thumb screws at each end of the Stock, can be easily adapted to plane circular work—either concave or convex.

☞ **Please insert opposite Page 32, Illustrated Catalogue, 1884.**

STANLEY'S LATERAL ADJUSTMENT.

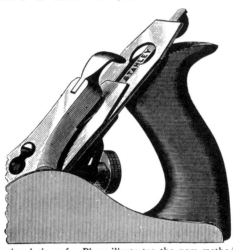

The above sectional view of a Plane illustrates the new method of Adjusting a Plane Iron, *sidewise*, to set the cutting edge exactly square with the face of the plane. A Lever, just under the Plane Iron, enables the workman thus to regulate his cutter.

This Lever, for sidewise adjustment, is attached to all BAILEY IRON PLANES, AND BAILEY WOOD PLANES, on pages 32 and 33 of this Catalogue—except Nos. 1, 9 and 11.

No. 10½. Carriage Makers' Rabbet Plane, 9 in., 2⅛ in. Cutter, $3 75.

Wood Planes.

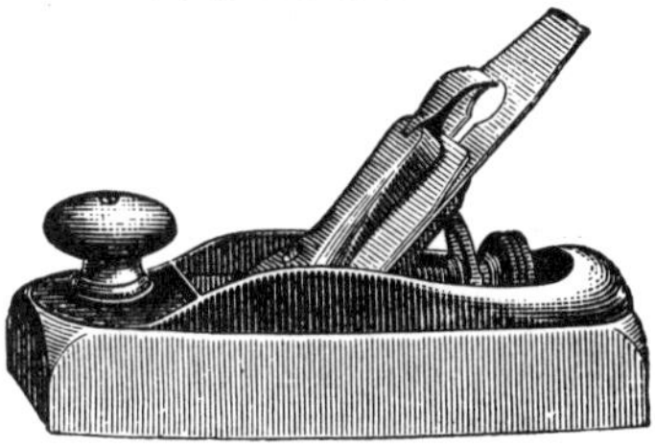

No.		Each.
21.	Smooth Plane, 7 inches in Length, 1¾ inch Cutter........	$2 00
22.	Smooth Plane, 8 inches in Length, 1¾ inch Cutter........	2 00
23.	Smooth Plane, 9 inches in Length, 1¾ inch Cutter........	2 00
24.	Smooth Plane, 8 inches in Length, 2 inch Cutter........	2 00
25.	Block Plane, 9½ inches in Length, 1¾ inch Cutter........	2 00

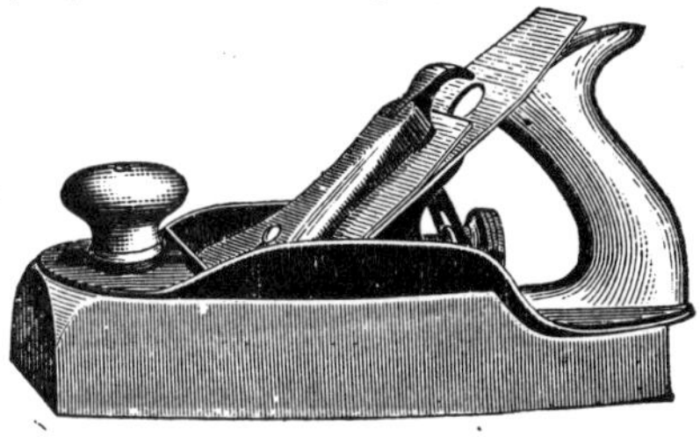

35.	Handle Smooth, 9 inches in Length, 2 inch Cutter.....	2 50
36.	Handle Smooth, 10 inches in Length, 2⅜ inch Cutter.....	2 75
37.	Jenny Smooth, 13 inches in Length, 2⅝ inch Cutter.....	3 00

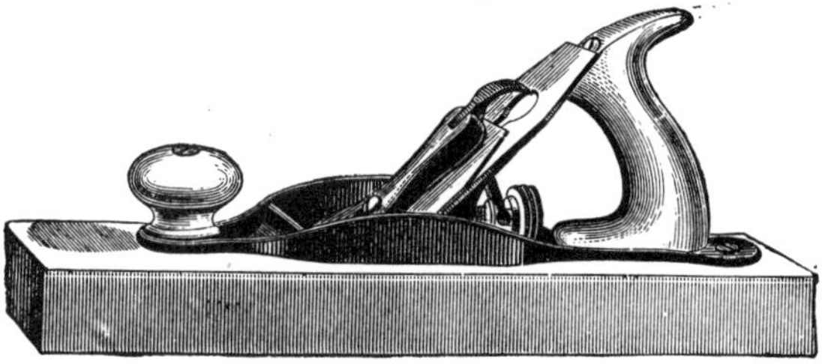

26.	Jack Plane, 15 inches in Length, 2 inch Cutter........	2 25
27.	Jack Plane, 15 inches in Length, 2⅛ inch Cutter........	2 50
28.	Fore Plane, 18 inches in Length, 2⅜ inch Cutter........	2 75
29.	Fore Plane, 20 inches in Length, 2⅜ inch Cutter........	2 75
30.	Jointer Plane, 22 inches in Length, 2⅜ inch Cutter........	3 00
31.	Jointer Plane, 24 inches in Length, 2⅜ inch Cutter........	3 00
32.	Jointer Plane, 26 inches in Length, 2⅝ inch Cutter........	3 25
33.	Jointer Plane, 28 inches in Length, 2⅝ inch Cutter........	3 25
34.	Jointer Plane, 30 inches in Length, 2⅝ inch Cutter........	3 50

☞ Extra Plane-woods of every style, can be supplied cheaply.

Bailey's Plane Irons.

	1¼	1⅝	1¾	2	2⅛	2⅜	2⅝ Inch.
SINGLE IRONS	20	25	28	30	33	37	40 cts.
DOUBLE IRONS.	40	45	50	55	60	65	70 cts.

☞ **Orders for Plane Irons should designate the No. of the Planes for which they are wanted.**

Bailey's Patent Adjustable Block Planes.

These Block Planes are adjusted by a Screw and Lever Movement, and the Mouth can be opened wide, or made close, as the nature of the work to be done may require. The handles to Nos. 9¾ and 15½ are secured to the stock by the use of an iron nut, and can be easily attached to or liberated from the Plane, as the convenience of the workman may require.

Each.
9½. Excelsior Block Plane, 6 inches Length, 1¾ inch Cutter...$1 50
15. Excelsior Block Plane, 7 inches Length, 1¾ inch Cutter... 1 60

9¾. Excelsior Block Plane, Rosewood Handle, 6 inches in Length, 1¾ in. Cutter................................. 1 75
15½. Excelsior Block Plane, Rosewood Handle, 7 inches in Length, 1¾ in. Cutter................................. 1 85
Cutters, warranted Cast Steel, per doz................... 2 40

Bailey's Adjustable Veneer Scraper.

12. 3-in. Cutter.. 3 50
CAST STEEL, HAND, VENEER SCRAPERS, 3 x 5 inch, per doz. 3 00

TESTIMONIALS.

"The best Plane now in use."—CHICKERING & SONS, Piano Forte Manufacturers, Boston, Mass.

"I have used a set of Bailey's Planes about six months, and would not use any other now."—EDWARD CARVILLE, Sacramento, Cal.

"The best plane ever introduced in carriage-making."—JAS. L. MORGAN, at J. B. Brewster & Co.'s Carriage Manufactory, New York.

"From the satisfaction they give would like to introduce them here."—J. ALEXANDER, at Bell's Melodeon Factory, Guelph, Canada.

"Bailey's planes are used in our factory. We can get only one verdict, and that is, the men would not be without them."—STEINWAY & SONS, Piano Forte Manufacturers, New York.

"I have owned a set of Bailey's Planes for six months, and can cheerfully say that I would not use any others, for I believe they are the best in use."—C. C. HARRIS, Stair Builder, St. Louis, Mo.

A Boston Mechanic says:—"I always tell my shopmates, when they wish to use my Plane, not to borrow a Bailey Plane unless they intend to buy one, as they will never be satisfied with any other Plane after using this."

THE STANLEY
ADJUSTABLE PLANES.
PATENTED.

These Planes are adjusted by the use of a Lever, and are equally well adapted to coarse or fine work. The Planes are thoroughly tested at our Factory, and are in perfect working order when sent into market. The Plane Irons are made of the best English Cast Steel, and are fully warranted.

Planes Nos. 104 and 105 have a Wrought Steel Stock, and are commended for their lightness and the ease with which they can be worked.

Steel Planes.

No.		Each.
104.	Smooth Plane, 9 inches in Length, $2\frac{1}{8}$ inch Cutter......	$2 75

| 105. | Jack Plane, 14 inches in Length, $2\frac{1}{8}$ inch Cutter....... | 3 50 |

Improved Adjustable Circular Plane.

No.		Each.
113.	Adjustable Circular Plane, $1\frac{3}{4}$ inch Cutter............	$4 00

This Plane has a Flexible Steel Face, which can be easily shaped to any required arc, either concave or convex, by turning the Knob on the front of the Plane. The Knob is attached to a double-acting screw, which moves two levers properly connected by gears, thus controlling accurately both ends of the flexible face. By the peculiar construction of the Plane, a smaller arc, either concave or convex, can be obtained by this Plane than by any other similar tool.

Wood Planes.

No.		Each.
122.	Smooth Plane, 8 inches in Length, $1\frac{3}{4}$ in. Cutter........	$1 50

135.	Handle Smooth, 10 inches in Length, $2\frac{1}{8}$ in. Cutter	2 00

127.	Jack Plane,	15 inches in Length, $2\frac{1}{8}$ in. Cutter.......	2 00
129.	Fore Plane,	20 inches in Length, $2\frac{3}{8}$ in. Cutter.......	2 25
132.	Jointer Plane,	26 inches in Length, $2\frac{5}{8}$ in. Cutter.......	2 50

☛ The STANLEY PLANE IRONS are the same prices, for corresponding widths, as per list of Bailey's Plane Irons—See page 37.

PATENT IMPROVED RABBET PLANES.

WITH STEEL CASE.

No.		Each.
80.	Rabbet Plane, Skew, 9 inches in Length, $1\frac{1}{2}$ in. Cutter.	$1 10
90.	Rabbet Plane, Skew, 9 inches in Length, with Spur, $1\frac{1}{2}$ in. Cutter..	1 25

The Stanley Iron Block Planes.

No.   Each.
101. Block Plane, 3½ inches in Length, 1 inch Cutter........$0 20
CUTTERS for above Block Plane, warranted Cast Steel, per doz.   75

102. Block Plane, 5½ inches in Length, 1¼ inch Cutter......   40
103. Block Plane, Adjustable, 5½ in. in Length, 1¼ in. Cutter.   60
CUTTERS for above Block Planes, warranted Cast Steel, per doz. 1 50

110. Block Plane, 7½ inches in Length, 1¾ inch Cutter.......   60
120. Block Plane, Adjustable, 7½ inches in Length, 1¼ inch Cutter...   85

130. Block Plane (Double-Ender), 8 inches in Length, 1¾ inch Cutter...   80

This Plane has two slots, or cutter seats; and can be used as a Block Plane, or, by reversing the position of the cutter and the clamping wedge, (see the dotted lines in the cut,) it can be used to plane close up into corners, or places difficult to reach with any other plane.

CUTTERS, for above Block Planes, warranted Cast Steel, per doz. 2 00

BULL-NOSE RABBET PLANE.

No.   Each.
75. Iron Stock, 4 inches in length, 1 inch Cutter........... $0 50

MILLER'S PATENT COMBINED
PLOW, FILLETSTER & MATCHING PLANE.

This tool embraces, in a most ingenious and successful combination, the common Carpenter's Plow, an adjustable Filletster, and a perfect Matching Plane. The entire assortment can be kept in smaller space, or made more portable, than an ordinary Carpenters' Plow.

☞ Each Plane is accompanied by a Slitting-tool, a Tonguing-tool (¼ inch), a Filletster Cutter, and eight Plow-bits (1-8, 3-16, 1-4, 5-16, 3-8, 7-16, 1-2 and 5-8 inches).

PRICES, including Plow Bits, Slitting and Tonguing Tools.
No. 41.   Iron Stock and Fence..$ 9 00
No. 42.   Gun Metal Stock and Fence.. 12 00

COMBINED.
PLOW AND MATCHING PLANE.

The above engraving represents the Tool adjusted for use as a Plow. With each Plow eight Bits (1-8, 3-16, 1-4, 5-16, 3-8, 7-16, 1-2 and 5-8) are furnished; also a Slitting-tool, and a Tonguing-tool (¼ inch); and by use of the latter, together with the ¼ inch Plow-bit for grooving, a perfect Matching Plane is made.

PRICES, including Plow Bits, Slitting and Tonguing Tools.
No. 43.   Iron Stock and Fence...$ 7 00
No. 44.   Gun Metal Stock and Fence.. 10 00

☞ Please insert opposite Page 38, Illustrated Catalogue, 1884.

STANLEY'S BULL-NOSE PLOW, ETC.   38A

STANLEY'S PATENT BULL-NOSE
PLOW, FILLETSTER AND MATCHING PLANE.

Two interchangeable front parts go with this Tool. The form shown above is that of a Bull-nose Plow; and the cutter will easily work up to, and into a ½ inch hole, or any larger size—as in Sash-fitting, Stair-work, etc. With the other front on, it takes the ordinary form of a Plow, and is adapted to all regular uses.

With each tool eight Plow Bits (1-8, 3-16, 1-4, 5-16, 3-8, 7-16, 1-2 and 5-8 inch), a Filletster Cutter and a Slitting Blade, are furnished; also a tonguing tool (1-4 inch).

Bull-Nose Plow, Filletster, and Matching Plane.
No. 141. Iron Stock and Fence..$7 00

Bull-Nose Plow, and Matching Plane.
No. 143. Iron Stock and Fence..$5 00

STANLEY'S
ADJUSTABLE CHAMFER PLANE.

The front section of this Plane is movable up and down. It can be firmly secured to the rear section at any desired point by means of a Thumb-screw. Without the use of any other tool, this Plane will do perfect chamfer, or stop chamfer, work of all ordinary widths.

For Beading, Reeding or Moulding a chamfer an additional section to this Plane is furnished (price $1 00) with six cutters, sharpened at both ends, including a large variety of ornamental forms.

No. 72.   Chamfer Plane, 9 inches in length, 1 5-8 in. Cutter.............. $2 00
No. 72½ Chamfer Plane, with Beading and Moulding Attachment............3 00

STANLEY'S PATENT
BIT AND SQUARE LEVEL.

The frame of this Level has three pairs of V slots on its back edges. The shank to a Bit will lie in these slots, either parallel with the bubble-glass, at an exact angle to it, or at an angle of 45 degrees. A Thumb Screw secures the Level to the Bit, in either position; and boring can be done with perfect accuracy as to perpendicular, horizontal, or angle of 45 degrees, by observing the bubble-glass while turning the bit.

This Level can with equal facility be attached to a Carpenter's Square, thus making an accurate Plumb, or Level, for all ordinary uses.

No. 44 Bit and Square Level, Brass Frame........................Per doz., $3 60

STANLEY'S
ADJUSTABLE CLAPBOARD MARKER.

This ingenious tool can be used with one hand, while the other is employed in holding a clapboard in position. The marking blade is properly slotted, so that the tool can be easily adjusted to any thickness of clapboard.

The sharp edge of the teeth on the marking blade, are just parallel with the outer edges of the legs when placed against the corner-board; and by moving the tool half an inch, it will mark a full line across the clapboard, exactly over and conformed to the edge of the corner-board. There is then no difficulty in sawing for a perfectly close joint.

No. 88. Iron Stock, with Wood Handle, Steel Blade..................Each, $0 50

DOUBLE ROUTERING IRON (1-8 & 1-4 in.)

FOR STANLEY'S HAND BEADER, NO. 66.

This tool, for doing all kinds of light Routering, has been added to the already liberal assortment of Cutters furnished with STANLEY'S UNIVERSAL HAND BEADER (see Catalogue, page 38¼), making *Seven* steel cutters with each Beader, and sold at the original price, $1.00 each. No more practical tool has ever been offered to Mechanics and Amateurs than this Hand Beader, and it is acknowledged to be a marvel of cheapness.

STANLEY'S PATENT FLOOR PLANE.

This tool will be found useful for planing Floors, Bowling Alleys, Skating Rinks, Decks of Vessels, etc. The construction of the Plane will enable the owner to do more work, and with less outlay of strength, than can be done with any other tool.

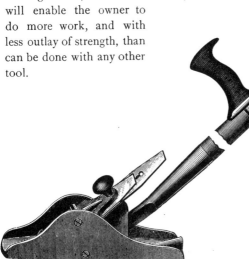

The weight of the Plane is about 10 lbs,, and the full length of the handle, 45 inches.

PRICE.
No. 74.   Floor Plane, 10½ inches in length, 2⅝ in. Cutter, $4.50.

STANLEY'S PATENT
UNIVERSAL HAND BEADER.

For Beading, Reeding or Fluting straight or irregular surfaces this tool is invaluable to woodworkers.

Six superior steel cutters go with each tool. Both ends are sharpened, thus embracing six ordinary sizes of Beads, four sets of Reeds and two Fluters.

The cutter is firmly clamped to the stock. A gauge with long straight bearing surfaces is used in ordinary work; and a gauge with oval bearing surfaces is used for curved or irregular forms of work.

Either gauge can be shifted from one side of the cutter to the other without separating it from the stock, and can be rigidly set at any required distance from either side of the cutter.

PRICE.
No. 66.   Iron Stock, with six Steel Cutters................$1.00
               Extra Cutters or blanks for same................. .05

STANLEY'S PATENT
IMPROVED CHAMFER PLANE.

The front section of this Plane, to which the cutter is attached, is movable up and down. It can be firmly secured to the rear section of the Plane at any desired point by means of a thumb-screw. Without the use of any other tool this Plane will do perfect chamfer or stop-chamfer work of all ordinary widths.

When the two sections are clamped together so as to form an even base-line the tool can be used as an ordinary bench plane.

PRICE.

No. 72. Iron Stock, 9 inches in length, 1⅝ in. Cutter.. .. $2.00

STANLEY'S PATENT
IMPROVED RABBET PLANE.

This Plane will lie perfectly flat on either side, and can be used with right or left hand equally well, while planing into corners or up against perpendicular surfaces.

PRICES.

No. 180. Iron Stock, 8 in. in length, 1½ in. wide..........$1.00
No. 181. Iron Stock, 8 in. in length, 1¼ in. wide.......... 1.00
No. 182. Iron Stock, 8 in. in length, 1 in. wide.......... 1.00
No. 190. Iron Stock, 8 in. in length, 1½ in. wide, with spur. 1.15
No. 191. Iron Stock, 8 in. in length, 1¼ in. wide, with spur. 1.15
No. 192. Iron Stock, 8 in. in length, 1 in. wide, with spur. 1.15

NOTE.—STANLEY'S ADJUSTABLE SCRAPER-PLANE (*No.* 112), *described on p.* 40½ *in this Catalogue, can be used equally well as a Tooth Plane.*

Price of Tooth Plane Irons, 3 inches wide, each...... ..... 35 cts.

TRAUT'S PATENT ADJUSTABLE
DADO, FILLETSTER, PLOW, ETC.

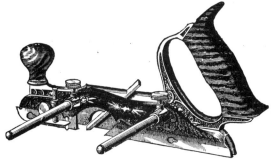

The Tool here represented consists of two Sections: A main stock, with two bars, or arms; and a sliding section, having its bottom, or face, level with that of the main stock.

It can be used as a Dado of any required width, by inserting the bit into the main stock, and bringing the sliding section snugly up to the edge of the bit. The two spurs, one on each section of the Plane, will thus be brought exactly in front of the edges of the bit. The gauge on the sliding section will regulate the depth to which the tool will cut.

By attaching the Guard-plate, shown above, to the sliding section, the Tool may be readily converted into a Plow, a Filletster, or a Matching Plane—as explained in the printed instructions which go with every tool.

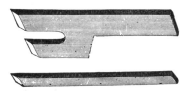

The Tool is accompanied by eight Plow-bits ($\frac{3}{16}$, $\frac{1}{4}$, $\frac{5}{16}$, $\frac{3}{8}$, $\frac{1}{2}$, $\frac{5}{8}$, $\frac{7}{8}$ and $1\frac{1}{4}$ inch), a Filletster Cutter, a Slitting-tool, and a Tonguing-tool. All these tools are secured in the main stock *on a skew*.

PRICE, including Plow-bits, Slitting and Tonguing Tools.
No. 46.   Iron Stock and Fence.. $7 00

ADJUSTABLE DADO.

It can be used as a Dado of any required width, by inserting the bit into the main stock, and bringing the sliding section snugly up to the edge of the bit. The two spurs, one on each section of the plane, will thus be brought exactly in front of the edges of the bit. The gauge on the sliding section will regulate the depth to which the tool will cut.

PRICE, including Bits ($\frac{3}{8}$, $\frac{1}{2}$, $\frac{5}{8}$, $\frac{7}{8}$ and $1\frac{1}{4}$ inch) **and Slitting-tool.**
No. 47.   Iron Stock and Fence.................................... $4 00

TRAUT'S PATENT ADJUSTABLE
BEADING, RABBET AND SLITTING PLANE.

This Plane embraces in a compact and practical form (1) Beading and Center Beading Plane ; (2) Rabbet and Filletster ; (3) Dado ; (4) Plow ; (5) Matching Plane ; and (6) a superior Slitting Plane.

In each of its several forms this Plane will do perfect work, even in the hands of an ordinary mechanic—its simplicity of construction and adaptation of parts (as described in the Directions which accompany each Tool) being easily understood.

☞ Each Plane is accompanied by seven Beading Tools (1-8, 3-16, 1-4, 5-16, 3-8, 7-16 and 1-2 inch), nine Plow and Dado Bits (1-8, 3-16, 1-4, 5-16, 3-8, 7-16, 1-2, 5-8 and 7-8 inch), a Slitting Tool and a Tonguing Tool.

Price, including Beading Tools, Bits, Slitting Tool, etc.,
No. 45. Iron Stock and Fence... $8 00

PATENT
ADJUSTABLE BEADING PLANE.

This Tool, for ordinary beading or for center-beading, cannot be surpassed. Two steel spurs are inserted in the stock so that they cut the grain of the wood just in advance of the beading tool, insuring a perfect edge to the bead. By adjustment of the fence, center-beading can be done up to three inches from the edge of a board.

Price, including Bits (1-8, 3-16, 1-4, 5-16, 3-8, 7-16 **and** 1-2 **inch**),
No. 50. Iron Stock and Fence... $4 00

☞ **Please insert opposite Page 40, Illustrated Catalogue, 1884.**

HOLLOWS AND ROUNDS, ETC.   40¼

HOLLOWS AND ROUNDS.
WITH CAST-STEEL BITS.

These additional pieces will be furnished, to order, and can be used with entire satisfaction in combination with Plane No. 45 (see opposite, page 40).

Insert the required form of Bit in the main stock of the Plane, and then secure the corresponding form of face to the arms. Superior work can thus be done with this Tool.

All parts of Plane No. 45 are made interchangeable, and these additional pieces for Hollows and Rounds, will fit any Plane of that style, so that mechanics can buy such sizes as they need, and at any time.

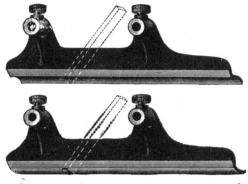

Nos.	6	8	10	12	
Works,	¾	1	1¼	1½	inch Circle.
Price,	$1.40	$1.40	$1.40	$1.40	In Pairs.
	.75	.75	.75	.75	Each.

Extra Short Arms (if wanted with Hollows and Rounds),   20 cts. Pairs.
Double Beading Tools (for Reeding), 1-8, 3-16, 1-4 inch,   20 cts. Each.

EXPLANATORY.
(See pages 38, 39 and 40.)

This is a representation of the Slitting-Tool added to Planes Nos. 41, 42, 43, 44, 45 and 46.

The improvement is an important one, as carpenters who have not the convenient use of circular saws, can with this Slitting-Tool rapidly slit up their stuff for fitting up doors and windows, or kindred work.

The Slitting-Tool is inserted into a slot on the right side of the main stock of the Plane, and just in front of the handle; a steel depth-gauge is placed over it on the same spindle, and both are fastened down by a brass thumb-screw.

The position of the Slitting-Tool being right under the hand of the workman, his full strength can be exerted, while the construction of the Plane renders it stiff enough to ensure perfect accuracy in working.

NOTE.—The several sets of Bits, etc., which accompany the Planes (Nos. 41, 42, 43, 44, 45 and 46), are now put up in wooden boxes, which protect the cutting edges, and keep the full assortment of tools always convenient for selection by the owner.

STANLEY'S PATENT DUPLEX
RABBET PLANE AND FILLETSTER.

The valuable features of this Plane can be seen by a glance at the illustration given above.
Remove the arm to which the fence is secured, and a Handled Rabbet Plane is had; and with two seats for the cutter, so that the tool can be used as a Bull-Nose Rabbet if required.
The construction of the stock is such that the Plane will lie perfectly flat on either side, and can be used with right or left hand equally well, while planing into corners or up against perpendicular surfaces.
The arm to which the fence is secured can be screwed into either side of the stock, thus making a superior right or left hand Filletster, with adjustable spur and depth gauge.

PRICE.
No. 78.—Iron Stock and Fence, $8\frac{1}{2}$ inches in Length, $1\frac{1}{2}$ inch Cutter....$1.50

WOODWORKER'S HANDY ROUTER PLANE.

This tool should be added to the kit of every skilled Carpenter, Cabinet Maker, Stair Builder, Pattern Maker or Wheelwright.
When in the form shown in the illustration it is perfectly adapted to smooth the bottom of grooves, panels, or all depressions below the general surface of any woodwork.
The Bits can also be clamped to the backside of the upright post, and outside of the stock. In this position they will plane into corners, or will smooth surfaces not easily reached with any other tool, as the form of the shank enables the Bit to work either parallel with the stock lengthwise, or at a right-angle with it.

PRICE.
No. 71.   Iron Stock, with Steel Bits ($\frac{1}{4}$ and $\frac{1}{2}$ inch) $1.50

STANLEY'S ADJUSTABLE SCRAPER PLANE.

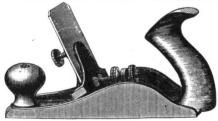

This Tool, by its peculiar construction, can be used for scraping and finishing Veneers, Cabinet Work, or Hard Woods in any form.

PRICE.
No. 112.   Adjustable Scraper Plane, 9 inches in Length, 3 in. Blade....$3.00

TONGUING AND GROOVING PLANE.
PATENTED.

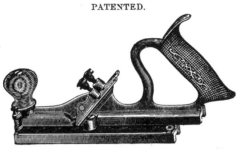

The stock of this tool is made of metal, and it has two cutters fastened into the stock by thumb screws. The guide, or fence, when set as shown in the above engraving, allows both of the cutters to act; and the cutters being placed at a suitable distance apart, a perfect Tonguing Plane is made. The guide, or fence, which is hung on a pivot at its centre, may be easily swung around, end for end; thus one of the cutters will be covered and the guide held in a new position, thereby converting the Tool into a Grooving Plane. A groove will be cut to exactly match the tongue which is made by the other adjustment of the Tool.

The ingenuity and simplicity of this Tool, together with its compact form and durability, will commend it to the favorable regard of all wood-working mechanics.

PRICE, including Tonguing and Grooving Tools.

No. 48. Iron Stock and Fence, for $\frac{3}{4}$ to $1\frac{1}{4}$ inch Boards...... $2 50
No. 49. Iron Stock and Fence, for $\frac{3}{8}$ to $\frac{3}{4}$ inch Boards...... 2 50

IMPROVED MITRE BOX.

The peculiar features which distinguish this Mitre Box above all others in market, are referred to below:

The Frame is made of a single casting, and is subject to no change of position, being finished accurately at first, it must always remain true. The slot in the back of the frame, through which the saw passes, is only three eighths of an inch wide, thereby obviating any liability to push short pieces of work through the slot, when the saw is in motion.

This Mitre Box can be used with a Back Saw or a Panel Saw, equally well. If a back saw is used, both links which connect the rollers, or guides, are left in the upper grooves, and the back of the saw is passed through under the links. If a Panel Saw is used, the link which connects the rollers on the back spindle is changed to the lower groove; and then the blade of the saw will be stiffly supported by both sets of rollers, and be made to serve as well as a Back Saw.

By slightly raising or lowering the spindles, when necessary, the leaden rolls at the bottom may be adjusted to stop the saw at the proper depth; and by the use of a set screw, the spindles on which the guides revolve, may be turned sufficiently to make the rollers bear firmly on the sides of a saw blade of any thickness.

If a narrow saw blade is used, or if the saw blade becomes narrower from use, the rollers may be lowered on the spindles by use of the brass collar and set screw under them.

PRICES.

Mitre Box, 20 inches..................................$7 00
Mitre Box, 20 inch, with 20 inch, Disston's Back Saw..........10 00

BAILEY'S IRON SPOKE SHAVES.

☞ The Spoke Shaves in the following List, are superior in style, quality and finish, to any in market. The Cutters are made of the best English Cast Steel, tempered and ground by an improved method, and are in perfect working order when sent from the factory.

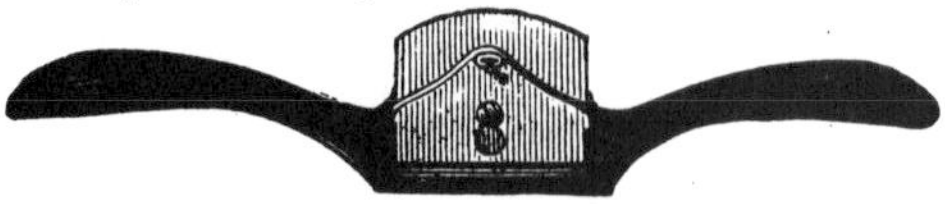

No.		Per Dozen.
51.	Double Iron, Raised Handle, 10 inch, 2⅛ in Cutter.......	$3 50
52.	Double Iron, Straight Handle, 10 inch, 2⅛ in. Cutter.....	3 50
53.	Adjustable, Raised Handle, 10 inch, 2⅛ in. Cutter........	4 50
54.	Adjustable, Straight Handle, 10 inch, 2⅛ in. Cutter........	4 50
55.	Model Double Iron, Hollow Face, 10 inch, 2⅛ in. Cutter..	3 00
56.	Coopers' Spoke Shave, 18 inch, 2⅝ in. Cutter............	7 00
56½.	Coopers' Spoke Shave (Heavy), 19 inch, 4 in. Cutter.....	9 00
57.	Coopers' Spoke Shave (Light), 18 inch, 2⅛ in. Cutter.....	4 50
58.	Model Double Iron, 10 inch, 2⅛ in. Cutter...............	3 00
59.	Single Iron (Pattern of No. 56), 10 inch, 2⅛ inch Cutter..	3 50
60.	Double Cutter, Hollow and Straight, 10 in., 1½ in Cutters.	4 50

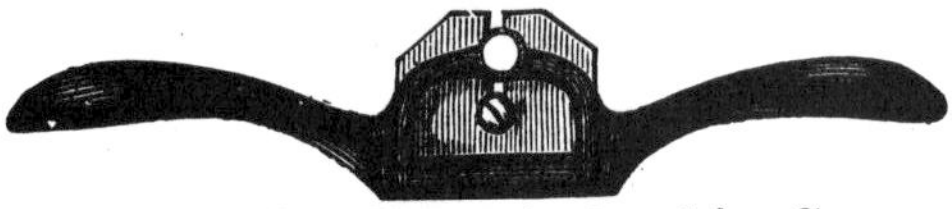

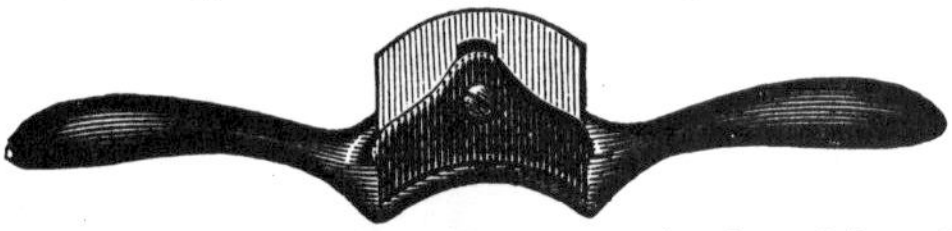

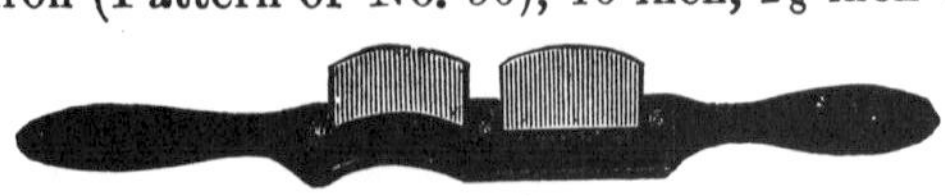

Price List of Spoke Shave Cutters

Nos. 51, 52, 53, 54, 55, 57, 58, 59 .... 1 00
No. 60 (in pairs), $1 50—No. 56, $1 50—No. 56½, $2 00.

IRON SPOKE SHAVES.

No.		Per Dozen.
64.	Straight Handle, 9 inch, 2⅛ inch Cutter (with Thumb Screw)........	$2 00
	Cast Steel Cutters.....	75

PATENT REVERSIBLE SPOKE SHAVE.

This Spoke Shave can be worked to and from the person using it, without changing position.

62.	Raised Handle, (Heavy), Double Cutter, 10 inch, 2⅛ inch Cutters.....	6 00
	Cast Steel Cutters............	1 00

PATENT CHAMFER SPOKE SHAVE.

This Tool can be easily adjusted by means of the Thumb-screws attached to the Guides; and will chamfer an edge any desired width up to 1½ inch.

65.	Raised Handle, 1¼ inch Cutter........	6 00
	Cast Steel Cutters...........	75

PATENT ADJUSTABLE BOX SCRAPER.

An excellent Box Scraper, and also well adapted for planing floors.

70.	Malleable Iron, 2 inch Steel Cutter............	6 00
	Cast Steel Cutters...........	1 50

WHEELER'S PATENT COUNTERSINK.
FOR WOOD.
OVER 75,000 HAVE BEEN SOLD.

The form of the tool between the cutter and the shank is that of a hollow half-cone, inverted, thus leaving ample space, just back of and above the cutter, for the free escape of shavings. This Countersink works equally well for every variety of screw, the pitch of the cone being the same as the taper given to the heads of all sizes of screws, thereby rendering only a single tool necessary for every variety of work. The Countersink cuts rapidly, and is easily sharpened by drawing a thin file lengthwise inside of the Cutter. The ingenious method of attaching a gauge to a Countersink will be observed by reference to the engraving. By fastening the gauge at a given point, any number of screws may be driven so as to leave the heads flush with the surface, or at a uniform depth below it. The gauge can be easily moved, or detached entirely, by means of the set screw. The Countersinks are sold with or without the gauge.

PRICES. Per dozen.

Countersinks, ½ dozen in a box......... .... $3 00
Countersinks, with Gauge, ¼ dozen in a box... .... 4 50

L. BAILEY'S PATENT ADJUSTABLE "VICTOR" PLANES.
Manufactured by Leonard Bailey & Co., Hartford, Conn.

The Tools comprising this list are made under the direct supervision of LEONARD BAILEY, the original inventor of L. Bailey's Patent Adjustable Iron and Wood Bench Planes. An improved device for fastening Plane Irons (Cutters) in the Stock (Patented Sept. 25, 1883,) has been recently applied to this full line of Planes, which is warranted superior to anything for the same purpose now in market. Circulars containing full illustrated description will be furnished on application. The Stanley Rule & Level Company are General Agents for the sale of these goods.

☞ The "Victor" Planes, have Iron Handles. They will be furnished with Wood Handles, if so ordered.

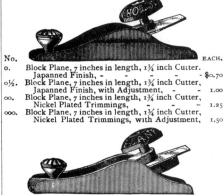

No.		EACH.
0.	Block Plane, 7 inches in length, 1¾ inch Cutter, Japanned Finish, - - - - -	$0.70
0½.	Block Plane, 7 inches in length, 1¾ inch Cutter, Japanned Finish, with Adjustment, -	1.00
00.	Block Plane, 7 inches in length, 1¾ inch Cutter, Nickel Plated Trimmings, -	1.25
000.	Block Plane, 7 inches in length, 1¾ inch Cutter, Nickel Plated Trimmings, with Adjustment,	1.50

No.		EACH.
3.	Smooth Plane, with Adjustment 8½ inches in length. 1¾ inch Cutter, Polished Trimmings,	$3.00
3½.	Smooth Plane, with Adjustment, 8½ inches in length 1¾ inch Cutter, Nickel Plated Trimmings. - - - - - - -	3.75
4.	Smooth Plane, with Adjustment, 9 inches in length, 2 inch Cutter, Polished Trimmings, -	3.25
4½.	Smooth Plane, with Adjustment, 9 inches in length, 2 inch Cutter, Nickel Plated Trimmings. - - - - - - -	4.00

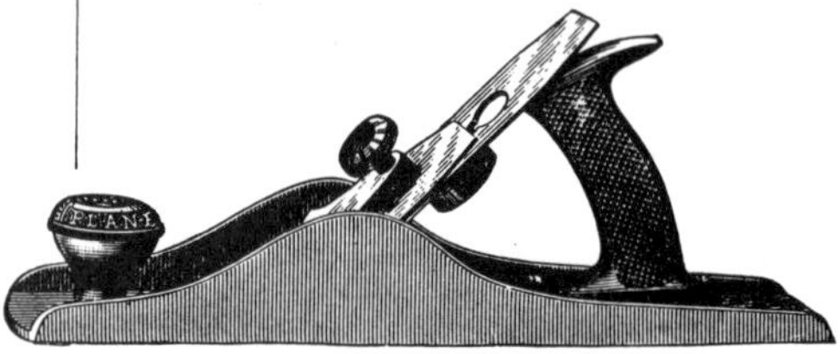

1.	Block Plane, adjustable Mouth and Cutter, 6 inches in length, 1¾ inch Cutter, Polished Trimmings, - - -	$1.50
1¾.	Block Plane, adjustable Mouth and Cutter, 6 inches in length, 1¾ inch Cutter, Nickel Plated Trimmings, -	1.75
2.	Block Plane, adjustable Mouth and Cutter, 7 inches in length. 1¾ inch Cutter, Polished Trimmings, -	1.75
2¾.	Block Plane, adjustable Mouth and Cutter, 7 inches in length, 1¾ inch Cutter, Nickel Plated Trimmings, -	2.00

5.	Jack Plane, with Adjustment, 14 inches in length, 2 inch Cutter, Polished Trimmings, -	$3.75
5½.	Jack Plane, with Adjustment, 14 inches in length, 2 inch Cutter, Nickel Plated Trimmings. -	4.50
6.	Fore Plane, with Adjustment, 18 inches in length, 2⅜ inch Cutter, Polished Trimmings,	4.75
6½.	Fore Plane, with Adjustment, 18 inches in length, 2⅜ inch Cutter, Nickel Plated Trimmings, -	5.50
7.	Jointer Plane, with Adjustment. 22 inches in length, 2⅜ inch Cutter, Polished Trimmings,	5.50
7½.	Jointer Plane, with Adjustment, 22 inches in length, 2⅜ inch Cutter, Nickel Plated Trimmings, -	6.25
8.	Jointer Plane, with Adjustment, 24 inches in length, 2⅝ inch Cutter, Polished Trimmings,	6.50
8½.	Jointer Plane, with Adjustment, 24 inches in length, 2⅝ inch Cutter, Nickel Plated Trimmings, -	7.25

1¼.	Block Plane, with Handle, adjustable Mouth and Cutter, 6 inches in length, 1¾ inch Cutter, Polished Trimmings, -	$1.75
1½.	Block Plane, with Handle, adjustable Mouth and Cutter, 6 inches in length, 1¾ inch Cutter, Nickel Plated Trimmings, -	2.00
2¼.	Block Plane, with Handle, adjustable Mouth and Cutter, 7 inches in length, 1¾ inch Cutter, Polished Trimmings, -	2.00
2½.	Block Plane, with Handle, adjustable Mouth and Cutter, 7 inches in length, 1¾ inch Cutter, Nickel Plated Trimmings, - - -	2.25

L. BAILEY'S POCKET BLOCK PLANE.

LITTLE VICTOR BLOCK PLANE.

No.		EACH.
12.	4½ inches in length, 1¼ inch Cutter, Japanned Finish, Polished Trimmings, - - -	$0.75
12½.	4½ inches in length, 1¼ inch Cutter, Japanned Finish, Nickel Plated Trimmings, -	1.00
12¾.	4½ inches in length, 1¼ inch Cutter, Full Nickel Plated, - - - - -	1.25

No		EACH.
50.	3¼ inches in length, 1 inch Cutter, complete Adjustment, Japanned, - - -	$0.45
50½.	3¼ inches in length, 1 inch Cutter, complete Adjustment, full Nickel Plated, -	.50
51.	3¼ inches in length. 1 inch Cutter, Screw Fastening, Japanned, -	.30
51½.	3¼ inches in length, 1 inch Cutter, Screw Fastening, full Nickel Plated, -	.45
52.	3¼ inches in length, 1 inch Cutter, Screw Eye Fastening, Japanned, - -	.25
Cutters,	- - - - - - per dozen,	1.25

ADJUSTABLE CIRCULAR PLANE.

No. 10. Circular Plane, with Adjustment, Flexible Steel Face, 1¾ inch Cutter, Nickel Plated Trimmings, - - - - - Each, $4.00.

IMPROVED
ADJUSTABLE CIRCULAR PLANE.

The above cut represents LEONARD BAILEY'S Improved Adjustable Plane, for working Concave or Convex Surfaces.

Its working capacity is 12½ inches inside circle. Both ends of the face of the Plane are moved simultaneously and precisely alike, by means of one screw. The simplicity of its operation is unequaled.

No. 20. Circular Plane, 9½ inches in length, 1¾ inch Cutter, Nickel Plated Trimmings. Each, $6.00.

L. Bailey's Victor Plane Irons.

The "Victor" Patent Adjustable Planes are supplied with this SUPERIOR COMPOUND PLANE IRON.

	1¾	2	2⅜	2⅝ inch.
SINGLE, - - -	38	42	48	50 cts.
PATENT DOUBLE,	62	68	80	84 cts.
" STEEL CAPS,	25	25	35	35 cts.

All Double Irons have Patent Steel Caps.
Block Plane Irons, 1¾ inch, 35 cents each.

These Plane Irons are made of the best English Steel, hardened and tempered by a new process. Each cutter is sharpened and put in perfect working order before leaving the factory, and fully warranted.

COMBINED
Smooth, Rabbet & Filletster Plane.

The Tool represented below consists of an Iron Smooth Plane, (same as our No. 4 Plane,) and is so constructed that by means of the attachments which accompany it, it can also be used as a Rabbet Plane or a Filletster.

This view of the tool shows the cutter on one side to be flush with the edge of the stock to the Plane, adapting it for use as a Rabbet plane; also, there is shown in this view the depth gauge, spur in edge of stock, and the fence, all of which belong to the tool when used as a Filletster.

This view of the tool shows the socket, which can be screwed on to the side of the stock, and through which a bar slides. The fence can thus be moved to any required distance from the edge of the stock, making an Adjustable Filletster of any desired width up to two inches.

No. 11. Combined Smooth, Rabbet and Filletster Plane, 9 in. long, 2 inch Cutter, Polished Trimmings, - - - - - - Each, $5.50

No. 11½. Combined Smooth, Rabbet and Filletster Plane, 9 in. long, 2 inch Cutter, Nickel Plated Trimmings, - - - Each, 6.00

L. BAILEY'S
Patent Combination Plane.

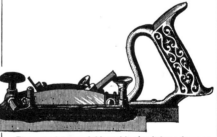

The main feature of this tool is, that it has a fence or guide, which is made to change from one side to the other as the nature of the work requires, same fence or guide also in itself being vertically self-adjusting. The same cutter is used for each special tool in the combination.

No. 14. Plow, Filletster, Back Filletster, Dado, Rabbet & Matching Plane combined, Each, $5.50

L. BAILEY'S
PATENT "VICTOR" REVERSIBLE BOX SCRAPER.

No. 48. Per Dozen......................................$7 00
Cutters, Per Dozen.................................... 1 50

DOUBLE IRON SPOKE SHAVE.

No. 41. Per Dozen......................................$5 00
Cutters, Per Dozen.................................... 1 50

ADJUSTABLE SPOKE SHAVE.

No. 43. Per Dozen......................................$6 00
Cutters, Per Dozen.................................... 1 50

L. BAILEY'S PATENT FLUSH T BEVEL.

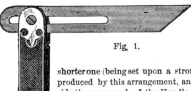

Fig. 1.

The method of securing the Blade at any angle desired is simple, yet effective. Fig. 2 shows the internal construction of the Handle. By moving the thumb-piece at the lower end of the Handle, the long lever acts upon the shorter one (being set upon a strong pivot). The strength of a compound lever is produced by this arrangement, and the short lever being attached to the Nut inside the upper end of the Handle, operates as a wrench to turn it upon the screw, and thus to fasten or release the Blade at the pleasure of the owner.

The Handle is made of Cast Iron. The Blade is fine quality Steel, spring temper, and with perfectly parallel edges.

Fig. 2.

Inches,	8	10
Per Dozen,	$14 00	16 00

½ dozen in a box.

IMPROVED TRAMMEL POINTS.

These Tools are used by Millwrights, Machinists, Carpenters, and all Mechanics having occasion to strike arcs, or circles, larger than can be conveniently done with ordinary Compass dividers. They may be used on a straight wooden bar of any length, and when secured in position by the thumbscrews, all circular work can be readily laid out by their use. They are made of Bronze Metal, and have Steel Points, either of which can be removed, and replaced by the pencil socket which accompanies each pair, should a pencil mark be preferred in laying out work.

PRICES.

No.		Per Pair.
1.	(Small) Br'ze Metal, Steel Points,	$1 00
2.	(Medium) " "	1 25
3.	(Large) " "	1 75

ADJUSTABLE PLUMB BOBS.

These Plumb Bobs are constructed with a reel at the upper end, upon which the line may be kept; and by dropping the bob with a slight jerk, while the ring is held in the hand, any desired length of line may be reeled off. A spring, which has its bearing on the reel, will check and hold the bob firmly at any point on the line. The pressure of the spring may be increased, or decreased, by means of the screw which passes through the reel. A suitable length of line comes already reeled on each Plumb Bob.

PRICES.

No.		Each.
1.	(Small) Bronze Metal, with Steel Point............	$1 50
2.	(Large) " " "	1 75
5.	(Large) Iron " " "	1 00

PATENT EXCELSIOR TOOL HANDLES.

DIRECTIONS.—Unscrew the Cap of the Handle, and select the Tool needed; with the thumb, the center bolt may be thrust down sufficiently to open the clamp at the small end of the Handle, into which the tool can be inserted; then in replacing the cover, and screwing it down to its place, the clamp will be closed and the Tool firmly secured for use.

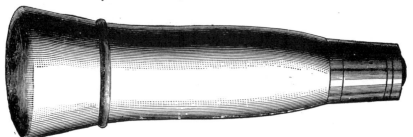

No.		Per Doz.
1.	Turkey Boxwood Handle, with Twenty Tools............	$7 50

1-2 dozen in a box.

2.	Iron Handle, with Twelve Tools.......................	4 00
3.	Iron Handle, with Twenty Tools.......................	5 50
22.	Iron Handle, Nickel Plated, with Twelve Tools..........	5 00
23.	Iron Handle, Nickel Plated, with Twenty Tools.........	6 50

½ dozen in a box.

CARPENTERS' TOOL HANDLES.

No.   Per Dozen.
8. Carpenter's Tool Handle, Steel Screw and Nut, with Wrench, 1 dozen in a box $0 90
8¼. Carpenter's Tool Handle, Steel Screw and Nut, with Wrench, and 10 Brad Awls, assorted sizes. One handle and awls, in a box, and 12 boxes in a package... 3 00

AWL HAFTS.

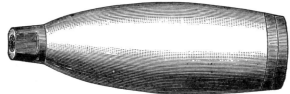

No.   Per Gross.
5. Hickory, Pegging Awl Haft, Plain Top, Steel Screw and Nut, with Wrench, 1 dozen in a box.............. 10 00
6. Hickory, Pegging Awl Haft, Leather Top, Steel Screw and Nut, with Wrench, 1 dozen in a box........... 12 00
7. Hickory, Pegging Awl Haft, Leather Top Extra Large, Riveted, Steel Screw and Nut, with Wrench, 1 dozen in a box...................................... 14 00

6½. Appletree Sewing Awl Haft, to hold any size Awl, Steel Screw and Nut, with Wrench, 1 dozen in a box..... 12 00

10. Common Sewing Awl Haft, Brass Ferrule, 3 doz. in a box   3 50
11. Common Pegging Awl Haft, Brass Ferrule, 3 doz. in a box   3 50

PATENT PEGGING AWLS.

PATENT PEGGING AWLS (short start, for Handle Nos. 5 and 6), assorted Nos. 00, 0, 1, 2, 3, 4, 5 and 6, 1 gross in a box   80

CHALK-LINE REELS, Etc.

	Per Dozen.
Chalk-Line Reels, 3 dozen in a box	$0 36
Chalk-Line Reels, with 60 feet best quality Chalk-Line, 1 dozen in a box	1 75
Chalk- Line Reels, with Steel Scratch Awls, 1 dozen in a box	0 95
Chalk-Line Reels, with Steel Scratch Awls, and 60 feet best quality Chalk-Line, 1 dozen in a box	2 25

HANDLED SCRATCH AWLS.

No.		Per Gross.
1.	Handled, Steel Scratch Awls, 1 dozen in a box	$7 00
2.	Handled, Steel Scratch Awls, Large, 1 dozen in a box	8 50

HANDLED BRAD AWLS.

	Per Gross.
Handled Brad Awls, assorted, 1 dozen in a box	$6 50
Handled Brad Awls, assorted, Large, 1 dozen in a box	7 00

Brad Awl Handles, Brass Ferrules, assorted, 3 dozen in a box,	3 50

SAW HANDLES.

All full sizes, of perfect timber, well seasoned, and every way superior and reliable goods.

No.		Per Dozen.
1.	Full size, Cherry, Varnished Edges......................	$1 65
2.	Full size, Beech, Varnished Edges.......................	1 40
3.	Full size, Beech, Plain Edges...........................	1 20
4.	Small Panel, Beech, Varnished Edges for 16 to 20 in. Saws	1 35
5.	Meat Saw, Beech, Varnished Edges.....................	1 35
6.	Compass Saw, Beech, Varnished Edges..................	1 25
7.	Back Saw, Beech, Varnished Edges..................	1 35

In paper boxes of 1 dozen each—packed 2 gross in a case.

PLANE HANDLES.

	Per Dozen.
Jack Plane Handles, 5 gross in a case......................	$0 42
Fore, or Jointer Handles, 3¾ gross in case...................	72

In packages of 1 dozen each.

MALLETS.—Mortised Handles.

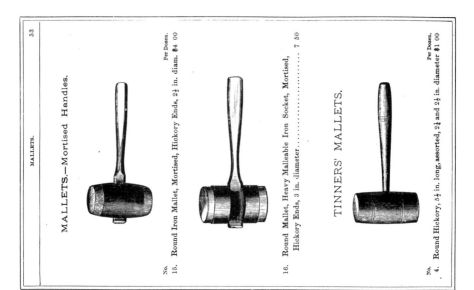

No.		Per Dozen
15.	Round Iron Mallet, Mortised, Hickory Ends, 2¼ in. diam.	$4 00
16.	Round Mallet, Heavy Malleable Iron Socket, Mortised, Hickory Ends, 3 in. diameter	7 50

TINNERS' MALLETS.

No.		Per Dozen
4.	Round Hickory, 5¼ in. long, assorted, 2¼ and 2¾ in. diameter	$1 00

MALLETS.—Mortised Handles.

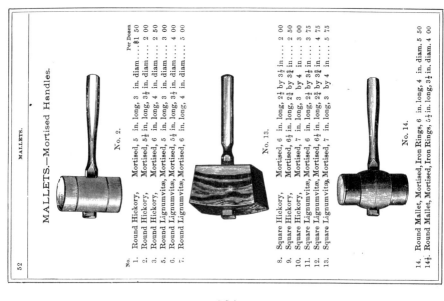

No.		Per Dozen
1.	Round Hickory, Mortised, 5 in. long, 3 in. diam.	$1 50
2.	Round Hickory, Mortised, 5½ in. long, 3½ in. diam.	2 00
3.	Round Hickory, Mortised, 6 in. long, 4 in. diam.	2 50
5.	Round Lignumvitæ, Mortised, 5 in. long, 3 in. diam.	3 00
6.	Round Lignumvitæ, Mortised, 5½ in. long, 3½ in. diam.	4 00
7.	Round Lignumvitæ, Mortised, 6 in. long, 4 in. diam.	5 00
8.	Square Hickory, Mortised, 6 in. long, 2½ by 3½ in.	2 00
9.	Square Hickory, Mortised, 6½ in. long, 2¾ by 3¾ in.	2 50
10.	Square Hickory, Mortised, 7 in. long, 3 by 4 in.	3 00
11.	Square Lignumvitæ, Mortised, 6 in. long, 2½ by 3½ in.	3 75
12.	Square Lignumvitæ, Mortised, 6½ in. long, 2¾ by 3¾ in.	4 75
13.	Square Lignumvitæ, Mortised, 7 in. long, 3 by 4 in.	5 75
14.	Round Mallet, Mortised, Iron Rings, 6 in. long, 4 in. diam.	5 50
14½.	Round Mallet, Mortised, Iron Rings, 5½ in. long, 3½ in. diam.	4 00

TACK HAMMERS, ETC.

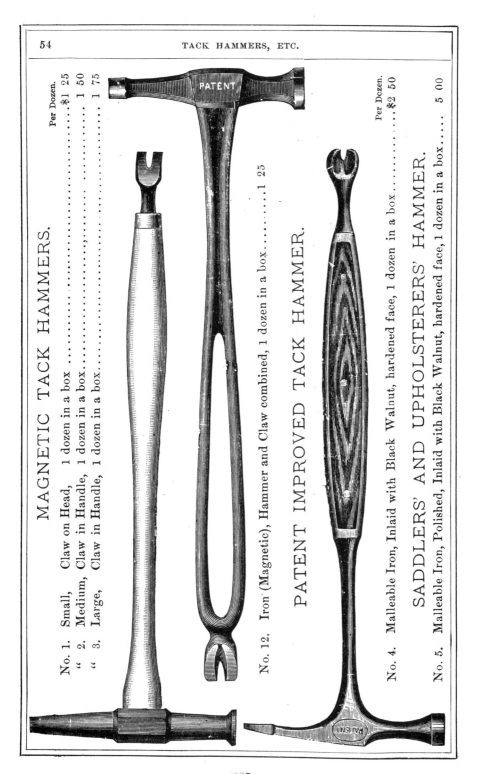

MAGNETIC TACK HAMMERS.

				Per Dozen.
No. 1.	Small,	Claw on Head,	1 dozen in a box	$1 25
" 2.	Medium,	Claw in Handle,	1 dozen in a box	1 50
" 3.	Large,	Claw in Handle,	1 dozen in a box	1 75

PATENT IMPROVED TACK HAMMER.

No. 12. Iron (Magnetic), Hammer and Claw combined, 1 dozen in a box........1 25

SADDLERS' AND UPHOLSTERERS' HAMMER.

		Per Dozen.
No. 4.	Malleable Iron, Inlaid with Black Walnut, hardened face, 1 dozen in a box.........	$2 50
No. 5.	Malleable Iron, Polished, Inlaid with Black Walnut, hardened face, 1 dozen in a box.....	5 00

STEAK HAMMERS, ETC. 55

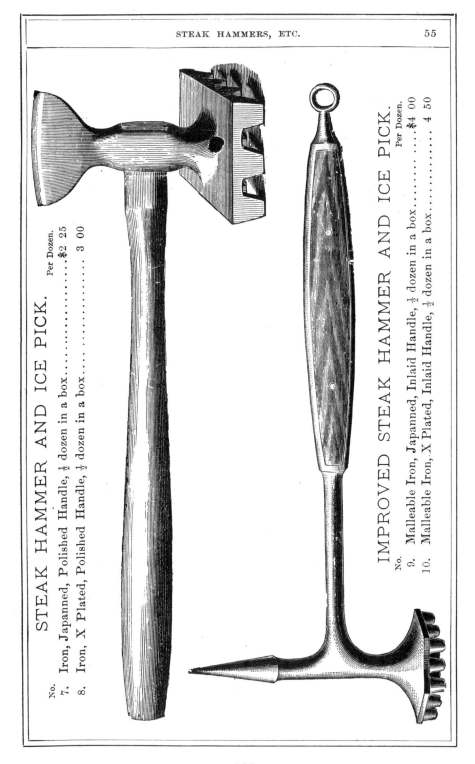

STEAK HAMMER AND ICE PICK.

Per Dozen.
No.
7. Iron, **Japanned**, Polished Handle, ½ dozen in a box........$2 25
8. Iron, X Plated, Polished Handle, ½ dozen in a box........ 3 00

IMPROVED STEAK HAMMER AND ICE PICK.

Per Dozen.
No.
9. Malleable Iron, **Japanned**, Inlaid Handle, ½ dozen in a box........$4 00
10. Malleable Iron, X Plated, Inlaid Handle, ½ dozen in a box........ 4 50

PATENT IMPROVED CATTLE TIE.

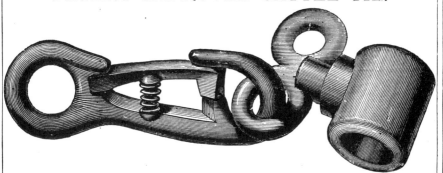

It is believed that this Cattle Tie in both its parts, embraces more truly valuable features than any similar article heretofore sold. The snap is properly proportioned for the greatest strength required, and is made of Malleable iron. The form of the hooked end prevents the snap from becoming unfastened by any accidental means; while the Improved Spiral Spring, which is made of brass, and not liable to rust from being wet, or to break in cold weather, acts with certainty, and is protected in its position under an independent lug, or bar, upon which all shocks or pressure must first come.

The other part of the Cattle Tie consists of an Iron socket, which may be secured at any desired position on a rope, by means of a Malleable thumb screw in one side of the socket; the thumb screw having a perforated head through which the snap is readily hooked. The socket may at once be changed from one position on the rope to another, and secured perfectly without the use of a screw-driver, or other Tool, thus adapting the length of the rope to the size of the neck or horns of any animal. The extreme simplicity of both parts insures great durability.

	Per Dozen.
Cattle Ties, Japanned, 1 dozen in a box	$1 20
Cattle Ties, Japanned, with Rope	3 60

SMITH'S SASH CORD IRON.

These Irons can be more easily put into the edge of a sash than any other, requiring only the boring of a round hole; and no screws are needed to keep them in position. When the windows are taken out for cleaning, or any other purpose, the Irons can be removed or replaced without the use of any Tool.

	Per Gross.
In boxes of 1 gross	$0 65

INDEX.

	Page.
Awl Hafts,	49
Awls, Patent Pegging,	49
Brad Awls, Handled,	50
Bevels, Sliding T,	27
Bevels, Patent Flush, Eureka,	27
Bevels, L. Bailey's Patent Flush,	46
Box Scraper, Adjustable,	43
" " "Victor" Adjustable,	46
Chalk-Line Reels and Awls	50
Carpenters' Tool Handles,	49
Cattle Ties,	56
Countersinks, Wheeler's Patent,	43
Dado, Filletster, Plow, Etc, Combined,	39
Dado, Adjustable,	39
Gauges,	28 to 30
Handles, Brad Awls,	50
" Plane,	51
" Saw,	51
" Screw Driver,	31
Hammers, Magnetic,	54
" Tack, No. 4,	54
" Steak,	55
" Upholsterers',	54
Level Glasses,	21
Mallets, Hickory and Lignumvitæ,	52 and 53
Mitre Box, Improved,	41
Mitre Squares, Improved,	24
Mitre Try Squares, Improved,	24
Miscellaneous Articles, Stanley's,	12 and 13

1892 STANLEY RULE & LEVEL Co. CATALOG and PRICE LIST, including Revision to 1897

At the date of the original issue of the 1892 Catalog, the new additions of products from the previous 1888 issue included the N° 62½ Extra Narrow, Two Foot, Four Fold Rule; N° 16 Duplex Plumb and Level; N° 30 Adjustable Patent Level and Level Sights; N° 82 Improved Rabbet and Butt Gauge; and the Dowel Sharpener. In addition, a few new illustrations of old products appeared in 1892.

However, soon after the 1892 Catalog and Price List was issued, several new products were added. These included: HAND-Y Plumb and Levels; Side Rabbet; Rabbet & Block; Core Box; and the famed N° 55 Universal Planes. As these products became available, supplementary sheets were added and in 1897 the 1892 Price List was re-issued showing all of these new tools.

Ken Roberts
1980

PRICE LIST

OF

U. S. STANDARD

Boxwood and Ivory

RULES,

PLUMBS AND LEVELS, TRY SQUARES,
BEVELS, GAUGES, MALLETS,
IRON AND WOOD ADJUSTABLE PLANES,
SPOKE SHAVES, SCREW DRIVERS,
AWL HAFTS, HANDLES, &c.

MANUFACTURED BY THE

STANLEY
Rule and Level Co.

NEW BRITAIN, CONN., U. S. A.

WAREROOMS:

No. 29 CHAMBERS STREET, NEW YORK.

ORDERS FILLED AT THE WAREROOMS, OR AT NEW BRITAIN.

JANUARY, 1892.

OFFICE OF THE

STANLEY RULE AND LEVEL CO.

NEW BRITAIN, CONN., U. S. A.

This Illustrated Catalogue presents the full line of

IMPROVED LABOR-SAVING

CARPENTERS' TOOLS

now manufactured by us. It will be interesting to Hardware Dealers, as containing genuine Goods, with an established reputation. Such Goods will sell themselves.

Mechanics require no explanation, or apology, from the Dealer who offers them STANLEY'S Tools; and the uniform report from those who use these Tools is, that they want more.

Our continued efforts are pledged to produce a grade of Tools acceptable to Mechanics, and worthy the ineffectual attempts made by other manufacturers to imitate them.

JANUARY 1, 1892.

PRICE LIST.

☞ All Rules embraced in the following Lists bear the Invoice Number, by which they are sold, and are graduated to correspond with the description given of each.

STANLEY'S BOXWOOD RULES.
ONE FOOT, FOUR FOLD, NARROW.

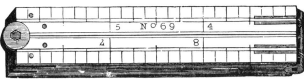

No.   Per Doz.
69. Round Joint, Middle Plates, 8ths and 16ths of inches.................................... $\tfrac{5}{8}$ in. wide, **$3 00**

65. Square Joint, Middle Plates, 8ths and 16ths of inches..................................... $\tfrac{5}{8}$ in. wide,   3 50
64. Square Joint, Edge Plates, 8ths and 16ths of inches..................................... $\tfrac{5}{8}$ "   5 00
65½. Square Joint, Bound, 8ths and 16ths of inches. $\tfrac{5}{8}$ "   11 00

55. Arch Joint, Middle Plates, 8ths and 16ths of inches..................................... $\tfrac{5}{8}$ in. wide,   4 00
56. Arch Joint, Edge Plates, 8ths and 16ths of inches..................................... $\tfrac{5}{8}$ "   6 00
57. Arch Joint, Bound, 8ths and 16ths of inches.. $\tfrac{5}{8}$ "   12 00

TWO FEET, FOUR FOLD, NARROW.

No.		Per Doz.
68.	Round Joint, Middle Plates, 8ths and 16ths of inches..................................1 in. wide,	$4 00
61.	Square Joint, Middle Plates, 8ths and 16ths of inches..................................1 "	5 00
63.	Square Joint, Edge Plates, 8ths, 10ths, 12ths and 16ths of inches, Drafting Scales.......1 "	7 00
84.	Square Joint, Half Bound, 8ths, 10ths, 12ths and 16ths of inches, Drafting Scales.......1 "	12 00
62.	Square Joint, Bound, 8ths, 10ths, 12ths and 16ths of inches, Drafting Scales ....1 "	15 00
51.	Arch Joint, Middle Plates, 8ths, 10ths, 12ths and 16ths of inches, Drafting Scales.......1 "	6 00
53.	Arch Joint, Edge Plates, 8ths, 10ths, 12ths and 16ths of inches, Drafting Scales..........1 "	8 00
53½.	Arch Joint, Edge Plates, 8ths, 10ths, 12ths and 16ths of inches, with inside Beveled Edges, and Architect's Drafting Scales...........1 " [See Engraving, p. 9.]	15 00
52.	Arch Joint, Half Bound, 8ths, 10ths, 12ths and 16ths of inches, Drafting Scales ....1 "	13 00
54.	Arch Joint, Bound, 8ths, 10ths, 12ths and 16ths of inches, Drafting Scales................1 "	16 00
59.	Double Arch Joint, Bitted, 8ths, 10ths, 12ths and 16ths of inches, Drafting Scales.......1 "	9 00
60.	Double Arch Joint, Bound, 8ths, 10ths, 12ths and 16ths of inches, Drafting Scales.......1 "	21 00

TWO FEET, FOUR FOLD, EXTRA NARROW.

61¼.	Square Joint, Middle Plates, 8ths and 16ths of inches..................................¾ in. wide,	5 50
63½.	Square Joint, Edge Plates, 8ths, 10ths and 16ths of inches........................¾ "	8 00
62½.	Square Joint, Bound, 8ths, 10ths, 12ths and 16ths of inches ....¾ "	15 00

STANLEY'S TWO FEET, FOUR FOLD, NARROW RULES.

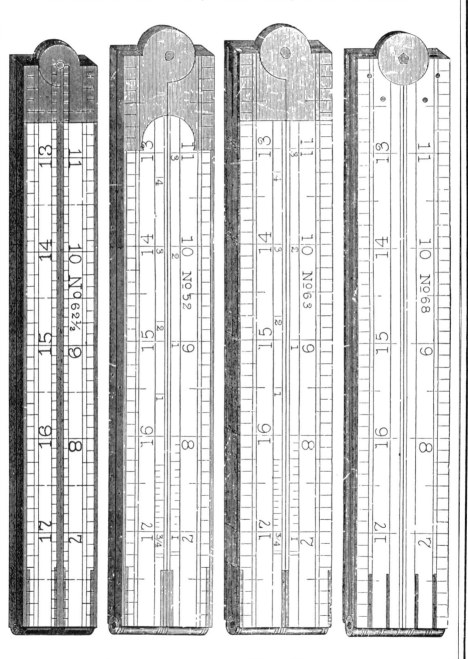

TWO FEET, FOUR FOLD, BROAD.

No.		Per Doz.
67.	Round Joint, Middle Plates, 8ths and 16ths of inches .... 1⅜ in. wide,	$5 00
70.	Square Joint, Middle Plates, 8ths and 16ths of inches, Drafting Scales .... 1⅜ "	7 00
72.	Square Joint, Edge Plates, 8ths, 10ths and 16ths of inches, Drafting Scales .... 1⅜ "	9 00
72½.	Square Joint, Bound, 8ths, 10ths and 16ths of inches, Drafting Scales .... 1⅜ "	18 00
73.	Arch Joint, Middle Plates, 8ths, 10ths and 16ths of inches, Drafting Scales .... 1⅜ "	9 00
75.	Arch Joint, Edge Plates, 8ths, 10ths and 16ths of inches, Drafting Scales .... 1⅜ "	11 00
76.	Arch Joint, Bound, 8ths, 10ths and 16ths of inches, Drafting Scales .... 1⅜ "	20 00
77.	Double Arch Joint, Bitted, 8ths, 10ths and 16ths of inches, Drafting Scales .... 1⅜ "	12 00
78.	Double Arch Joint, Half Bound, 8ths, 10ths and 16ths of inches, Drafting Scales ...... 1⅜ "	20 00
78½.	Double Arch Joint, Bound, 8ths, 10ths and 16ths of inches, Drafting Scales .... 1⅜ "	24 00
83.	Arch Joint, Edge Plates, Slide, 8ths, 12ths and 16ths of inches, 100ths of a foot, and Octagonal Scales .... 1⅜ "	14 00

BOARD MEASURE, TWO FEET, FOUR FOLD.

79.	Square Joint, Edge Plates, 12ths and 16ths of inches, Drafting Scales .... 1⅜ in. wide,	11 00
81.	Arch Joint, Edge Plates, 12ths and 16ths of inches, Drafting Scales .... 1⅜ "	13 00
82.	Arch Joint, Bound, 12ths and 16ths of inches, Drafting Scales .... 1⅜ "	22 00

THREE FEET, FOUR FOLD RULES.

66.	Arch Joint, Middle Plates, Four Fold, 16ths of inches outside, and Yard Stick on inside ... 1 in. wide,	8 00
66½.	Arch Joint, Middle Plates, Four Fold, 8ths and 16ths of inches .... 1 "	8 00

CARRIAGE MAKERS', FOUR FEET, FOUR FOLD.

94.	Arch Joint, Bound, 8ths and 16ths of inches. 1½ in. wide,	48 00

STANLEY'S TWO FEET, FOUR FOLD, BROAD RULES.

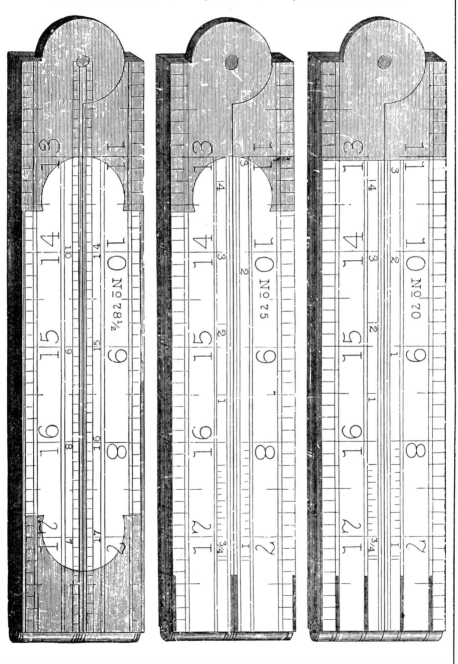

TWO FEET, TWO FOLD.

No.			Per Doz.
29.	Round Joint, 8ths and 16ths of inches ...... 1⅜ in. wide,		$3 50
18.	Square Joint, 8ths and 16ths of inches 1½	"	5 00
22.	Square Joint, Bitted, Board Measure, 10ths, 12ths and 16ths of inches, Octagonal Scales. 1½	"	8 00
1.	Arch Joint, 8ths and 16ths of inches, Octagonal Scales 1½	"	7 00
2.	Arch Joint, Bitted, 8ths, 10ths and 16ths of inches, Octagonal Scales 1½	"	8 00
4.	Arch Joint, (plates on outside of wood) Bitted, Extra Thin, 8ths and 16ths of inches, Drafting and Octagonal Scales 1½	"	10 00
5.	Arch Joint, Bound, 8ths, 10ths and 16ths of inches, Drafting and Octagonal Scales 1½	"	16 00

TWO FEET, TWO FOLD, SLIDE.

26.	Square Joint, Slide, 8ths, 10ths and 16ths of inches, Octagonal Scales 1½	"	9 00
27.	Square Joint, Bitted, Gunter's Slide, 8ths, 10ths and 16ths of inches, 100ths of a foot, Drafting and Octagonal Scales 1½	"	12 00
12.	Arch Joint, Bitted, Gunter's Slide, 8ths, 10ths and 16ths of inches, 100ths of a foot, Drafting and Octagonal Scales 1½	"	14 00
15.	Arch Joint, Bound, Gunter's Slide, 8ths, 10ths and 16ths of inches, Drafting and Octagonal Scales 1½	"	24 00
6.	Arch Joint, Bitted, Gunter's Slide, Engineering, 8ths, 10ths and 16ths of inches, 100ths of a foot, Octagonal Scales 1½	"	18 00
16.	Arch Joint, Bound, Gunter's Slide, Engineering, 8ths, 10ths and 16ths of inches, Octagonal Scales 1½	"	28 00

SHIP CARPENTERS' BEVELS.

42.	Boxwood, Double Tongue, 8ths and 16ths of inches,	6 00
43.	Boxwood, Single Tongue, 8ths and 16ths of inches,	6 00

☞ With recently constructed machinery, we can furnish, to order, Rules marked with Spanish graduations, or with Metric graduations.

STANLEY'S SLIDE RULES, ARCHITECTS' RULES, ETC.   9

STANLEY'S TWO FEET, TWO FOLD, SLIDE RULES.

No. 12.

☞ We have an improved Treatise on the Gunter's Slide and Engineer's Rules, showing their utility, and containing full and complete instructions, enabling Mechanics to make their own calculations. It is also particularly adapted to the use of persons having charge of cotton or woolen machinery, surveyors, and others. 200 pages, bound in cloth. Price, $1.00, net. Sent by mail, postpaid, on receipt of the price.

SHIP CARPENTERS' BEVELS.

No. 42.

ARCHITECTS' RULES, WITH BEVELED EDGES.

Nos. 53½ and 86½.

☞ For description of No. 53½ (Boxwood) see page 4, and for description of No. 86½ (Ivory) see page 11.

BOXWOOD CALIPER RULES, SIX INCH.

No.		Per Doz.
36. Square Joint, Two Fold, 8ths, 10ths, 12ths and 16ths of inches .... 7/8 in. wide,		**$7 00**
13. Square Joint, Two Fold, 8ths and 16ths inches, 1 1/8 "		**10 00**
13½. Square Joint, Two Fold, 8ths and 16ths inches, 1½ "		**12 00**

CALIPER, ONE FOOT, FOUR FOLD.

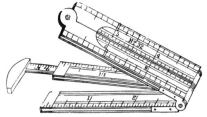

32. Arch Joint, Edge Plates, 8ths, 10ths, 12ths and 16ths of inches .... 1 in. wide, **12 00**
32½. Arch Joint, Bound, 8ths, 10ths, 12ths and 16ths of inches .... 1 " **20 00**

CALIPER, ONE FOOT, TWO FOLD.

36½. Square Joint, Two Fold, 12 inch, 8ths, 10ths, 12ths and 16ths of inches .... 1⅜ " **12 00**

TWO FEET, SIX FOLD RULES.

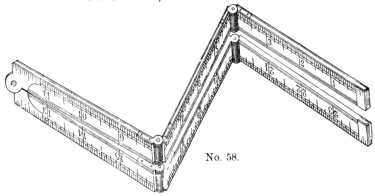

No. 58.

58. Arch Joint, Edge Plates, 8ths, 10ths, 12ths and 16ths of inches .... ¾ in. wide, **13 00**
58½. Arch Joint, Bound, 8ths, 10ths, 12ths and 16ths of inches .... ¾ " **36 00**

STANLEY'S IVORY RULES.

No.	IVORY CALIPER, SIX INCH.		Per Doz.
38.	Square Joint, German Silver, Two Fold, 8ths, 10ths, 12ths and 16ths of inches .... $\frac{7}{8}$ in. wide,		$15 00
40½.	Square Joint, German Silver, Bound, Two Fold, 8ths and 16ths of inches .... $\frac{5}{8}$	"	24 00

IVORY CALIPER, ONE FOOT, FOUR FOLD.

39.	Square Joint, German Silver, Edge Plates, 8ths, 10ths, 12ths and 16ths of inches ..... $\frac{7}{8}$ in. wide,		38 00
40.	Square Joint, German Silver, Bound, 8ths and 16ths of inches .... $\frac{5}{8}$	"	44 00

IVORY, ONE FOOT, FOUR FOLD.

90.	Round Joint, Brass, Middle Plates, 8ths and 16ths of inches		10 00
92½.	Square Joint, German Silver, Middle Plates, 8ths and 16ths of inches................ $\frac{5}{8}$	"	14 00
92.	Square Joint, German Silver, Edge Plates, 8ths and 16ths of inches .... $\frac{5}{8}$	"	17 00
88½.	Arch Joint, German Silver, Edge Plates, 8ths and 16ths of inches .... $\frac{5}{8}$	"	21 00
88.	Arch Joint, German Silver, Bound, 8ths and 16ths of inches .... $\frac{5}{8}$	"	32 00
91.	Square Joint, German Silver, Edge Plates, 8ths, 10ths, 12ths and 16ths of inches ..... $\frac{3}{4}$	"	23 00

IVORY, TWO FEET, FOUR FOLD.

85.	Square Joint, German Silver, Edge Plates, 8ths, 10ths, 12ths and 16ths of inches ..... $\frac{7}{8}$ in. wide,		54 00
86.	Arch Joint, Ger. Silv., Ed. Plates, 8ths, 10ths, 12ths, 16ths in., 100ths foot, Draft'g Scales, 1	"	64 00
86½.	Arch Joint, German Silver, Edge Plates, 8ths, 10ths, 12ths and 16ths of inches, with Inside Beveled Edges, and Architects' Dftg. Scales, 1 [See Engraving, p. 9.]	"	96 00
87.	Arch Joint, German Silver, Bound, 8ths, 10ths, 12ths and 16ths of inches, Drafting Scales .1	"	80 00
89.	Double Arch Joint, German Silver, Bound, 8ths, 10ths, 12ths and 16ths of in. Drafting Scales, 1	"	92 00
95.	Arch Joint, German Silver, Bound, 8ths, 10ths 12ths and 16ths of inches, Drafting Scales, 1$\frac{3}{8}$	"	102 00
97.	Double Arch Joint, German Silver, Bound, 8ths, 10ths, 12ths and 16ths of in. Drafting Scales, 1$\frac{3}{8}$	"	116 00

MISCELLANEOUS ARTICLES.

BENCH RULES.

No.			Per Doz.
34.	Bench Rule, Maple, Brass Tips .... 2 feet,		$3 00
35.	Bench Rule, Board Measure, Brass Tips .... 2 "		6 00

BOARD, LOG AND WOOD MEASURES.
[For Explanation, see opposite page.]

46.	Board Stick, Octagon, Brass Caps, 8 to 23 feet....2 feet,	8 00
46½.	Board Stick, Square, Brass Caps, 8 to 23 feet....2 "	8 00
47.	Board Stick, Octagon, Brass Caps, 8 to 23 feet....3 "	12 00
47½.	Board Stick, Square, Brass Caps, 8 to 23 feet....3 "	12 00
43½.	Board Stick, Flat, Hickory, Cast Brass Head and Tip, 6 Lines, 12 to 22 feet ....3 "	12 00
49.	Board Stick, Flat, Hickory, Steel Head, Brazed, Extra Strong, 6 Lines, 12 to 22 feet..........3 "	26 00
48.	Walking Cane, Board Measure, Octagon, Hickory, Cast Brass Head and Tip, 8 Lines, 9 to 16 feet..3 "	12 00
48½.	Walking Cane, Log Measure (Doyle's Revised), Octagon, Hickory, Cast Brass Head and Tip...3 "	15 00
71.	Wood Measure, Brass Caps, 8ths of inches and 10ths of a foot............................4 "	8 00
49½.	Forwarding Stick, Cast Brass (T) Head ....5 "	24 00

YARD STICKS.

33.	Yard Stick, Polished	2 00
41.	Yard Stick, Brass Tips, Polished	3 50
50.	Yard Stick, Hickory, Brass Cap'd Ends, Polished.	4 50

WANTAGE AND GAUGING RODS.
[For Explanation, see opposite page.]

44.	Wantage Rod, 8 Lines	5 00
37.	Wantage Rod, 12 Lines	7 00
45.	Gauging Rod, 120 gallons....................3 feet,	7 00
45½.	Gauging Rod, 180 gallons, and Wantage Tables..4 "	18 00

SADDLERS' RULE.

80.	Saddlers' Rule, Maple, Capped Ends, 8ths and 16ths of inches ....1½ in. wide, 36 inches,	9 00

PATTERN MAKERS' SHRINKAGE RULES.

30.	Pattern Makers' Shrinkage Rule, Boxwood, 8ths and 16ths of inches.................24¼ inches,	15 00
31.	Pattern Makers' Shrinkage Rule, Two Fold, Boxwood, Triple Plated Edge Plates, 8ths and 16ths of inches ....24¼ "	18 00

BOARD AND LOG MEASURES.

NOTE.—Nos. 43½, 46, 46½, 47, 47½, 48 and 49 give the contents in Board Measure of 1 inch Boards.

DIRECTIONS.—Place the stick across the flat surface of the Board, bringing the inside of the Cap snugly to the edge of the same; then follow with the eye the Column of figures in which the length of the Board is given as the first figure under the Cap, and at the mark nearest the opposite edge of the Board will be found the contents of the Board in feet.

No. 41.

No. 43½

No. 47½

No. 47.

No. 49.

No. 48.

NOTE.—The Log Measure (No. 48½) has Doyle's Revised Tables, and gives the number of feet of one inch square edged boards which can be sawed from a log of any size, from 12 to 36 inches in diameter, and of any length. The figures immediately under the head of the Cane are for the lengths of Logs in feet. Under these figures, on the same line at the mark nearest the diameter of the Log, will be found the number of feet the Log will make.

No. 45.

No. 45 Gauging Rod is graduated in Wine Measure, up to 120 gallons, on one side, and in inches and parts of inches on the other sides.

DIRECTIONS.—To ascertain the capacity of a Barrel insert the Rod in the bung-hole, in a slanting direction to the chime, note point on the rod which comes exactly in the middle of the bung-hole, on a line with the under side of the stave ; then reverse the process, running the point of the rod to the chime at the other end of the barrel ; and if the bung-hole is exactly in the middle of the barrel, the result will be the same as before, and the capacity of the barrel will be shown. If the measurements differ, add them together and divide by two, and you have the number of gallons the barrel will hold.

No. 44.

No. 44 Wantage Rod has 8 Tables, or Lines, for Barrels of 16, 23, 32, 42, 48, 84, 110 and 120 Gallons.

DIRECTIONS.—Having found the capacity of the Barrel by use of the Gauge Rod, insert the Wantage Rod perpendicularly in the bung-hole, holding it so that the brass lip points toward the head of the barrel ; lower it slowly until the lip comes just under the inner side of the stave, then withdraw it, being careful not to let the rod go any further into the barrel ; and the mark where the rod is wet, on the line which has the full capacity of the barrel at the top, shows the number of gallons that are wanting to fill it.

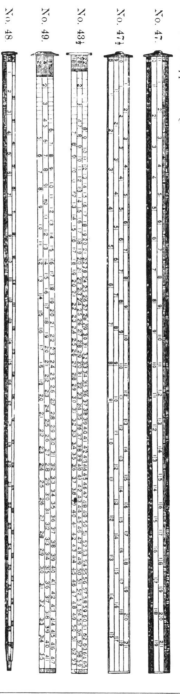

STEARNS' IVORY RULES.

Formerly Manufactured in Brattleboro', Vt., by C. L. Mead, successor to E. A. Stearns & Co.

IVORY, ONE FOOT, FOUR FOLD.

		Per Dozen,	
No.		Unbound.	Bound.
51.	Arch Joint, Edge Plates, German Silver, 8ths, 10ths, 12ths and 16ths of inches (100ths of a foot on edges of unbound), 13-16ths in. wide,	$33 00	$44 00
52.	Square Joint, Edge Plates, German Silver, 8ths, 10ths, 12ths and 16ths of inches (100ths of a foot on edges of unbound), 13-16ths in. wide,	30 00	40 00

IVORY, ONE FOOT, FOUR FOLD, CALIPER.

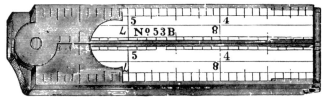

53.	Arch Joint, Edge Plates, German Silver, 8ths, 10ths, 12ths and 16ths of inches (100ths of a foot on edges of unbound), 13-16ths in. wide,	38 00	48 00
54.	Square Joint, Edge Plates, German Silver, 8ths, 10ths, 12ths and 16ths of inches (100ths of a foot on edges of unbound), 13-16ths in. wide,	35 00	45 00

IVORY, SIX INCH, TWO FOLD, CALIPER.

55.	Square Joint, German Silver, 8ths and 16ths of inches ....13-16ths in. wide,	17 00
55½.	Square Joint, German Silver Case, Spring Caliper, 8ths and 16ths of inches, 13-16th in. wide,	21 00

IVORY, ONE FOOT, FOUR FOLD.

57.	Square Joint, Edge Plates, German Silver, 8ths and 16ths of inches....5-8ths in. wide,	20 00	30 00

IVORY, TWO FEET, FOUR FOLD, BROAD.

		Per Dozen.	
No.		Unbound.	Bound.
47.	Arch Joint, Triple Plated Edge Plates, German Silver, 8ths, 10ths, 12ths and 16ths of inches, 100ths of a foot, Drafting and Octagonal Scales .... 1½ in. wide,	$100 00	$120 00

IVORY, TWO FEET, FOUR FOLD, MEDIUM.

48.	Arch Joint, Edge Plates, German Silver, 8ths, 10ths, 12ths and 16ths of inches (100ths of a foot on edges of unbound), Drafting Scales .... 1⅛ in. wide,	80 00	95 00

IVORY, TWO FEET, FOUR FOLD, NARROW.

50.	Square Joint, Edge Plates, German Silver, 8ths, 10ths, 12ths and 16ths of inches (100ths of a foot on edges of unbound), Drafting Scales .... 1 in. wide,	72 00	84 00

IVORY, TWO FEET, FOUR FOLD, EXTRA NARROW.

56.	Arch Joint, Edge Plates, German Silver, 8ths, 10ths, 12ths and 16ths of inches (100ths of a foot on edges of unbound) .... ¾ in. wide,	65 00	80 00

IVORY, TWO FEET, SIX FOLD, EXTRA NARROW.

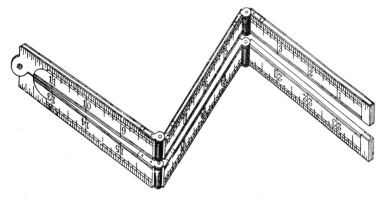

60.	Arch Joint, Edge Plates, German Silver, 8ths, 10ths, 12ths and 16ths of inches (100 of a foot on edges of unbound) .... ¾ in. wide,	80 00	95 00

16 STANLEY'S PATENT HAND-Y PLUMB AND LEVEL, ETC.

STANLEY'S PATENT
HAND-Y PLUMB AND LEVEL.

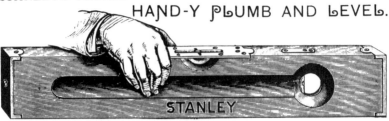

This Level can be used in horizontal position, or can be brought to perpendicular for ascertaining a plumb, with remarkable ease, as the shallow grooves along the two sides afford an excellent grip on the Tool. This Level will be especially useful in House-framing, Bridge-building and general out-of-door work.

No. Per Doz.
16. Patent Adjustable Hand-y Plumb and Level, Cherry, Arch Top Plate, Two Side Views, Polish'd, Tipped, Ass'td, 24 to 30 inches, **$15 00**

STANLEY'S PATENT
DUPLEX PLUMBS AND LEVELS.

These Levels have the ordinary form of leveling glass, set in the top surface of the Stock. For any uses where an observation of the glass, *sidewise*, may be found convenient, an additional leveling glass is set in the side, at the opposite end from the Plumb. Both glasses are protected by Brass Discs, can be seen from either side, and are inserted in the Level with the least possible removal of wood from the Stock.

No. Per Doz.
25. Patent Adjustable Plumb and Level, Mahogany, Arch Top Plate, Improved Duplex Side Views, Polished, Tip'd, Ass'td, 24 to 30 in., **$24 00**
30. Patent Adjustable Plumb and Level, Cherry, Arch Top Plate, Improved Duplex Side Views, Polished, Tip'd, Ass'td, 24 to 30 in., **18 00**
50. Patent Adjustable Plumb and Level, Cherry, Triple Stock, Arch Top Plate, Improved Duplex Side Views, Polished and Tipped, Assorted, 24 to 30 inches.................................... **24 00**

STANLEY'S PATENT ADJUSTABLE PLUMB AND LEVEL.

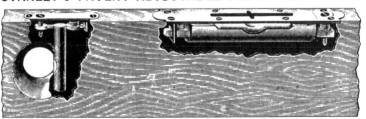

The Spirit-glass (or bubble tube), in both the Level and Plumb, is set in a metallic case attached rigidly to the Brass Top-plate above; and should it be necessary from any cause to adjust either glass, it can be done by means of the screw designated "Adjusting Screw."
Plumbs and Levels Nos. 01, 02, 03, etc., in this Price List, correspond exactly with the Nos. 1, 2, 3, etc., except that the former are non-adjustable and the latter adjustable.
The improved slots now made in all our Top-plates afford a ready means for detecting the movement of the bubble, to or from the center of the Glass.

PLUMBS AND LEVELS. 17

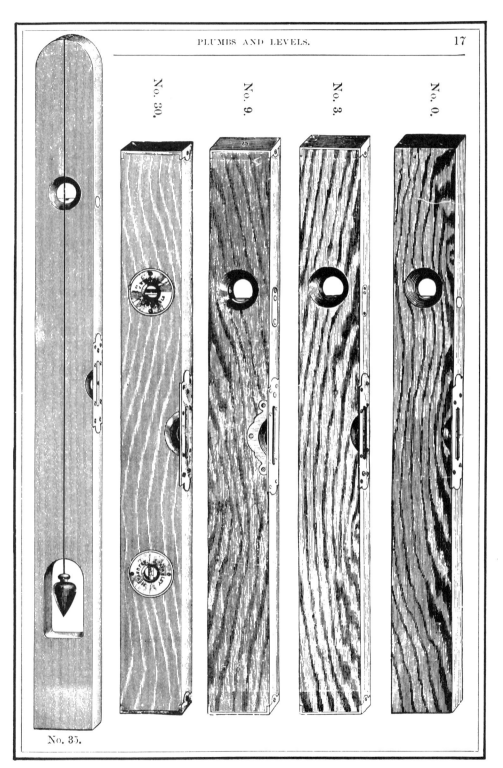

PLUMBS AND LEVELS.

No.		Per Doz.
102.	Levels, Arch Top Plate, Two Side Views, Polished, Assorted..................10 to 16 in.,	**$9 00**
103.	Levels, Arch Top Plate, Two Side Views, Polished, Assorted..................18 to 24 "	**12 00**
104.	Plumb and Level, Arch Top Plate, Two Side Views, Polished, Assorted............12 to 18 "	**14 00**
1½.	Mahogany Plumb and Level, Arch Top Plate, Two Side Views, Polished, Assorted,18 to 24 "	**16 50**
1¾.	Mahogany Plumb and Level, Arch Top Plate, Two Brass Lipped Side Views, Polished and Tipped, Assorted.........12 to 18 "	**27 00**
00.	Plumb and Level, Arch Top Plate, Two Side Views, Polished, Assorted............18 to 24 "	**16 00**
0.	Plumb and Level, Arch Top Plate, Two Side Views, Polished, Assorted.........24 to 30 " [See Engraving, p. 17.]	**18 00**
01.	Mahogany Plumb and Level, Arch Top Plate, Two Side Views, Polished, Assorted,24 to 30 "	**22 50**
02.	Plumb and Level, Arch Top Plate, Two Brass Lip'd Side Views, Polished, Ass'td,24 to 30 "	**24 00**
03.	Plumb and Level, Arch Top Plate, Two Side Views, Polished and Tipped, Assorted..24 to 30 "	**28 00**
04.	Plumb and Level, Arch Top Plate, Two Brass Lipped Side Views, Polished and Tipped, Assorted.....................24 to 30 "	**35 00**
06.	Mahogany Plumb and Level, Arch Top Plate, Two Brass Lipped Side Views, Polished, Assorted..................24 to 30 "	**30 00**
7.	Masons' Plumb and Level, Arch Top Plate, Two Plumbs, Two Side Views, Polished and Tipped........................36 "	**36 00**
7½.	Masons' Plumb and Level, Arch Top Plate, Two Plumbs, Two Side Views, Polished.36 "	**30 00**
8.	Masons' Plumb and Level, Arch Top Plate, Two Plumbs, Two Side Views, Polished.42 "	**36 00**

PATENT IMPROVED
ADJUSTABLE PLUMBS AND LEVELS.

No.		Per Doz.
1.	Patent Adjustable Mahogany Plumb and Level, Arch Top Plate, Two Side Views, Polished, Assorted .... 24 to 30 in.,	$27 00
2.	Patent Adjustable Plumb and Level, Arch Top Plate, Two Brass Lipped Side Views, Polished, Assorted .... 24 to 30 "	27 00
3.	Patent Adjustable Plumb and Level, Arch Top Plate, Two Side Views, Polished and Tipped, Assorted .... 24 to 30 "	30 00

[See Engraving, p. 17.]

NOTE.—No. 3 Plumb and Level, we also keep in stock, assorted 18 to 24 inches, one-half dozen in a box. The regular assortment 24 to 30 inches will always be sent, unless otherwise ordered.

4.	Patent Adjustable Plumb and Level, Arch Top Plate, Two Brass Lipped Side Views, Polished and Tipped, Assorted .... 24 to 30 "	39 00
5.	Patent Adjustable Plumb and Level, Triple Stock, Arch Top Plate, Two Ornamental Brass Lipped Side Views, Polished and Tipped, Assorted .... 24 to 30 "	48 00
6.	Patent Adjustable Mahogany Plumb and Level, Arch Top Plate, Two Brass Lipped Side Views, Polished, Assorted .... 24 to 30 "	33 00
9.	Patent Adjustable Mahogany Plumb and Level, Arch Top Plate, Two Ornamental Brass Lipped Side Views, Polished and Tipped, Assorted .... 24 to 30 "	48 00

[See Engraving, p. 17.]

10.	Patent Adjustable Mahogany Plumb and Level, Triple Stock, Two Ornamental Brass Lipped Side Views, Arch Top Plate, Polished and Tipped, Assorted .... 24 to 30 "	60 00
11.	Patent Adjustable Rosewood Plumb and Level, Arch Top Plate, Two Ornamental Brass Lipped Side Views, Polished and Tipped, Assorted .... 24 to 30 "	90 00
32.	Patent Adjustable Mahogany Graduated Plumb and Level (easily adjusted to work at any angle or elevation required) 28	" 100 00
35.	Patent Adjustable Masons' Plumb and Level, adapted also for Plumb Bob and Line 42	" 40 00

[See Engraving, p. 17.]

NICHOLSON'S PATENT
METALLIC PLUMBS AND LEVELS.

No.		Per Doz.
13.	14 inches long	$18 00
14.	20 inches long	21 00
15.	24 inches long	27 00

STANLEY'S PATENT
BIT AND SQUARE LEVEL.

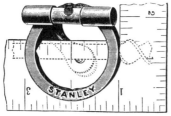 

The frame of this Level has three pairs of V slots on its back edges. The shank to a Bit will lie in these slots, either parallel with the bubble-glass, at an exact angle to it, or at an angle of 45 degrees.

A Thumb-Screw secures the Level to the Bit, in either position; and boring can be done with perfect accuracy as to perpendicular, horizontal or angle of 45 degrees, by observing the bubble-glass while turning the bit.

The frame can also be attached to a Carpenter's Square. Two shoulders rest on the top of the horizontal leg to the square, thus making it an accurate Spirit-level. The upright leg of the square will indicate an exact Plumb-line.

No. 44. Bit and Square Level, Brass Frame, in a box, each..**$0 30**

STANLEY'S IMPROVED
LEVEL SIGHTS.

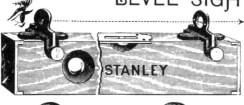

By the use of these ingenious devices, which can be attached to any Level, the owner has a convenient and accurate means for leveling, from one given point to another at a long distance away. When not in use the Level Sights are easily detached, and can be packed away in a small space for future use.

No. 1. Improved Level Sights, per pair, in a box............**$0 75**

POCKET LEVELS.

No. 41.

No.		Per Doz.
40.	Iron, Iron Top Plate, Japanned, 1 dozen in a box	$2 50
41.	Iron, Brass Top Plate, 1 dozen in a box	3 00
42.	Brass, Brass Top Plate, 1 dozen in a box	8 00

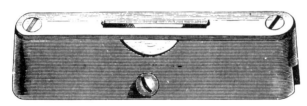

46.	Iron, Brass Top Plate, a superior article	3 50

MACHINISTS' IRON LEVELS.

38.	Iron Level, Brass Top Plate	4 inch,	3 50
39.	Iron Level, Brass Top Plate	6 "	4 50

LEVEL GLASSES.

			Per Gross.
Level Glasses, packed 1 dozen in a box	1¾ inch,	$9 50
Level Glasses, packed 1 dozen in a box	2 "	10 00
Level Glasses, packed 1 dozen in a box	2½ "	10 50
Level Glasses, packed 1 dozen in a box	3 "	11 50
Level Glasses, packed 1 dozen in a box	3½ "	13 00
Level Glasses, packed 1 dozen in a box	4 "	14 50
Level Glasses, packed 1 dozen in a box	4½ "	16 00
Assorted, 1¾, 3 and 3½ inch, 1 dozen in a box		12 00

PATENT
IMPROVED TRY SQUARES.

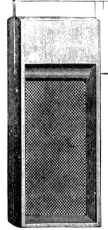

No. 12.

Iron Handle, Graduated Steel Blade,
Square Inside and out.

Inches......	2	4	6	8	10	12
Per Doz.....	$2 25	2 75	3 50	4 50	5 50	7 00

One-half dozen in a box.

PATENT
ADJUSTABLE TRY SQUARES.

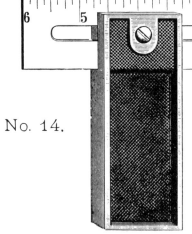

No. 14.

The Blade of this Square can be firmly and accurately secured in its seat, at any point. When the Blade is carried fully to the front of the Handle, it is like an ordinary Try Square in all its uses, and the moving of the Blade admits of making the Try Square equally perfect down to ¼ inch Blade, or less. This Tool will be found invaluable in putting on Butts or Locks, and doing short work about windows, doors, etc.

Iron Handle, Graduated Steel Blade.

Inches........	4	6
Per Doz........	$3 75	4 50

One-half dozen in a box.

COMBINED TRY AND MITRE SQUARE.

WINTERBOTTOM'S PATENT COMBINED
TRY AND MITRE SQUARE.

This Tool can be used with equal convenience and accuracy as a **Try Square** or a **Mitre Square**. By simply changing the position of the handle and bringing the mitred face at the top of the handle against one edge of the work in hand, a perfect mitre, or angle of 45 degrees, can be struck from either edge of the blade.

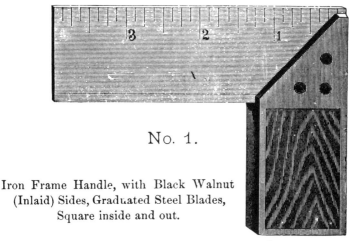

No. 1.

Iron Frame Handle, with Black Walnut (Inlaid) Sides, Graduated Steel Blades, Square inside and out.

Inches.......	4	6	8
Per Doz......	**$6 00**	**7 50**	**9 00**

One-half dozen in a box.

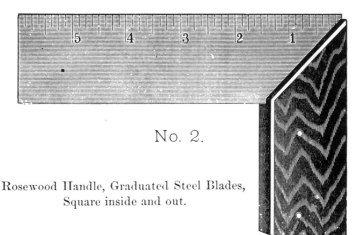

No. 2.

Rosewood Handle, Graduated Steel Blades, Square inside and out.

Inches.......	4½	6	7½	9	12
Per Doz......	**$4 50**	**5 00**	**6 00**	**7 00**	**9 00**

One-half dozen in a box.

IMPROVED MITRE TRY SQUARE.

No. 15.

Cast Brass Handle, 7½ in. blade, **$12 00** Per Doz.
One-half dozen in a box.

No. 16.

IMPROVED MITRE SQUARES.

Iron Frame Handle, with Black Walnut (Inlaid) Sides.

Inches	8	10	12
Per Doz.	$7 00	8 00	9 00

One-half dozen in a box.

INLAID TRY SQUARES.

No. 10.

Iron Frame Handle, with Rosewood (Inlaid) Sides, Steel Blades, Square inside and out.

Inches....	3	4	6	8	9	10	12
Per Doz...	$5 50	6 50	8 50	11 00	12 00	15 00	18 00

One-half dozen in a box.

PATENT COMBINATION
TRY SQUARE AND BEVEL.

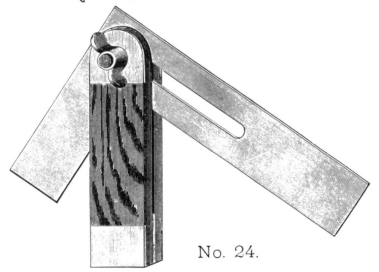

No. 24.

Inches.....	4	6	8	10	12
Per Doz....	$5 00	6 00	7 50	9 00	10 50

One-half dozen in a box.

TRY SQUARES.

GRADUATED STEEL BLADES.

Handles full size, and Brass Face-plates put on with Screws.

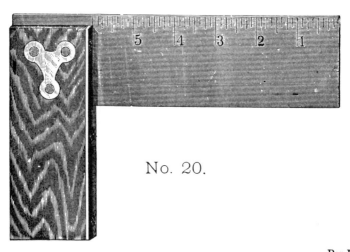

No. 20.

                                                                                   Per Doz.

Rosewood, 1 dozen in a box .... 3 inch, **$3 00**
Rosewood, ½ dozen in a box .... 4½ inch, **3 75**
Rosewood, ½ dozen in a box .... 6 inch, **5 00**
Rosewood, ½ dozen in a box .... 7½ inch, **5 75**
Rosewood, ½ dozen in a box .... 9 inch, **6 50**
Rosewood, ½ dozen in a box ....10 inch, **7 25**
Rosewood, ½ dozen in a box ....12 inch, **8 50**
Rosewood, with Rest, ½ dozen in a box ....15 inch, **12 50**
Rosewood, with Rest, ½ dozen in a box ....18 inch, **15 50**

PLUMB AND LEVEL TRY SQUARES.

A very convenient tool for Mechanics. A spirit-glass being set in the inner edge of the Try Square Handle, constitutes the Handle a Level; and when the handle is brought to an exact level, the blade of the square will be upright, and become a perfect Plumb.

No. 22.

7½ inch, Rosewood, ½ dozen in a box **$7 50**

SLIDING T BEVELS.

ROSEWOOD HANDLE, WITH BRASS LEVER (FLUSH).

6 inch....per doz., **$5 50**
8 inch....per doz., **6 00**
10 inch....per doz., **6 50**
12 inch....per doz., **7 00**
14 inch....per doz., **7 50**

One-half dozen in a box.

No. 25.

The Bevel Blade can be made fast or loose by moving the Lever with the thumb of the hand which grasps the Handle, thus leaving the other hand of the workman free.

☞ Bevels with Thumb Screws will be furnished if so ordered.

No. 18.

PATENT EUREKA FLUSH T BEVELS.

IRON HANDLE, STEEL BLADE, WITH PARALLEL EDGES.

The Blade is easily secured at any angle by turning the Thumb Screw at the lower end of the Handle.

Inches,	6	8	10
Per doz.,	$6 00	6 50	7 50

One-half dozen in a box.

GAUGES.

☞ All Marking Gauges, excepting Nos. 0, 61 and 61½, have an Adjusting Point of finely tempered steel, which may be readily removed and replaced if it needs sharpening. The Point can be thrust down as it wears away; or if by any means it be broken off, it can be easily repaired.

All Gauges with Brass Thumb Screws have also a brass shoe inserted in the head, under the end of the Thumb Screw. This shoe protects the gauge-bar from being dented by the action of the screw; and the broad surface of the shoe being in contact with the bar, the head is held more firmly in position than by any other method, and with less wear of the screw threads.

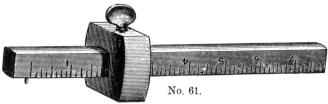

No. 61.

No.		Per Doz.
0.	Marking Gauge, Beechwood, Boxwood Thumb Screw, Marked, Steel Point, 1 dozen in a box	$0 75
61.	Marking Gauge, Beechwood, Boxwood Thumb Screw, Oval Bar, Marked, Steel Point, 1 dozen in a box ...	1 00
61½.	Marking Gauge, Beechwood, Boxwood Thumb Screw, Oval Head and Bar, Marked, Steel Point, 1 dozen in a box	1 25

No. 65.

62.	Patent Marking Gauge, Beechwood, Polished, Boxwood Thumb Screw, Oval Bar, Marked, Adjusting Steel Point, 1 dozen in a box	2 00
64.	Patent Marking Gauge, Polished, Plated Head, Boxwood Thumb Screw, Oval Bar, Marked, Adjusting Steel Point, 1 dozen in a box....................	2 75
64½.	Patent Marking Gauge, Polished, Oval Plated Head, Brass Thumb Screw and Shoe, Oval Bar, Marked, Adjusting Steel Point, ½ dozen in a box	4 50
65.	Patent Marking Gauge, Boxwood, Polished, Plated Head, Brass Thumb Screw and Shoe, Oval Bar, Marked, Adjusting Steel Point, ½ dozen in a box ...	5 00
66.	Patent Marking Gauge, Rosewood, Oval Plated Head and Bar, Brass Thumb Screw and Shoe, Oval Bar, Marked, Adjusting Steel Point, ½ dozen in a box ...	6 00

GAUGES.

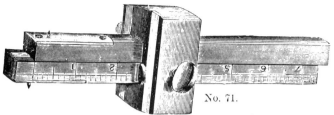

No. 71.

No.		Per Doz.
71.	Patent Double Gauge (Marking and Mortise Gauge combined), Beechwood, Polished, Plated Head and Bars, Brass Thumb Screws and Shoes, Oval Bars, Marked, Steel Points, ½ dozen in a box	$8 00
72.	Patent Double Gauge (Marking and Mortise Gauge combined), Beechwood, Polished, Boxwood Thumb Screws, Oval Bars, Marked, Steel Points, ½ doz. in box,	4 00
74.	Patent Double Gauge (Marking and Mortise Gauge combined), Boxwood, Polished, Full Plated Head and Bars, Brass Thumb Screws and Shoes, Oval Bars, Marked, Steel Points, ½ dozen in a box	14 00
70.	Cutting Gauge, Mahogany, Polished, Plated Head, Boxwood Thumb Screw, Oval Bar, Marked, Steel Cutter, 1 dozen in a box	4 00

No. 73.

73.	Patent Mortise Gauge, Boxwood, Polished, Plated Head, Brass Slide, Brass Thumb Screw and Shoe, Oval Bar, Marked, Steel Points, ½ dozen in a box	8 00
76.	Patent Mortise Gauge, Boxwood, Polished, Plated Head, Screw Slide, Brass Thumb Screw and Shoe, Oval Bar, Marked, Steel Points, ½ dozen in a box	11 00
80.	Patent Mortise Gauge, Boxwood, Full Plated Head, Plated Bar, Screw Slide, Brass Thumb Screw and Shoe, Marked, Steel Points, ½ dozen in a box	18 00
67.	Mortise Gauge, Adjustable Wood Slide, Boxwood Thumb Screw, Oval Bar, Marked, Steel Points, ½ dozen in a box	4 00
68.	Mortise Gauge, Plated Head, Adjustable Wood Slide, Brass Thumb Screw and Shoe, Oval Bar, Marked, Steel Points, ½ dozen in a box	6 00

GAUGES.

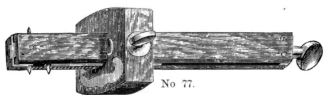

No 77.

No.		Per Doz.
77.	Patent Mortise and Marking Gauge, Rosewood, Plated Head, Improved Screw Slide, Brass Thumb Screw and Shoe, Oval Bar, Marked, Steel Points, ½ dozen in a box,	$10 00
78.	Patent Mortise Gauge, Rosewood, Plated Head, Screw Slide, Brass Thumb Screw and Shoe, Oval Bar, Marked, Steel Points, ½ dozen in a box	11 00
79.	Patent Mortise Gauge, Rosewood, Plated Head and Bar, Screw Slide, Brass Thumb Screw and Shoe, Marked, Steel Points, ½ dozen in a box	13 00
83.	Handled Slitting Gauge, 17-inch Bar, Marked, ½ dozen in a box	9 00
84.	Handled Slitting Gauge, with Roller, 17-inch Bar, Marked, ½ dozen in a box	10 00
85.	Panel Gauge, Beechwood, Boxwood Thumb Screw, Oval Bar, Steel Points, 1 dozen in a box	3 20
85½.	Panel Gauge, Rosewood, Plated Head and Bar, Brass Thumb Screw and Shoe, Steel Point, ½ dozen in a box,	18 00

IRON GAUGES.

The engraving below (No. 60), shows an Iron Marking and Cutting Gauge; No. 60½ is the same in form, but has a Reversible Brass Slide slotted into the face of the bar. When a Mortise Gauge is required, the Brass Slide may be turned over in the bar. The point in the Brass Slide may be moved to any position, and the slide will be secured by a single turn of the screw which fastens the head of the Gauge.

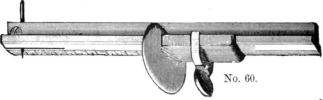

No. 60.

No.		Per Doz.
60.	Patent Iron Marking and Cutting Gauge, Oval bar, Marked, Adjusting Steel Point, equally well adapted for use on Metals or Wood, ½ dozen in a box	$5 00
60½.	Patent Iron Reversible Gauge, Mortise, Marking and Cutting combined, Brass Slide, Oval Bar, Marked, Adjusting Steel Points, equally well adapted for use on Metals or Wood, ½ dozen in a box	8 00

STANLEY'S PATENT
IMPROVED GAUGES.

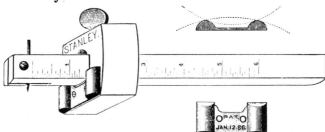

The Brass Face, with two ribs or protections, attached to one side of the Gauge-head (see Engraving), will enable the owner to run a gauge-line with perfect steadiness and accuracy around curves of any degree, and either concave or convex. The Gauge-head is reversible, and the flat side can be used for all ordinary work.

Any style of Gauge found in this Catalogue, pages 28, 29 and 30—with this valuable improvement included—may be ordered as Nos. 161, 162, etc., instead of Nos. 61, 62, etc.

The following numbers are carried in stock, and have the most liberal sale:

No.	MARKING GAUGES.	Per Doz.
161.	Beechwood, Boxwood Thumb Screw, Oval Bar, Steel Point	$2 00
162.	Beechwood, Boxwood Thumb Screw, Adjusting Steel Point	3 00
164.	Beechwood, Boxwood Thumb Screw, Plated Head, Adjusting Steel Point	3 75
165.	Boxwood, Brass Thumb Screw and Shoe, Plated Head, Adjusting Steel Point	6 00
166.	Rosewood, Brass Thumb Screw and Shoe, Oval Plated Head, Adjusting Steel Point	7 00

MORTISE GAUGES.

No.		Per Doz.
171.	Double Gauge (Marking and Mortise), Beechwood, Plated Heads and Bars, Brass Thumb Screws and Shoes	9 00
172.	Double Gauge (Marking and Mortise), Beechwood, Boxwood Thumb Screws	5 00
173.	Boxwood, Brass Slide, Plated Head, Brass Thumb Screw and Shoe	9 00
177.	Rosewood, Screw Slide, Plated Head, Brass Thumb Screw and Shoe	11 00

☞ The price of any Improved Gauge, not mentioned in above List, may be ascertained by adding $1.00 per dozen to the price given on the corresponding Gauge—without the improvement—in our regular List of Gauges.

BAILEY'S PATENT ADJUSTABLE PLANES.

Manufactured only by the Stanley Rule & Level Co., under the original Patents.

Over 1,500,000 Already Sold.

These Planes meet with universal approbation from the best Mechanics, as their extensive sale abundantly testifies. For beauty of style and finish they are unequaled, and the superior methods for adjusting them readily in all their parts render them economical to the owner.

☞ Each Plane is thoroughly tested at our Factory and put in perfect working order before being sent into the market. A printed description of the method of adjustment accompanies each plane. The Plane Irons are by us fully warranted.

IRON PLANES.

No.			Each.
1.	Smooth Plane,	$5\frac{1}{2}$ inches in Length, $1\frac{1}{4}$ inch Cutter	$2 25
2.	Smooth Plane,	7 inches in Length, $1\frac{5}{8}$ inch Cutter	2 75
3.	Smooth Plane,	8 inches in Length, $1\frac{3}{4}$ inch Cutter	3 00
4.	Smooth Plane,	9 inches in Length, 2 inch Cutter	3 25
$4\frac{1}{2}$.	Smooth Plane,	10 inches in Length, $2\frac{3}{8}$ inch Cutter	3 75

5.	Jack Plane,	14 inches in Length, 2 inch Cutter	3 75
6.	Fore Plane,	18 inches in Length, $2\frac{3}{8}$ inch Cutter	4 75
7.	Jointer Plane,	22 inches in Length, $2\frac{3}{8}$ inch Cutter	5 50
8.	Jointer Plane,	24 inches in Length, $2\frac{5}{8}$ inch Cutter	6 50
9.	Cabinet Makers' Block Plane, 10 in. Length, 2 in. Cutter,		6 00

$10\frac{1}{2}$.	Carriage Makers' Rabbet Plane, 9 in. Length, $2\frac{1}{8}$ in. Cutter,	3 75
10.	Carriage Makers' Rabbet Plane, 13 in. Length, $2\frac{1}{8}$ in. Cutter,	4 50
11.	Belt Makers' Plane, $2\frac{3}{8}$ inch Cutter	3 00

BAILEY'S PATENT ADJUSTABLE PLANES.

TORN CATALOGUES.—Every article in this Book is described either by figure, number, or name. We beg our customers not to cut or tear out parts of pages as this destroys it for reference.

STANLEY'S PATENT
LATERAL ADJUSTMENT.

The above sectional view of a Plane illustrates the new method of adjusting a Plane Iron, *sidewise*, to set the cutting edge exactly square with the face of the plane.

At the lower end of the Lever, a revolving (anti-friction) Disc fits into the slot in the Plane Iron; thus furnishing an easy sidewise adjustment, entirely independent of the forward and backward adjustment of the Cutter.

This Lever for sidewise adjustment, as shown above, is attached to all BAILEY IRON PLANES, AND BAILEY WOOD PLANES, on pages 32, 34 and 35 of this Catalogue (except Nos. 1, 9 and 11). It will also be found attached to Block Planes, Nos. 16, 17, 18 and 19, on page 36.

BAILEY'S ADJUSTABLE CABINET MAKERS' PLANE.

No. **9.** Cabinet Makers' Plane, 10 in. Length, 2 in. Cutter ... **$6 00**

This Plane is used by Piano Forte Makers, Cabinet Makers and kindred trades, where an extra-fine Tool is required in finishing hard woods, etc.

A Metallic Handle, with Slot and Set-screw, is furnished with each Plane. This Handle can be attached to the top of the Plane, on either edge. The Plane, turned on its side, will then work perfectly on a shooting-board, for planing mitres, etc.

BAILEY'S PATENT ADJUSTABLE PLANES.

TORN CATALOGUES.—Every article in this Book is described either by figure, number, or name. We beg our customers not to cut or tear out parts of pages as this destroys it for reference.

BAILEY'S ADJUSTABLE BELT PLANE.

No. **11.** Belt Makers' Plane, 2⅜ inch Cutter................$3 00

This Tool is used by Belt Makers, for chamfering down the laps of a Belt, before fastening them together. It is equally well adapted to use in repairing Belts in all manufacturing establishments.

BAILEY'S ADJUSTABLE VENEER SCRAPER.

No. **12.** Adjustable Veneer Scraper, 3 inch Cutter..........$3 50
CAST STEEL, HAND VENEER SCRAPERS, 3 x 5 inch, per dozen .. 3 00

BAILEY'S ADJUSTABLE CIRCULAR PLANE.

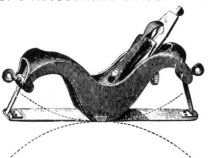

No. **13.** Circular Plane, 1¾ inch Cutter $4 00

This Plane has a Flexible Steel Face, and by means of the thumb screws at each end of the Stock, can be easily adapted to plane circular work—either concave or convex.

BAILEY'S ADJUSTABLE WOOD PLANES.

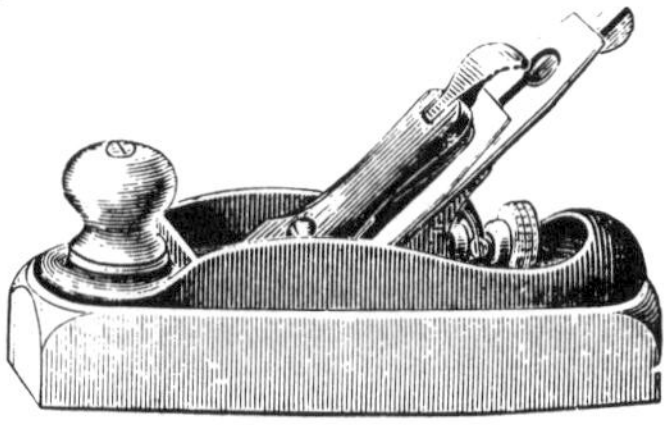

No.				Each.
21.	Smooth Plane,	7 inches in Length,	1¾ inch Cutter......	$2 00
22.	Smooth Plane,	8 inches in Length,	1¾ inch Cutter......	2 00
23.	Smooth Plane,	9 inches in Length,	1¾ inch Cutter......	2 00
24.	Smooth Plane,	8 inches in Length,	2 inch Cutter......	2 00
25.	**Block Plane,**	9½ inches in Length,	1¾ inch Cutter......	2 00

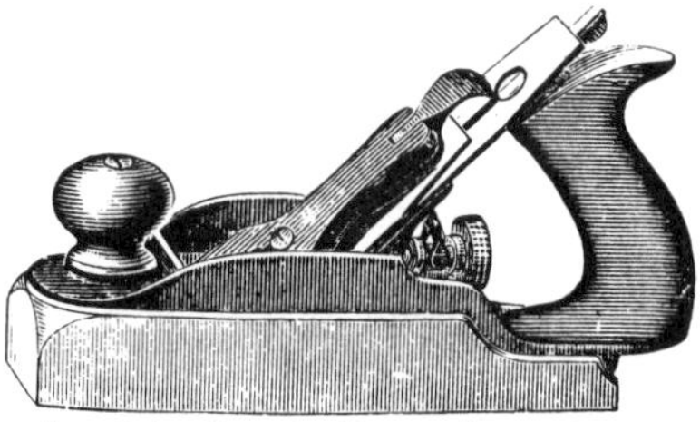

35.	Handle Smooth,	9 inches in Length,	2 inch Cutter.....	2 50
36.	Handle Smooth,	10 inches in Length,	2⅜ inch Cutter.....	2 75
37.	Jenny Smooth,	13 inches in Length,	2⅝ inch Cutter.....	3 00

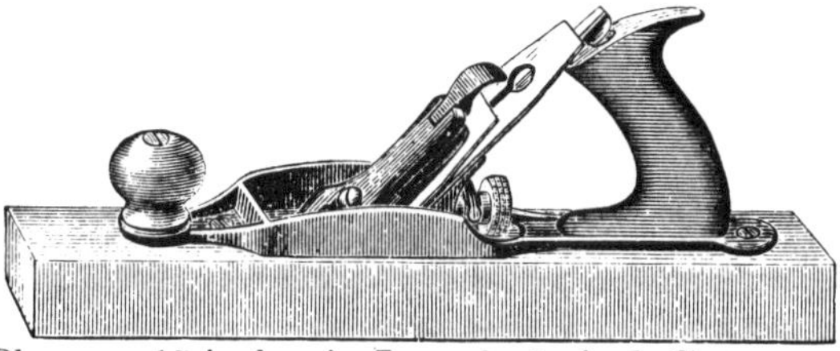

26.	Jack Plane,	15 inches in Length,	2 inch Cutter	2 25
27.	Jack Plane,	15 inches in Length,	2⅛ inch Cutter	2 50
28.	Fore Plane,	18 inches in Length,	2⅜ inch Cutter	2 75
29.	Fore Plane,	20 inches in Length,	2⅜ inch Cutter	2 75
30.	Jointer Plane,	22 inches in Length,	2⅜ inch Cutter	3 00
31.	Jointer Plane,	24 inches in Length,	2⅜ inch Cutter	3 00
32.	Jointer Plane,	26 inches in Length,	2⅝ inch Cutter	3 25
33.	Jointer Plane,	28 inches in Length,	2⅝ inch Cutter	3 25
34.	Jointer Plane,	30 inches in Length,	2⅝ inch Cutter	3 50

☞ Extra Plane-woods of every style can be supplied cheaply.

BAILEY'S PLANE IRONS.

Packed, one-half dozen in a box.

	1¼	1⅝	1¾	2	2⅛	2⅜	2⅝ Inch.
SINGLE IRONS,	20	25	28	30	33	37	40 cts. Each.
DOUBLE IRONS,	40	45	50	55	60	65	70 cts. Each.

☞ Orders for Plane Irons should designate the No. of the Planes for which they are wanted.

BAILEY'S ADJUSTABLE BLOCK PLANES.

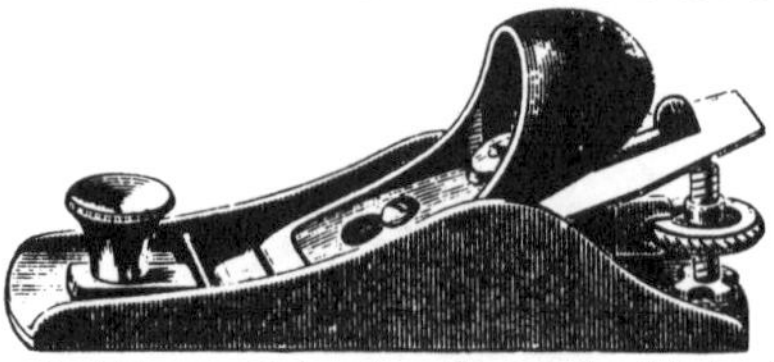

No.		Each.
9½.	Excelsior Block Plane, 6 inches Length, 1¾ inch Cutter..	$1 50
15.	Excelsior Block Plane, 7 inches Length, 1¾ inch Cutter..	1 60

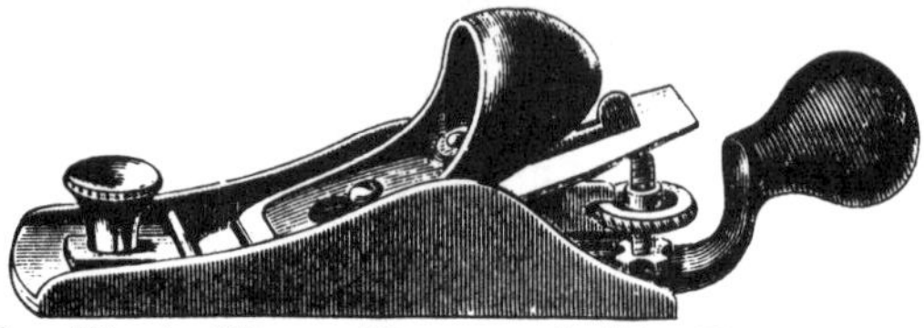

9¾.	Excelsior Block Plane, Rosewood Handle, 6 inches in Length, 1¾ inch Cutter	1 75
15½.	Excelsior Block Plane, Rosewood Handle, 7 inches in Length, 1¾ inch Cutter	1 85

The handles to Nos. 9¾ and 15½ are secured to the stock by the use of an iron nut, and can be attached to or liberated from the plane at convenience.

CUTTERS for above Block Planes, warranted Cast Steel, per doz. 2 40

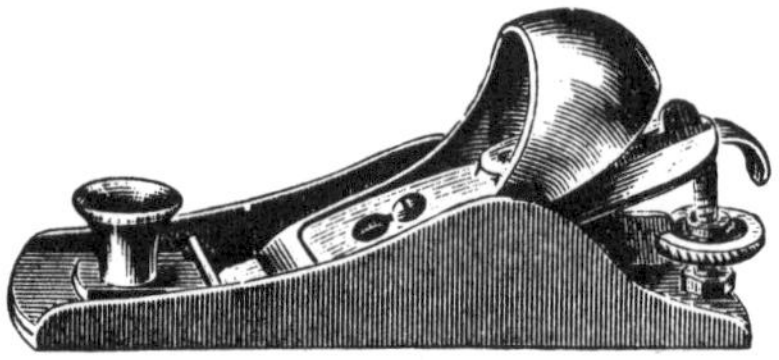

16.	Excelsior Block Plane, 6 inches Length, 1¾ in. Cutter, with Stanley's Lateral Adjustment, Nickel Plated Trimmings,	1 65
17.	Excelsior Block Plane, 7 inches Length, 1¾ in. Cutter, with Stanley's Lateral Adjustment, Nickel Plated Trimmings,	1 75

IMPROVED BLOCK PLANES.

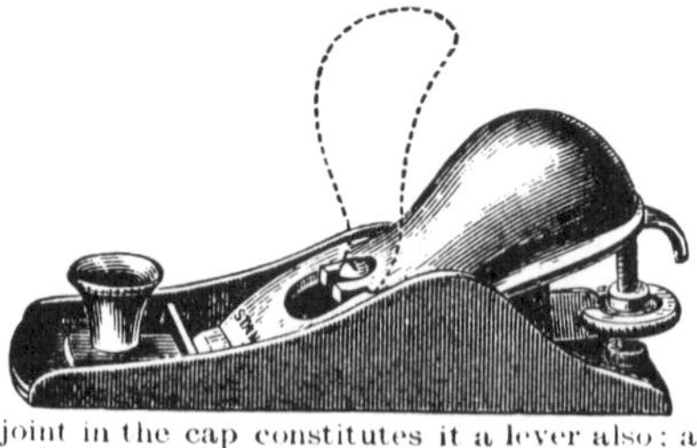

The knuckle-joint in the cap constitutes it a lever also; and the single movement of putting the cap in position clamps the cutter firmly in its seat.

18.	Knuckle-Joint Block Plane, 6 inches Length, 1¾ in. Cutter, Stanley's Lateral Adjustment, Nickel Plated Trimmings,	1 75
19.	Knuckle-Joint Block Plane, 7 inches Length, 1¾ in. Cutter, Stanley's Lateral Adjustment, Nickle Plated Trimmings,	1 85

CUTTERS for above Block Planes, warranted Cast Steel, per doz. 2 40

STANLEY PATENT ADJUSTABLE PLANES.

STANLEY ADJUSTABLE PLANES.
PATENTED.

STEEL PLANES.

Planes Nos. 104 and 105 have a Wrought Steel Stock, and are adjusted by the use of a Lever. These Planes are commended for their lightness and the ease with which they can be worked.

No.			Each.
104.	Smooth Plane,	9 inches in Length, 2⅛ inch Cutter	$2 75
105.	Jack Plane,	14 inches in Length, 2⅜ inch Cutter	3 50

STANLEY ADJUSTABLE SCRAPER PLANE.

This Tool, by its peculiar construction, can be used for scraping and finishing Veneers, Cabinet Work or Hard Woods in any form.

By use of an Extra Cutter, specially prepared, a superior Toothing Plane is made; and in addition to the ordinary uses of such a Plane, this one will do excellent work in scraping off old paint and glue.

No.		Each.
112.	Adjustable Scraper Plane, 9 in. in Length, 3 in. Cutter	$3 00
	CUTTERS, for Veneer Scraping	25
	CUTTERS, for Toothing, Nos. 22, 28, 32 (22, 28 or 32 teeth per in.)	35

STANLEY ADJUSTABLE CIRCULAR PLANE.

No.		Each.
113.	Adjustable Circular Plane, 1¾ inch Cutter	$4 00

This Plane has a Flexible Steel Face, which can be easily shaped to any required arc, either concave or convex, by turning the Knob on the front of the Plane.

STANLEY ADJUSTABLE WOOD PLANES.

These Planes are adjusted by a Lever, and are especially adapted for working on soft woods.

No.   Each.
122. Smooth Plane, 8 inches in Length, $1\frac{3}{4}$ inch Cutter **$1 50**

135. Handle Smooth, 10 inches in Length, $2\frac{1}{8}$ inch Cutter ... **2 00**

127. Jack Plane,   15 inches in Length, $2\frac{1}{8}$ inch Cutter..... **2 00**
129. Fore Plane,   20 inches in Length, $2\frac{3}{8}$ inch Cutter..... **2 25**
132. Jointer Plane, 26 inches in Length, $2\frac{3}{8}$ inch Cutter..... **2 50**

☞ Extra Plane-woods of every style can be supplied cheaply.

STANLEY PLANE IRONS.

Packed, one-half dozen in a box.

	$1\frac{3}{4}$	$2\frac{1}{8}$	$2\frac{3}{8}$	$2\frac{5}{8}$ Inch.
SINGLE IRONS	28	33	37	40 cts. each.
DOUBLE IRONS	50	60	65	70 cts. each.

☞ Orders for Plane Irons should designate the Number of the Planes for which they are wanted.

STANLEY IRON BLOCK PLANES.

No.		Each.
101.	Block Plane, 3½ inches in Length, 1 inch Cutter	$0 20
Cutters for above Block Plane, warranted Cast Steel, per doz.		75

102.	Block Plane, 5½ inches in Length, 1¼ inch Cutter	40
103.	Block Plane, Adjustable, 5½ in. in Length, 1¼ in. Cutter.	60
Cutters for above Block Planes, warranted Cast Steel, per doz.		1 50

110. Block Plane, 7½ inches in Length, 1¾ inch Cutter 60

120. Block Plane, Adjustable, 7½ in. in Length, 1¾ in. Cutter, 85

130. Block Plane, (Double-Ender), 8 in. in Length, 1¾ in. Cutter, 80
This Plane has two slots, and two cutter seats. It can be used as a Block Plane, or, by reversing the position of the cutter and the clamping wedge (see dotted lines in the engraving), it can be used to plane close up into corners, or places difficult to reach with any other plane.

Cutters for above Block Planes, warranted Cast Steel, per doz. 2 00

BULL-NOSE RABBET PLANE.

75. Iron Stock, 4 inches in Length, 1 inch Cutter 50

PATENT ADJUSTABLE
BEADING, RABBET AND SLITTING PLANE.

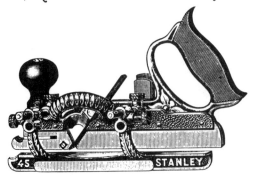

This Plane embraces in a compact and practical form: (1) Beading and Center Beading Plane; (2) Rabbet and Filletster; (3) Dado; (4) Plow; (5) Matching Plane; and (6) a superior Slitting Plane.

In each of its several forms this Plane will do perfect work, even in the hands of an ordinary mechanic—its simplicity of construction and adaptation of parts (as described in the Directions which accompany each Tool) being easily understood.

☞ Each Plane is accompanied by seven Beading Tools ($\frac{1}{8}$, $\frac{3}{16}$, $\frac{1}{4}$, $\frac{5}{16}$, $\frac{3}{8}$, $\frac{7}{16}$ and $\frac{1}{2}$ inch), nine Plow and Dado Bits ($\frac{1}{8}$, $\frac{3}{16}$, $\frac{1}{4}$, $\frac{5}{16}$, $\frac{3}{8}$, $\frac{7}{16}$, $\frac{1}{2}$, $\frac{5}{8}$ and $\frac{7}{8}$ inch), a Slitting Tool and a Tonguing Tool.

Price, including Beading Tools, Bits, Slitting Tool, etc.

No. **45.** Iron Stock and Fence.......................**$8 00**

[Samples of Tools embraced in the set of Eighteen Tools sold with Plane No. 45.]

A Southern Carpenter writes concerning this Tool:
"A first-class Mechanic's pet. Worth its weight in silver."

A Western Carpenter writes concerning this Tool:
"I have finished one house, on which it paid for itself."

REEDING TOOLS.

☞ Additional Tools for REEDING, consisting of two, three, four or five Beads each, will be furnished to order, at prices given below. They can only be used with Plane No. 45.

[In ordering, always state size and number of Beads wanted.]

Sizes.	2	3	4	5 Beads.
$\frac{1}{8}$ inch,	20	30	40	50 cents each.
$\frac{3}{16}$ inch,	20	30	40	50 cents each.
$\frac{1}{4}$ inch,	20	30	40	50 cents each.

PRICES OF
HOLLOWS AND ROUNDS.
[To be used only with Plane No. 45.]

These additional parts, for working Hollows and Rounds, can be attached to any Plane No. 45; but they are not included in the regular assortment of parts sent out with the Plane.

The full width of each Cutter is indicated in the following list; and, also the full diameter to the circle of which each Cutter will work only an arc. The Numbers correspond with the regular lists of common Hollows and Rounds. Order always by Numbers, and thus avoid confusion.

☞ The extra cost of these Hollows and Rounds, including Cast Steel Bits, is given below.

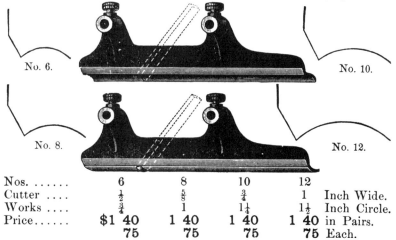

No. 6.   No. 10.

No. 8.   No. 12.

Nos.	6	8	10	12	
Cutter	$\frac{1}{2}$	$\frac{5}{8}$	$\frac{3}{4}$	1	Inch Wide.
Works	$\frac{3}{4}$	1	$1\frac{1}{4}$	$1\frac{1}{2}$	Inch Circle.
Price	$1 40	1 40	1 40	1 40	in Pairs.
	75	75	75	75	Each.

NOSING TOOL, FOR PLANE No. 45.
[Attach same as Hollows and Rounds.]

No. **5.** Nosing Tool, $1\frac{1}{4}$ inch Cutter $1 00

EXPLANATORY.
[See pages 40, 42 and 43.]

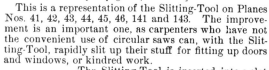

This is a representation of the Slitting-Tool on Planes Nos. 41, 42, 43, 44, 45, 46, 141 and 143. The improvement is an important one, as carpenters who have not the convenient use of circular saws can, with the Slitting-Tool, rapidly slit up their stuff for fitting up doors and windows, or kindred work.

The Slitting-Tool is inserted into a slot on the right side of the main stock of the Plane, and just in front of the handle; a steel depth-gauge is placed over it on the same spindle, and both are fastened down by a brass thumb-screw.

The position of the Slitting-Tool being right under the hand of the workman, his full strength can be exerted while the construction of the Plane renders it stiff enough to secure perfect accuracy in working.

NOTE.—The several sets of Bits, etc., which accompany the Planes (Nos. 41, 42, 43, 44, 45, 46, 141 and 143) are put up in wooden boxes, which protect the cutting edges and keep the full assortment of tools always convenient for selection by the owner.

TRAUT'S PATENT ADJUSTABLE
DADO, FILLETSTER, PLOW, ETC.

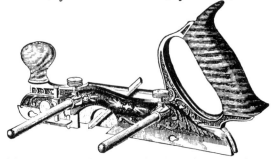

The Tool here represented consists of two sections: a main stock, with two bars, or arms; and a sliding section, having its bottom, or face, level with that of the main stock.

It can be used as a Dado of any required width, by inserting the bit into the main stock and bringing the sliding section snugly up to the edge of the bit. The two spurs, one on each section of the Plane, will thus be brought exactly in front of the edges of the bit. The gauge on the sliding section will regulate the depth to which the tool will cut.

By attaching the Guard-plate, shown above, to the sliding section, the Tool may be readily converted into a Plow, a Filletster or a Matching Plane—as explained in the printed instructions which go with every tool.

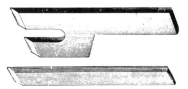

The Tool is accompanied by eight Plow and Dado Bits ($\frac{3}{16}, \frac{1}{4}, \frac{5}{16}, \frac{3}{8}, \frac{1}{2}, \frac{5}{8}, \frac{7}{8}$ and $1\frac{1}{4}$ inch), a Filletster Cutter, a Slitting-tool and a Tonguing-tool. All except the Slitting Blade are secured in the main stock on a *skew*.

Price, including Plow-bits, Slitting and Tonguing Tools.

No. **46.** Iron Stock and Fence, each.....................$7 00

ADJUSTABLE DADO.

This Tool has the same stock and arms as shown above (No. 46)—but it can be used as a Dado only. Insert the bit into the main stock, and bring the sliding section snugly up to the edge of the bit. The two spurs, one on each section of the plane, will thus be brought exactly in front of the edges of the bit. The gauge on the sliding section will regulate the depth to which the tool will cut. The tools are secured in the stock *on a skew*.

Price, including Bits ($\frac{3}{8}, \frac{1}{2}, \frac{5}{8}, \frac{7}{8}$ and $1\frac{1}{4}$ inch), and Slitting-tool.

No. **47.** Iron Stock and Fence, each.....................$4 00

MILLER'S PATENT COMBINED
PLOW, FILLETSTER AND MATCHING MACHINE.

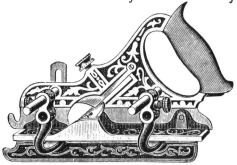

This Tool embraces, in a most ingenious and successful combination, the common Carpenter's Plow, an adjustable Filletster and a perfect Matching Plane. The entire assortment can be kept in a smaller space, or made more portable, than an ordinary Carpenter's Plow.

With each Tool eight forged Plow Bits (⅛, 3-16, ¼, 5-16, ⅜, 7-16, ½ and ⅝ inch), a Filletster Cutter and a Slitting Blade are furnished; also a Tonguing Tool (¼ inch), and by use of the latter, with the ¼ inch Plow Bit for grooving, a perfect Matching Plane is made.

No.	Combined Plow, Filletster and Matching Plane.	Each.
41.	Iron Stock and Fence	$9 00
42.	Gun Metal Stock and Fence	12 00

Combined Plow and Matching Plane—without Filletster.

43.	Iron Stock and Fence	7 00
44.	Gun Metal Stock and Fence	10 00

PATENT BULL-NOSE
PLOW, FILLETSTER AND MATCHING PLANE.

Two interchangeable front parts go with this Tool. The form shown above is that of a Bull-nose Plow; and the Cutter will easily work up to, and into a ½ inch hole, or any larger size—as in Sash-fitting, Stair-work, etc. With the other front on, it takes the ordinary form of a Plow, and is adapted to all regular uses.

With each tool eight Plow Bits (⅛, 3-16, ¼, 5-16, ⅜, 7-16, ½ and ⅝ inch), a Filletster Cutter and a Slitting Blade are furnished; also a Tonguing Tool (¼ inch).

No.	Bull-Nose Plow, Filletster and Matching Plane.	Each.
141.	Iron Stock and Fence	$7 00

Bull-Nose Plow and Matching Plane—without Filletster.

143.	Iron Stock and Fence	5 00

PATENT
TONGUING AND GROOVING PLANE.

The Plane has two separate cutters, a suitable distance apart. When the guide, or fence, is set as shown below, both cutters work and a tongue can be made.

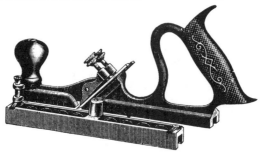

This fence is hung on a pivot, and can be swung around, end for end. This movement covers one of the cutters, and also furnishes a guide for grooving an exact match for the tongue.

Prices, including Tonguing and Grooving Tools.

No.		Each.
48.	Iron Stock and Fence, for $\frac{3}{4}$ to $1\frac{1}{4}$ inch Boards	$2 50
49.	Iron Stock and Fence, for $\frac{3}{8}$ to $\frac{3}{4}$ inch Boards	2 50

PATENT
ADJUSTABLE BEADING PLANE.

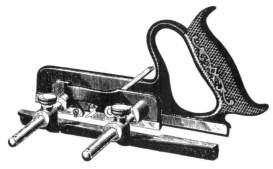

This Tool, for ordinary beading or for center beading, cannot be surpassed. By adjustment of the fence, center beading can be done up to five inches from the edge of a board. Except for working across the grain, the spurs need not be used.

Price, including Bits ($\frac{1}{8}$, $\frac{3}{16}$, $\frac{1}{4}$, $\frac{5}{16}$, $\frac{3}{8}$, $\frac{7}{16}$ and $\frac{1}{2}$ inch).

No.		Each.
50.	Iron Stock and Fence	$4 00

PATENT
IMPROVED RABBET PLANE.

This Plane will lie perfectly flat on either side, and can be used with right or left hand equally well, while planing into corners or up against perpendicular surfaces.

No.		Each.
180.	Iron Stock, 8 in. in length, 1½ in. wide	$1 00
181.	Iron Stock, 8 in. in length, 1¼ in. wide	1 00
182.	Iron Stock, 8 in. in length, 1 in. wide	1 00
190.	Iron Stock, 8 in. in length, 1½ in. wide, with Spur	1 15
191.	Iron Stock, 8 in. in length, 1¼ in. wide, with Spur	1 15
192.	Iron Stock, 8 in. in length, 1 in. wide, with Spur	1 15

PATENT DUPLEX
RABBET PLANE AND FILLETSTER.

The valuable features of this Plane can be seen by a glance at the illustration given above.

Remove the arm to which the fence is secured, and a Handled Rabbet Plane is had; and with two seats for the cutter, so that the tool can be used as a Bull-Nose Rabbet if required.

The construction of the stock is such that the Plane will lie perfectly flat on either side, and can be used with right or left hand equally well, while planing into corners or up against perpendicular surfaces.

The arm to which the fence is secured can be screwed into either side of the stock, thus making a superior right or left hand Filletster, with adjustable spur and depth gauge.

No.		Each.
78.	Iron Stock and Fence, 8½ in. in length, 1½ in. Cutter	$1 50

PATENT
ADJUSTABLE CHAMFER PLANE.

The front section of this Plane is movable up and down. It can be firmly secured to the rear section at any desired point by means of a thumb-screw. Without the use of any other tool, this Plane will do perfect chamfer or stop-chamfer work of all ordinary widths.

For Beading, Reeding or Moulding a chamfer, an additional section to this Plane is furnished (price $1.00) with six cutters, sharpened at both ends, including a large variety of ornamental forms.

No. 72. Chamfer Plane, 9 inches in length, $1\frac{5}{8}$ inch Cutter..............$2 00
No. 72½. Chamfer Plane, with Beading and Moulding Attachment........ 3 00

WOODWORKERS' HANDY ROUTER PLANE.

This Tool should be added to the kit of every skilled Carpenter, Cabinet Maker, Stair Builder, Pattern Maker or Wheelwright. It is perfectly adapted to smooth the bottom of grooves, panels or all depressions below the general surface of any woodwork.

The Bits can also be clamped to the backside of the upright post, and outside of the stock. In this position they will plane into corners, will router out mortises for Sash-frame Pulleys, or will smooth surfaces not easily reached with any other tool.

No. 71. Iron Stock, with Steel Bits (¼ and ½ inch).....................$1 50

"VICTOR" ADJUSTABLE CIRCULAR PLANE.

The working capacity is 12½ inches inside circle. Both ends of the face of the Plane are moved simultaneously, and precisely alike, by means of one screw. The simplicity of its operation is unequaled.

No. 20. Circular Plane, 1¾ inch Cutter, Nickel Plated Trimmings.......$6 00

STANLEY'S PATENT
UNIVERSAL HAND BEADER.

For Beading, Reeding or Fluting straight or irregular surfaces, and for all kinds of light Routering. With a square gauge for straight, and an oval gauge for curved work.

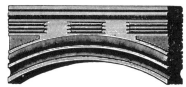

Seven superior steel cutters go with each Tool. Both ends are sharpened, thus embracing six ordinary sizes of Beads, four sets of Reeds, two Fluters and a double Router Iron (⅛ and ¼ inch).

No.		Each.
66.	Iron Stock, with seven Steel Cutters	$1 00
	Extra Cutters or blanks for same	05

PATENT FLOOR PLANE.

This Tool will be found useful for planing Floors, Bowling Alleys, Skating Rinks, Decks of Vessels, etc. The construction of the Plane will enable the owner to do more work, and with less outlay of strength, than can be done with any other tool.

The weight of the Plane is about 10 pounds, and the full length of the handle, 45 inches.

No.		Each.
74.	Floor Plane, 10¼ inches in length, 2⅝ inch Cutter	$4 50

STANLEY'S RULE TRAMMEL POINTS.

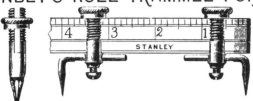

A practical form of Trammel Points, adapted for convenient use on a **Carpenters' Rule**. They can be attached to Folding Rules of any ordinary width; **and** on many kinds of work will take the place of regular Trammel Points, **Calipers** or Dividers.

A complete set consists of two Brass Trammel Heads, with movable **Steel Points**, and one Head with a Pencil Socket.

No.		Per Set.
99.	Rule Trammel Points, per set of three, in a box.........	$0 50

STANLEY'S ADJUSTABLE CHISEL GAUGE.

Attach to a ¼ inch Chisel (with beveled edge up) and a shaving of **any desired** thickness can be raised, for blind-nailing or for inlaying wood strips in **ornamental** surface work.

No.		Per Doz.
96.	Chisel Gauge, Steel Stock, ½ dozen in a box.............	$2 40

IMPROVED MITRE BOX.

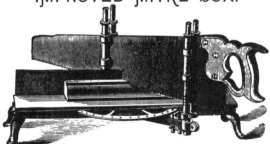

This Mitre Box can be used with a Back Saw or a Panel Saw equally **well**. If a Back Saw is used, both links which connects the rollers, or guides, **are left in** the upper grooves, and the back of the saw is passed through under the links.

If a Panel Saw is used, the link which connects the rollers on the **back spindle** is changed to the lower groove; and then the blade of the saw will be stiffly supported by both sets of rollers, and work like a Back Saw.

No.		Each.
50.	Mitre Box, 20 inches................................	$7 00
60.	Mitre Box, 20 inch, with 20 inch Disston's Back Saw	10 00

STANLEY'S "ODD-JOBS."

20,000 Already Sold.

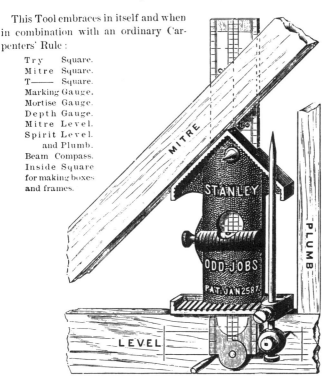

This Tool embraces in itself and when in combination with an ordinary Carpenters' Rule:

    Try Square.
    Mitre Square.
    T——— Square.
    Marking Gauge.
    Mortise Gauge.
    Depth Gauge.
    Mitre Level.
    Spirit Level.
      and Plumb.
    Beam Compass.
    Inside Square for making boxes and frames.

A Mechanic who has this Tool to use on his rule, can do all **ordinary Jobs** with only a Saw, a Hammer and a Plane, in addition.

Attach the Tool to a Rule, making a try-square, or a T-square, with **long or short tongue**; also, a right or left-hand mitre-square. The pointed steel rod **is a** scratch-awl.

A marking-gauge is made by setting the point, or pencil, at **any required** distance from the square end of the stock. A mortise-gauge, also, by **inserting an** additional point, or pencil, in the angle at head of the Rule. A graduated depth-gauge is furnished by extending the Rule down from the square end of the stock.

The steel point at mitred end of the tool forms a center, from which a circle can be swung (as with a beam compass) 1½ to 13 inches in diameter. A circle of 25 inches diameter can be made, if the Rule alone is used, with a pencil in the angle at its head.

No.		Each.
1.	Odd Jobs, Nickel Plated, with Level	$0 75

STANLEY'S ADJUSTABLE
CLAPBOARD (SIDING) MARKER.

This ingenious Tool can be used with one hand, while the other is employed in holding a clapboard in position.

The marking blade is easily adjusted to any thickness of clapboard, or siding.

The sharp edges of the teeth are just parallel with the legs when placed against the corner-board, or window casing.

By moving the tool half an inch, it will mark a full line across the clapboard, exactly over and conformed to the edge of the corner-board.

There is then no difficulty in sawing for a perfectly close joint.

No.		Each.
88.	Metal Stock, with Wood Handle, Steel Blade	**$0 50**

One in a box.

STANLEY'S ADJUSTABLE
CLAPBOARD (SIDING) GAUGE.

A simple and practical Clapboard Gauge, or Holder, is needed by Carpenters; and is offered to them in this Tool.

Two thin Steel Blades, which form a part of the base of the Tool, will slide under the last clapboard already laid (see broken corner in engraving).

When the bottom of the Gauge is brought firmly up to the lower edge of this clapboard, press the handle over sidewise, and this will force another thin blade down into the next lower clapboard, rendering the Tool immovable.

The clapboard to be laid can be held any width to the weather, by means of the graduated scale on the Tool; and after the Tool is released, the mark left is so slight that painting alone will fill it.

☞ A full set of three Gauges is recommended, for laying long clapboards, or siding.

No.		Each.
89.	Metal Stock, with Wood Handle, Steel Blades	**$0 50**

Three in a box.

STANLEY'S PATENT ROOFING BRACKET.

15,000 Already Sold.

The parts are of spring steel, and firmly riveted together. Push the beveled ends up under two layers of shingles already nailed down; the Bracket will then have two separate bearings on the roof, and is so formed that any increase of pressure from above, increases its stability. Two steel spurs project above the horizontal surface of the Bracket, to secure the staging boards.

One dozen per minute can be placed in position, or removed; and great economy in lumber and nails will be found. There are no loose parts to get lost; and no nail-holes are made in the roof. In constant use these Brackets will last a life-time.

DIRECTIONS.—To set the Bracket: grasp the back standard with the fingers through the center part, and spring the bow open enough for the front spurs to clear the shingle butt; then press the two beveled ends up under the shingles already laid, until the front shoulder strikes the butt of the upper course. To remove the Bracket: grasp with the fingers under the front of the bow, thus lifting the spurs until the Bracket is released.

No.		Per Doz.
1.	Roofing Brackets, 8 inch, ½ dozen in box	$3 00

STANLEY'S BUTT AND RABBET GAUGE.
For Hanging Doors, Mortising, Marking, etc.

This Gauge has two bars, both of brass, one movable within the other; also three blades or markers. The two steel blades or markers at the extreme end of the inner bar, can be moved to any position by means of the thumb-screw at opposite end of the Gauge.

A Rosewood Head on the outer bar makes the ordinary Marking, or Slitting Gauge. At one end of the outer bar a steel plate or stop is attached; and, in connection with the Rosewood Head set at the proper distance from the plate, a double head for Mortising is made.

When the Head and Plate have been adjusted to the desired width for the mortise, both markers can still be moved by means of the thumb-screw, and the mortise lines set further in from the edge of the work in hand, or brought out nearer to the edge, without disturbing the width of the mortise.

A superior Butt Gauge, for hanging doors, is made by bringing the marker on the upper side of the bar back of, and to any required distance from the outer surface of the steel plate.

Turn the Gauge over, and place this outer surface against the back of the Rabbet in the door-casing. The owner can then gauge correctly for the back leaf of the Butt; and, without changing any of the parts, the inner surface of the steel plate will act as a guide for the other marker in gauging on the edge of the door, for the front leaf of the Butt. The thickness of the plate (1-16th inch), will set off the door sufficiently to clear the jamb, in opening and shutting.

The Rosewood Head should be secured at the end of the bar nearest the thumb-screw, when the Tool is being used as a Butt Gauge. By placing one edge of a Butt between the surface of this Head and the small blade which stands up back of it, the exact thickness of the Butt is ascertained, and can be accurately gauged on both the door-post and the door. The movable Steel Plate affords a perfect guide for gauging in, or near, Rabbets of all kinds.

No.		Per Doz.
92.	Improved Butt and Rabbet Gauge, ½ dozen in a box	$15 00

IMPROVED TRAMMEL POINTS.

These Tools are used by Millwrights, Machinists, Carpenters, and all Mechanics having occasion to strike arcs, or circles, larger than can be conveniently done with ordinary Compass dividers. They may be used on a straight wooden bar of any length, and when secured in position by the thumbscrews, all circular work can be readily laid out by their use. They are made of Bronze Metal, and have Steel Points, on either of which a pencil socket (which accompanies each pair) can be firmly clamped close up to the main stock. The pencil will thus be secured at a point nearer the work than is possible by any other method.

No.			Per Pair.
1. (Small) Br'ze Metal, Steel Points,			$1 00
2. (Medium)	"	"	1 25
3. (Large)	"	"	1 75

ADJUSTABLE PLUMB BOBS.

These Plumb Bobs are constructed with a reel at the upper end, upon which the line may be kept; and by dropping the bob with a slight jerk, while the ring is held in the hand, any desired length of line may be reeled off. A spring, which has its bearing on the reel, will check and hold the bob firmly at any point on the line. The pressure of the spring may be increased, or decreased, by means of the screw which passes through the reel. A suitable length of line comes already reeled on each Plumb Bob.

No.					Each.
1. (Small) Bronze Metal, with Steel Point				....	$1 50
2. (Large)	"	"	"	....	1 75
5. (Large) Iron	"	"	....	1 00

SOLID CAST STEEL SCREW DRIVERS. 53

PATENT IMPROVED
SOLID CAST STEEL SCREW DRIVERS.

The Blades of both brands of our Screw Drivers are made from the same superior quality of Cast Steel, and are tempered with great care. They are ground down to a correct taper, and shaped at the end by special machinery; thus procuring perfect uniformity in size and strength, while the peculiar form of the point gives it unequaled firmness in the screw-head when in use.

The Handles are of the most approved pattern, the Brass Ferrules, of the thimble form, extra heavy, and closely fitted.

The shanks of the blades to No. 64 Screw Drivers are properly slotted to receive a patent metallic fastening, which secures them permanently in the handles.

☞ Our Screw Drivers are fully **WARRANTED.**

No. 64. Varnished Handles.
[With Patent Metallic Fastening.]

Sizes,	1½	2	3	4	5	6	7	8	10	12	Inches.
	$1 00	1 50	2 00	2 50	3 00	3 50	4 00	4 75	6 00	8 00	Per Doz.

One-half dozen in a box.

No. 86. Polished Handles.

Sizes,	2	3	4	5	6	7	8	10	12	Inches.
	1 50	2 00	2 50	3 00	3 50	4 00	4 75	6 00	8 00	Per Doz.

One-half dozen in a box.

SCREW DRIVER HANDLES.

No.			Per Doz.
7.	Assorted Sizes,	1 dozen in a box	$1 00
8.	Large Sizes,	1 dozen in a box	1 25
9.	Extra Large,	1 dozen in a box	1 50

BAILEY'S IRON SPOKE SHAVES.

☞ The Spoke Shaves in the following List are superior in style, quality and finish to any in market. The Cutters are made of the best English Cast Steel, tempered and ground by an improved method, and are in perfect working order when sent from the factory.

No.		Per Doz.
51.	Double Iron, Raised Handle, 10 inch, $2\frac{1}{8}$ inch Cutter	$3 50
52.	Double Iron, Straight Handle, 10 inch, $2\frac{1}{8}$ inch Cutter	3 50
53.	Adjustable, Raised Handle, 10 inch, $2\frac{1}{8}$ inch Cutter	4 50
54.	Adjustable, Straight Handle, 10 inch, $2\frac{1}{8}$ inch Cutter	4 50
55.	Model Double Iron, Hollow Face, 10 inch, $2\frac{1}{8}$ inch Cutter	3 00
56.	Coopers' Spoke Shave, 18 inch, $2\frac{5}{8}$ inch Cutter	7 00
56½.	Coopers' Spoke Shave (Heavy), 19 inch, 4 inch Cutter	9 00
57.	Coopers' Spoke Shave (Light), 18 inch, $2\frac{1}{8}$ inch Cutter	4 50
58.	Model Double Iron, 10 inch, $2\frac{1}{8}$ inch Cutter	3 00
59.	Single Iron (Pattern of No. 56), 10 inch, $2\frac{1}{8}$ inch Cutter	3 50
60.	Double Cutter, Hollow and Straight, 10 inches, $1\frac{1}{2}$ inch Cutter	4 50

Price List of Spoke Shave Cutters.

Nos. 51, 52, 53, 54, 55, 57, 58, 59 .... 1 00
No. 60 (in pairs), $1 50; No. 56, $1 50; No. 56½, $2 00

IRON SPOKE SHAVE.

No.   Per Doz.
64. Straight Handle, 9 inch, 2⅛ inch Cutter (with Thumb Screw)........$2 00
     Cast Steel Cutters 75

PATENT REVERSIBLE SPOKE SHAVE.

This Spoke Shave can be worked to and from the person using it, without changing position.

62. Raised Handle (Heavy), Double Cutter, 10 inch, 2⅛ inch Cutters...... 6 00
     Cast Steel Cutters 1 00

PATENT CHAMFER SPOKE SHAVE.

This Tool can be easily adjusted by means of the thumb-screws attached to the Guides; and will chamfer an edge any desired width up to 1½ inch.

65. Raised Handle, 1¼ inch Cutter..................................... 6 00
     Cast Steel Cutters 75

PATENT ADJUSTABLE BOX SCRAPER.

An excellent Box Scraper, and also well adapted for planing floors.

70. Malleable Iron, 2 inch Steel Cutter................ .... 6 00
     Cast Steel Cutters 1 50

WHEELER'S PATENT COUNTERSINK.
FOR WOOD.

This Countersink works equally well for every variety of screw. The Countersink cuts rapidly, and is easily sharpened by drawing a thin file lengthwise inside of the cutter. By fastening the Gauge at a given point, any number of screws may be driven so as to leave the heads flush with the surface, or at a uniform depth below it.

No.   Per doz.
18. Countersinks, ½ dozen in a box $3 00
20. Countersinks, with Gauge, ¼ dozen in a box ... 4 50

STANLEY'S IMPROVED
DOWEL SHARPENER.

This Tool is the reverse of the Countersink, and can be sharpened with equal ease. By its use Dowel Pins can be sharpened rapidly and uniformly.

No.   Per Doz.
22. Dowel Sharpener, ½ dozen in a box $3 00

PATENT EXCELSIOR TOOL HANDLES.

DIRECTIONS.—Unscrew the Cap of the Handle and select the Tool needed; with the thumb, the center bolt may be thrust down sufficiently to open the clamp at the small end of the Handle, into which the Tool can be inserted; then, in replacing the cover and screwing it down to its place, the clamp will be closed and the Tool firmly secured for use.

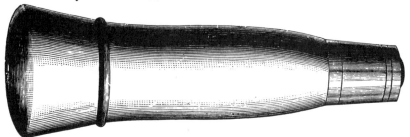

No.		Per Doz.
1.	Turkey Boxwood Handle, with Twenty Tools	$7 50

One-half dozen in a box.

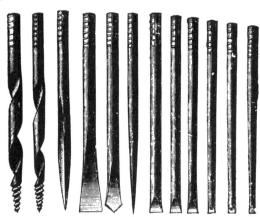

2.	Iron Handle, with Twelve Tools	4 00
3.	Iron Handle, with Twenty Tools	5 50
22.	Iron Handle, Nickel Plated, with Twelve Tools	5 00
23.	Iron Handle, Nickel Plated, with Twenty Tools	6 50

One-half dozen in a box.

CARPENTERS' TOOL HANDLES.

No.		Per Doz.
8.	Carpenter's Tool Handle, Steel Screw and Nut, with Wrench, 1 dozen in a box..	$0 90
8½.	Carpenter's Tool Handle, Steel Screw and Nut, with Wrench, and 10 Brad Awls, assorted sizes. One handle and awls, in a box, and 12 boxes in a package...	3 00

AWL HAFTS.

No.		Per Gross.
5.	Hickory, Pegging Awl Haft, Plain Top, Steel Screw and Nut, with Wrench, 1 dozen in a box	10 00
6.	Hickory Pegging Awl Haft, Leather Top, Steel Screw and Nut, with Wrench, 1 dozen in a box	12 00
7.	Hickory, Pegging Awl Hafts, Leather Top, Extra Large, Riveted, Steel Screw and Nut, with Wrench, 1 dozen in a box.............	14 00

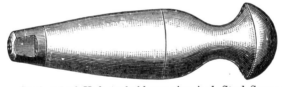

6½.	Appletree, Sewing Awl Haft, to hold any size Awl, Steel Screw and Nut, with Wrench, 1 dozen in a box	12 00

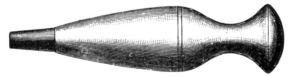

10.	Common Sewing Awl Haft, Brass Ferrule, 3 dozen in a box	3 50
11.	Common Pegging Awl Haft, Brass Ferrule, 3 dozen in a box	3 50

PATENT PEGGING AWLS.

PATENT PEGGING AWLS (short start, for Handle Nos. 5 and 6), Assorted, 1 gross in a box	80

CHALK-LINE REELS, ETC.

No.		Per Doz.
12.	Chalk-Line Reels, 3 dozen in a box	$0 36
13.	Chalk-Line Reels, with 60 feet best quality Chalk-Line, 1 dozen in a box	1 75
14.	Chalk-Line Reels, with Steel Scratch Awls, 1 doz. in a box	95
15.	Chalk-Line Reels, with Steel Scratch Awls, and 60 feet best quality Chalk-Line, 1 dozen in a box.................	2 25

HANDLED SCRATCH AWLS.

No.		Per Gross.
1.	Handled, Steel Scratch Awls, 1 dozen in a box...........	7 00
2.	Handled, Steel Scratch Awls, Large, 1 dozen in a box.....	8 50

HANDLED BRAD AWLS.

3.	Handled Brad Awls, assorted, 1 dozen in a box	6 50
4.	Handled Brad Awls, assorted, Large, 1 dozen in a box	7 00

5.	Brad Awl Handles, Brass Ferrules, assorted, 3 doz. in a box	3 50

SAW HANDLES.

All full sizes, of perfect timber, well seasoned, and every way superior and reliable goods.

No.		Per Doz.
1.	Full Size, Cherry, Varnished Edges	$1 65
2.	Full Size, Beech, Varnished Edges	1 40
3.	Full Size, Beech, Plain Edges	1 20
4.	Small Panel, Beech, Varnished Edges, 16 to 20 in. Saws	1 35
5.	Meat Saw, Beech, Varnished Edges	1 35
6.	Compass Saw, Beech, Varnished Edges	1 25
7.	Back Saw, Beech, Varnished Edges	1 35

In paper boxes of one dozen each—packed two gross in a case.

PLANE HANDLES.

No. 10.

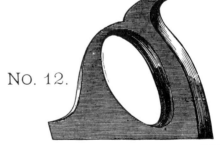

No. 12.

	Per Doz.
Jack Plane Handles, 5 gross in a case	$0 42
Fore, or Jointer Handles, 3¾ gross in case	75

In packages of one dozen each.

MALLETS.

MORTISED HANDLES.

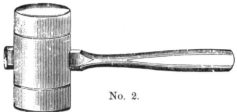

No. 2.

No.			Per Doz.
1.	Round Hickory,	Mortised, 5 in. long, 3 in. diam...	$1 50
2.	Round Hickory,	Mortised, 5½ in. long, 3½ in. diam...	2 00
3.	Round Hickory,	Mortised, 6 in. long, 4 in. diam...	2 50
5.	Round Lignumvitæ,	Mortised, 5 in. long, 3 in. diam...	3 00
6.	Round Lignumvitæ,	Mortised, 5½ in. long, 3½ in. diam...	4 00
7.	Round Lignumvitæ,	Mortised, 6 in. long, 4 in. diam...	5 00

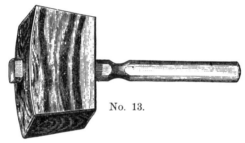

No. 13.

8.	Square Hickory,	Mortised, 6 in. long, 2½ by 3½ in...	2 00
9.	Square Hickory,	Mortised, 6½ in. long, 2¾ by 3¾ in...	2 50
10.	Square Hickory,	Mortised, 7 in. long, 3 by 4 in...	3 00
11.	Square Lignumvitæ,	Mortised, 6 in. long, 2½ by 3½ in...	3 75
12.	Square Lignumvitæ,	Mortised, 6½ in. long, 2¾ by 3¾ in...	4 75
13.	Square Lignumvitæ,	Mortised, 7 in. long, 3 by 4 in...	5 75

No. 14.

14.	Round Mallet, Mortised, Iron Rings, 6 in. long, 4 in. diam.	5 50
14½.	Round Mallet, Mortised, Iron Rings, 5½ in. long, 3½ in. diam.	4 00

MALLETS.

MORTISED HANDLES.

No.   Per Doz.
15. Round Iron Mallet, Mortised, Hickory Ends, $2\frac{1}{2}$ in. diam. **$4 00**

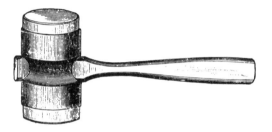

16. Round Mallet, Heavy Malleable Iron Socket, Mortised, Hickory Ends, 3 inch diameter **7 50**

TINNERS' MALLETS.

No.   Per Doz.
4. Round Hickory, $5\frac{1}{2}$ in. long, assorted, $2\frac{1}{4}$ and $2\frac{1}{2}$ in. diameter, **$1 00**

TACK HAMMERS, ETC.

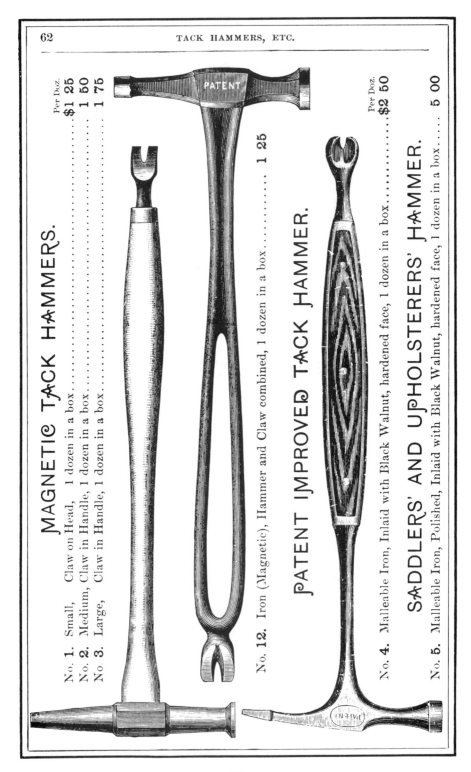

MAGNETIC TACK HAMMERS.

Per Doz.
- No. 1. Small, Claw on Head, 1 dozen in a box $1 25
- No. 2. Medium, Claw in Handle, 1 dozen in a box 1 50
- No. 3. Large, Claw in Handle, 1 dozen in a box 1 75

PATENT IMPROVED TACK HAMMER.

No. 12. Iron (Magnetic), Hammer and Claw combined, 1 dozen in a box 1 25

SADDLERS' AND UPHOLSTERERS' HAMMER.

Per Doz.
- No. 4. Malleable Iron, Inlaid with Black Walnut, hardened face, 1 dozen in a box $2 50
- No. 5. Malleable Iron, Polished, Inlaid with Black Walnut, hardened face, 1 dozen in a box 5 00

STEAK HAMMERS, ETC.

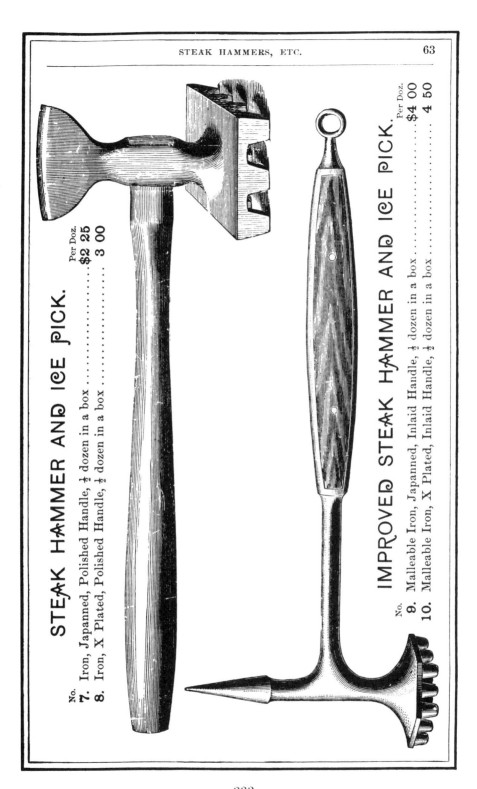

STEAK HAMMER AND ICE PICK.
Per Doz.
No.
7. Iron, Japanned, Polished Handle, ½ dozen in a box $2 25
8. Iron, X Plated, Polished Handle, ½ dozen in a box 3 00

IMPROVED STEAK HAMMER AND ICE PICK.
Per Doz.
No.
9. Malleable Iron, Japanned, Inlaid Handle, ½ dozen in a box $4 00
10. Malleable Iron, X Plated, Inlaid Handle, ½ dozen in a box 4 50

INDEX.

	Page.
Awl Hafts	57
Awls, Patent Pegging	57
Beader, Stanley's Universal	47
Bevels, Sliding T	27
Bevels, Patent Flush, Eureka	27
Bit and Square Level	20
Box Scraper, Adjustable	55
Brad Awls, Handled	58
Carpenters' Tool Handles	57
Chalk-Line Reels and Awls	58
Chisel Gauge	48
Clapboard Gauge	50
Clapboard Marker	50
Countersinks, Wheeler's Patent	55
Dado, Filletster, Plow, etc., combined	42
Dado, Adjustable	42
Dowel Sharpener	55
Gauges	28 to 30
Gauges, with Improved Face Plate	31
Gauges, Butt and Rabbet	51
Handles, Brad Awl	58
Handles, Plane	59
Handles, Saw	59
Handles, Screw Driver	53
Hammers, Magnetic	62
Hammers, Tack, No. 4	62
Hammers, Steak	63
Hammers, Upholsterers'	62
Hollows and Rounds, for Plane No. 45	41
Level, Bit and Square	20
Level Glasses	21
Level Sights	20
Levels, Machinists' Iron	21
Mallets, Hickory and Lignumvitæ	60 and 61
Mitre Box, Improved	48
Mitre Squares, Improved	24
Mitre Try Squares, Improved	24
Odd Jobs	49
Planes, Bailey's Patent Adjustable	32 to 36
Planes, Stanley Patent Adjustable	37 to 39
Planes, "Victor" Circular	46
Plane Irons	35 and 38

INDEX.

	Page.
Planes, Belt	34
Planes, Beading	44
Planes, Beading, Rabbet and Slitting	40
Planes, Bull-Nose Rabbet	39
Planes, Chamfer	46
Planes, Dado, Filletster, etc.	42
Planes, Floor	47
Planes, Knuckle-Joint	36
Planes, Nosing	41
Planes, Rabbet	45
Planes, Router	46
Planes, Tonguing and Grooving	44
Plow, Bull-Nose Matching Plane, etc.	43
Plow, Filletster and Matching Plane	43
Plow and Matching Plane	43
Plumbs and Levels, Non-Adjustable	18
Plumbs and Levels, Patent Adjustable	19
Plumbs and Levels, Nicholson's Patent	20
Plumbs and Levels, Hand-y	16
Plumbs and Levels, Duplex	16
Plumb Bobs, Adjustable	52
Pocket Levels	21
Roofing Brackets	51
Rules, Stanley's Boxwood	3 to 10
Rules, Stanley's Ivory	11
Rules, Stanley's Miscellaneous	12 and 13
Rules, Stearns' Ivory	14 and 15
Siding Gauge	50
Siding Marker	50
Scratch Awls, Handled	58
Screw Drivers	53
Spoke Shaves, Bailey's	54
Spoke Shaves, Iron	55
Spoke Shaves, Patent Reversible	55
Spoke Shaves, Patent Chamfering	55
Spoke Shave Cutters	54
Trammel Points	52
Trammel Points, for Rules	48
Tool Handles and Tools, Excelsior	56
Try Squares	26
Try Squares, Adjustable	22
Try Squares, Improved Iron Handle	22
Try Squares, Inlaid	25
Try Squares, Plumb and Level	26
Try Square and Bevel, Patent Combination	25
Try and Mitre Squares, Winterbottom's Patent	23
Veneer Scrapers	34 and 37

Attach to our PRICE LIST of January, 1892.

DISCOUNT SHEET.

STANLEY RULE AND LEVEL CO.

JANUARY 1, 1892.

Catalogue Pages.		Discount per cent.
57.	AWL HAFTS	50
57.	Awls, Patent Pegging	50
47.	BEADER, Stanley's Universal	20
27.	BEVELS, Sliding T	60
27.	Bevels, Patent Flush Eureka	30
20.	BIT and SQUARE LEVEL	20
55.	Box Scraper, Adjustable	30
58.	Brad Awls, Handled	30
51.	BUTT and RABBET GAUGE	20
50.	CLAPBOARD MARKER	20
50.	Clapboard Gauge	20
58.	Chalk-lines, Reels and Awls	30
57.	Carpenters' Tool Handles	40
48.	CHISEL GAUGE	20
55.	Countersinks, Wheeler's Patent	40
55.	DOWEL SHARPENER	40
28 to 30.	GAUGES	60
31.	Gauges, with Improved Face-Plate	60
51.	Gauges, Butt and Rabbet	20
58.	Handles, Brad Awl	30
59.	Handles, Plane	40
59.	Handles, Saw	40
53.	Handles, Screw Driver	50
62.	Hammers, Magnetic	30
62.	Hammers, Tack, No. 4	30
63.	Hammers, Steak	30
62.	Hammers, Upholsterers'	30
41.	Hollows & Rounds, for Plane No. 45	25
20.	LEVEL, Bit and Square	20
21.	Levels, Machinists' Iron	30
21.	Level Glasses	70
20.	LEVEL SIGHTS	20
60 & 61.	MALLETS	25
48.	Mitre Boxes	20
24.	Mitre Squares	30
24.	Mitre Try Squares	30
49.	ODD-JOBS, Stanley's	20
18.	PLUMBS and LEVELS, Non-Adjustable	75
19.	Plumbs and Levels, Patent Adjustable	75
16.	Plumbs and Levels, HAND-Y	20
16.	Plumbs and Levels, DUPLEX	20
20.	Plumbs and Levels, Nicholson's Patent	30
21.	Pocket Levels	70

Catalogue Pages.		Discount per cent.
32 to 34.	PLANES, Bailey's Adjustable—Iron.	50
35.	Planes, Bailey's Adjustable—Wood.	50
36.	Planes, Bailey's Adjustable—Block.	50
37.	Planes, Stanley Adjustable—Iron.	50
38.	Planes, Stanley Adjustable—Wood.	50
39.	Planes, Stanley Adjustable—Block.	50
46.	Planes, Victor Circular	50
35 & 38.	Plane Irons	50
	MISCELLANEOUS PLANES.	
44.	Planes, Beading	25
40.	Planes, Beading, Rabbet and Slitting	25
46.	Planes, Chamfer	25
42.	Planes, Dado	25
47.	Planes, Floor	25
41.	Planes, Nosing, for Plane No. 45	25
45.	Planes, Rabbet	25
45.	Planes, Rabbet and Filletster	25
46.	Planes, Router	25
44.	Planes, Tonguing and Grooving	25
42.	PLOW, Dado, Filletster, etc.	25
43.	Plow, Filletster and Matching Plane	25
43.	Plow and Matching Plane, Bull-Nose	25
52.	PLUMB BOBS, Adjustable	30
51.	ROOFING BRACKETS	20
3 to 10.	RULES. Boxwood	Stanley's 80
11.	Rules, Ivory	Stanley's 50
14 & 15.	Rules, Ivory	Stearn's 50
12 & 13.	Rules, Miscellaneous	Stanley's 60
58.	Scratch Awls Handled	30
53.	SCREW DRIVERS, No. 64, Var. Hdls.	65
53.	Screw Drivers, No. 86	70
54 & 55.	SPOKE SHAVES, Bailey's	40
52.	Spoke Shave Cutters, Bailey's	40
52.	TRAMMEL POINTS	30
48.	Trammel Points, for Rules	20
56.	Tool Handles and Tools, Excelsior	40
26.	TRY SQUARES, No. 20	60
22.	Try Squares, Adjustable, No. 14	30
22.	Try Squares, Iron Handle, No. 12	30
25.	Try Squares, Inlaid, No. 10	30
26.	Try Squares, Plumb and Level	30
25.	Try Square and Bevel, Combination	30
23.	Try and Mitre Square, Winterbottom's	30
34 & 37.	Veneer Scrapers	50

DISCOUNT for Cash (IF PAID WITHIN THIRTY DAYS), 10 per Cent.

☞ *TERMS CASH.—Payable in current New York or Boston funds. Invoices remaining unpaid after Thirty Days from their date are subject to our Draft, payable at sight.*

PRICE LIST

OF

U. S. STANDARD

BOXWOOD AND IVORY

RULES,

PLUMBS AND LEVELS, TRY SQUARES,
BEVELS, GAUGES, MALLETS,
IRON AND WOOD ADJUSTABLE PLANES,
SPOKE SHAVES, SCREW DRIVERS,
AWL HAFTS, HANDLES, &c.

MANUFACTURED BY THE

STANLEY
RULE AND LEVEL CO.

NEW BRITAIN, CONN., U. S. A.

WAREROOMS:

No. 29 CHAMBERS STREET, NEW YORK.

ORDERS FILLED AT THE WAREROOMS, OR AT NEW BRITAIN.

ABRIDGMENT   Revised to 1897

JANUARY, 1892.

OFFICE OF THE

STANLEY RULE AND LEVEL CO.

NEW BRITAIN, CONN., U. S. A.

This Illustrated Catalogue presents the full line of

IMPROVED LABOR-SAVING

CARPENTERS' TOOLS

now manufactured by us. It will be interesting to Hardware Dealers, as containing genuine Goods, with an established reputation. Such Goods will sell themselves.

Mechanics require no explanation, or apology, from the Dealer who offers them STANLEY'S Tools; and the uniform report from those who use these Tools is, that they want more.

Our continued efforts are pledged to produce a grade of Tools acceptable to Mechanics, and worthy the ineffectual attempts made by other manufacturers to imitate them.

JANUARY 1, 1892.

The Illustration of the No. 62-1/2 Rule has been added.

STANLEY'S BOXWOOD RULES.

STANLEY'S TWO FEET, FOUR FOLD, NARROW RULES.

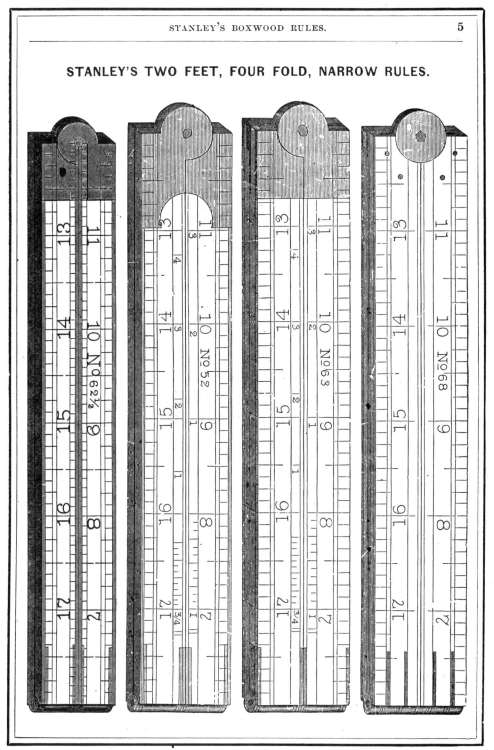

Illustration of Ship Carpenters' Bevel has been added.

STANLEY'S SLIDE RULES, ARCHITECTS' RULES, ETC.

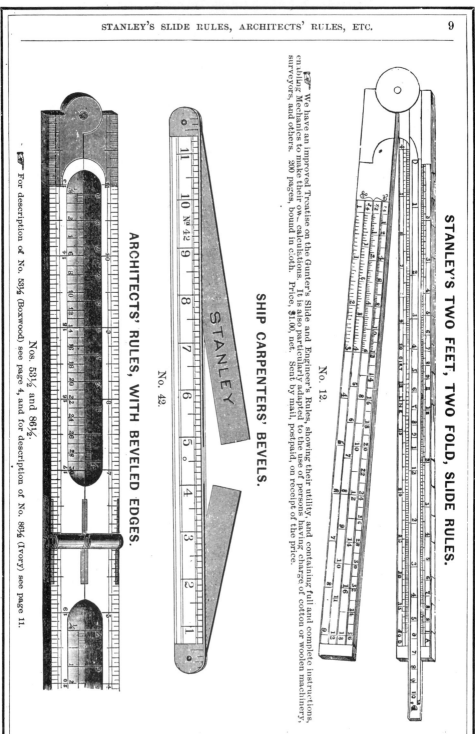

STANLEY'S TWO FEET, TWO FOLD, SLIDE RULES.

No. 12.

☞ We have an improved Treatise on the Gunter's Slide and Engineer's Rules, showing their utility, and containing full and complete instructions enabling Mechanics to make their own calculations. It is also particularly adapted to the use of persons having charge of cotton or woolen machinery, surveyors, and others. 200 pages, bound in cloth. Price, $1.00, net. Sent by mail, postpaid, on receipt of the price.

SHIP CARPENTERS' BEVELS.

No. 42.

ARCHITECTS' RULES, WITH BEVELED EDGES.

☞ For description of No. 53½ (Boxwood) see page 4, and for description of No. 86½ (Ivory) see page 11.

Nos. 53½ and 86½.

Illustration of No. 36 Caliper Rule has been added.

STANLEY'S BOXWOOD RULES.

BOXWOOD CALIPER RULES, SIX INCH.

No.		Per Doz.
36.	Square Joint, Two Fold, 8ths, 10ths, 12ths and 16ths of inches .... ⅞ in. wide,	$7 00
13.	Square Joint, Two Fold, 8ths and 16ths inches, 1⅛ "	10 00
13½.	Square Joint, Two Fold, 8ths and 16ths inches, 1½ "	12 00

CALIPER, ONE FOOT, FOUR FOLD.

32.	Arch Joint, Edge Plates, 8ths, 10ths, 12ths and 16ths of inches ....1 in. wide,	12 00
32½.	Arch Joint, Bound, 8ths, 10ths, 12ths and 16ths of inches ....1 "	20 00

CALIPER, ONE FOOT, TWO FOLD.

36½.	Square Joint, Two Fold, 12 inch, 8ths, 10ths, 12ths and 16ths of inches ....1⅜ "	12 00

TWO FEET, SIX FOLD RULES.

No. 58.

58.	Arch Joint, Edge Plates, 8ths, 10ths, 12ths and 16ths of inches........................¾ in. wide,	13 00
58½.	Arch Joint, Bound, 8ths, 10ths, 12ths and 16ths of inches........................¾ "	36 00

Illustration of Ivory Caliper Rule has been added.

STANLEY'S IVORY RULES.

No.		Per Doz.
	IVORY CALIPER, SIX INCH.	
38.	Square Joint, German Silver, Two Fold, 8ths, 10ths, 12ths and 16ths of inches .... $\frac{7}{8}$ in. wide,	$15 00
40½.	Square Joint, German Silver, Bound, Two Fold, 8ths and 16ths of inches .... $\frac{5}{8}$ "	24 00
	IVORY CALIPER, ONE FOOT, FOUR FOLD.	
39.	Square Joint, German Silver, Edge Plates, 8ths, 10ths, 12ths and 16ths of inches ..... $\frac{7}{8}$ in. wide,	38 00
40.	Square Joint, German Silver, Bound, 8ths and 16ths of inches .... $\frac{5}{8}$ "	44 00
	IVORY, ONE FOOT, FOUR FOLD.	
90.	Round Joint, Brass, Middle Plates, 8ths and 16ths of inches..........	10 00
92½.	Square Joint, German Silver, Middle Plates, 8ths and 16ths of inches.......... $\frac{5}{8}$ "	14 00
92.	Square Joint, German Silver, Edge Plates, 8ths and 16ths of inches .... $\frac{5}{8}$ "	17 00
88½.	Arch Joint, German Silver, Edge Plates, 8ths and 16ths of inches .... $\frac{5}{8}$ "	21 00
88.	Arch Joint, German Silver, Bound, 8ths and 16ths of inches .... $\frac{5}{8}$ "	32 00
91.	Square Joint, German Silver, Edge Plates, 8ths, 10ths, 12ths and 16ths of inches ..... $\frac{3}{4}$ "	23 00
	IVORY, TWO FEET, FOUR FOLD.	
85.	Square Joint, German Silver, Edge Plates, 8ths, 10ths, 12ths and 16ths of inches ..... $\frac{7}{8}$ in. wide,	54 00
86.	Arch Joint, Ger. Silv., Ed. Plates, 8ths, 10ths, 12ths, 16ths in., 100ths foot, Draft'g Scales,1 "	64 00
86½.	Arch Joint, German Silver, Edge Plates, 8ths, 10ths, 12ths and 16ths of inches, with Inside Beveled Edges, and Architects' Dftg. Scales,1 " [See Engraving, p. 9.]	96 00
87.	Arch Joint, German Silver, Bound, 8ths, 10ths, 12ths and 16ths of inches, Drafting Scales .1 "	80 00
89.	Double Arch Joint, German Silver, Bound, 8ths, 10ths, 12ths and 16ths of in. Drafting Scales,1 "	92 00
95.	Arch Joint, German Silver, Bound, 8ths, 10ths 12ths and 16ths of inches, Drafting Scales,1$\frac{3}{8}$ "	102 00
97.	Double Arch Joint, German Silver, Bound, 8ths, 10ths, 12ths and 16ths of in. Drafting Scales,1$\frac{3}{8}$ "	116 00

Illustrations of No.45 & No.44 Gauging Rods have been added.

STANLEY'S BOARD AND LOG MEASURES, ETC.

BOARD AND LOG MEASURES.

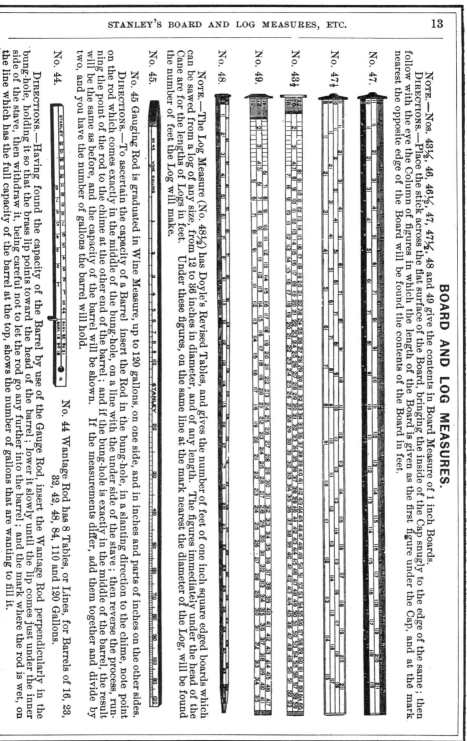

Note.—Nos. 43½, 46, 46½, 47, 47½, 48 and 49 give the contents in Board Measure of 1 inch Boards.

Directions.—Place the stick across the flat surface of the Board, bringing the inside of the Cap snugly to the edge of the same; then follow with the eye the Column of figures in which the length of the Board is given as the first figure under the Cap, and at the mark nearest the opposite edge of the Board will be found the contents of the Board in feet.

No. 47.

No. 47½.

No. 43½.

No. 49.

No. 48.

Note.—The Log Measure (No. 48½) has Doyle's Revised Tables, and gives the number of feet of one inch square edged boards which can be sawed from a log of any size, from 12 to 36 inches in diameter, and of any length. The figures immediately under the head of the Cane are for the lengths of Logs in feet. Under these figures, on the same line at the mark nearest the diameter of the Log, will be found the number of feet the Log will make.

No. 45.

No. 45 Gauging Rod is graduated in Wine Measure, up to 120 gallons, on one side, and in inches and parts of inches on the other sides.

Directions.—To ascertain the capacity of a Barrel insert the Rod in the bung-hole, in a slanting direction to the chime, note point on the rod which comes exactly in the middle of the bung-hole, on a line with the under side of the stave; then reverse the process, running the point of the rod to the chime at the other end of the barrel; and if the bung-hole is exactly in the middle of the barrel, the result will be the same as before, and the capacity of the barrel will be shown. If the measurements differ, add them together and divide by two, and you have the number of gallons the barrel will hold.

No. 44.

No. 44 Wantage Rod has 8 Tables, or Lines, for Barrels of 16, 28, 32, 42, 48, 84, 110 and 120 Gallons.

Directions.—Having found the capacity of the Barrel by use of the Gauge Rod, insert the Wantage Rod perpendicularly in the bung-hole, holding it so that the brass lip points toward the head of the barrel; lower it slowly until the lip comes just under the inner side of the stave, then withdraw it, being careful not to let the rod go any further into the barrel; and the mark where the rod is wet, on the line which has the full capacity of the barrel at the top, shows the number of gallons that are wanting to fill it.

Illustrations of HAND-Y and DUPLEX PLUMBS & LEVELS added.

16   STANLEY'S PATENT HAND-Y PLUMB AND LEVEL, ETC.

STANLEY'S PATENT
HAND-Y PLUMB AND LEVEL.

This Level can be used in horizontal position, or can be brought to perpendicular for ascertaining a plumb, with remarkable ease, as the shallow grooves along the two sides afford an excellent grip on the Tool. This Level will be especially useful in House-framing, Bridge-building and general out-of-door work.

No.   Per Doz.

16. Patent Adjustable Hand-y Plumb and Level, Cherry, Arch Top Plate, Two Side Views, Polish'd, Tipped, Ass'td, 24 to 30 inches, **$15 00**

STANLEY'S PATENT
DUPLEX PLUMBS AND LEVELS.

These Levels have the ordinary form of leveling glass, set in the top surface of the Stock. For any uses where an observation of the glass, *sidewise*, may be found convenient, an additional leveling glass is set in the side, at the opposite end from the Plumb. Both glasses are protected by Brass Discs, can be seen from either side, and are inserted in the Level with the least possible removal of wood from the Stock.

No.   Per Doz.

25. Patent Adjustable Plumb and Level, Mahogany, Arch Top Plate, Improved Duplex Side Views, Polished, Tip'd, Ass'td, 24 to 30 in., **$24 00**

30. Patent Adjustable Plumb and Level, Cherry, Arch Top Plate, Improved Duplex Side Views, Polished, Tip'd, Ass'td, 24 to 30 in., **18 00**

50. Patent Adjustable Plumb and Level, Cherry, Triple Stock, Arch Top Plate, Improved Duplex Side Views, Polished and Tipped, Assorted, 24 to 30 inches.. **24 00**

STANLEY'S PATENT ADJUSTABLE PLUMB AND LEVEL.

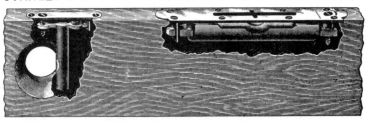

The Spirit-glass (or bubble tube), in both the Level and Plumb, is set in a metallic case attached rigidly to the Brass Top-plate above; and should it be necessary from any cause to adjust either glass, it can be done by means of the screw designated "Adjusting Screw."

Plumbs and Levels Nos. 01, 02, 03, etc., in this Price List, correspond exactly with the Nos. 1, 2, 3, etc., except that the former are non-adjustable and the latter adjustable.

The improved slots now made in all our Top-plates afford a ready means for detecting the movement of the bubble, to or from the center of the Glass.

☞ **Please insert over Page 17, Illustrated Catalogue, 1892,**
and remove p. 16½, as Levels Nos. 20 and 36 are discontinued.

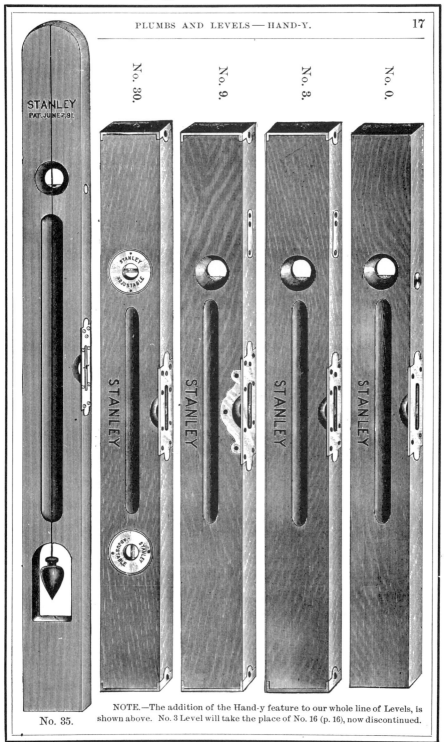

New Illustration of No. 30 Duplex Plumb and Level.

345

PROVED LEVEL GLASSES.

. Made of extra thick tubing. By a patented process, each Level Glass receives an indelible mark at its highest, or crowning point in the center; and the owner can thus easily set the glass accurately in its proper position.

PRICES.

Proved Level Glasses, packed 1 doz. in a box........1¾ inch,	$9	50
Proved Level Glasses, packed 1 doz. in a box........2 "	10	00
Proved Level Glasses, packed 1 doz. in a box........2½ "	10	50
Proved Level Glasses, packed 1 doz. in a box........3 "	11	50
Proved Level Glasses, packed 1 doz. in a box........3½ "	13	00
Proved Level Glasses, packed 1 doz. in a box........4 "	14	50
Proved Level Glasses, packed 1 doz. in a box........4½ "	16	00
Proved, Assorted, 1¾, 3 and 3½ inch, 1 doz. in a box	12	00

SIDE RABBET PLANE.

(See page 40½)

A convenient tool for side-rabbeting and trimming dados, mouldings and grooves of all sorts. A reversible nose-piece will give the tool a form by which it will work close up into corners when required.

Each.

No. **98.** Side Rabbet Plane, 4 inches in length, Right Hand, **$0.90**
No. **99.** Side Rabbet Plane, 4 inches in length, Left Hand, .90

☞ The liberal sale of this excellent Tool has developed a demand for a Left Handed pattern, which we now furnish, as above, No. 99.

(17½)

Side Rabbet Plane Patented by J.A.Traut
January 29, 1895 US Pat. No. 533,329

☞ Please insert opposite Page 20, Illustrated Catalogue, 1892.

HAND-Y PLUMBS AND LEVELS.

WITH GROUND GLASSES.

No.		Per Doz.
60.	Mahogany Plumb and Level, Arch Top Plate, Two Brass Lipped Side Views, with Ground Glasses, 24 to 30 inches,	**$35.00**
90.	Mahogany Plumb and Level, Arch Top Plate, Two Brass Lipped Side Views, Tipped, with Ground Glasses..................24 to 30 inches,	**39.00**

No. 95.

95.	Brass Bound, Mahogany Plumb and Level, Two Brass Lipped Side Views, with Ground Glasses, 24 to 30 inches,	**72.00**
96.	Brass Bound, Rosewood Plumb and Level, Two Brass Lipped Side Views, with Ground Glasses, 24 to 30 inches,	**80.00**
98.	Brass Bound, Rosewood Plumb and Level, Two Brass Lipped Side Views, with Ground Glasses, 12 inches,	**48.00**

☞ Packed separately, in paste-board boxes. In ordering, dealers should designate the lengths of the Levels wanted—24, 26, 28 or 30 inch.

HAND-Y PLUMBS AND LEVELS.

FOR MASONS.

The two Plumb Glasses are set in separate holes ; but each on the right-hand side. This will enable the workman to plumb work below where he stands, or above him, without turning the Level Stock end for end ; which is found to be a great convenience.

70.	Mahogany Plumb and Level, Arch Top Plate, Two Brass Lipped Side Views, Two Plumbs, with Ground Glasses ....36 inches,	**45.00**
80.	Mahogany Plumb and Level, Arch Top Plate, Two Brass Lipped Side Views, Two Plumbs, with Ground Glasses ....42 inches,	**50.00**

GROUND GLASSES.

The inside surfaces of these Glasses are ground perfectly smooth ; and thus the bubble is made extremely sensitive.

HEXAGON POCKET LEVELS.

WITH DETACHABLE BASE PIECE.

No.		Each.
33.	Nickel Plated, with Ground Glasses, complete.........	$1.25

One in a box.

33½. Hexagon Pocket Level, Nickel Plated, 2½ inch, with Ground Glasses,................................. .75

One in a box.

Base-pieces, 3½ inch, if ordered separately............ .50

ECLIPSE LEVELS.

A CONVENIENT TOOL FOR MACHINISTS, ELECTRICIANS AND EXPERT MECHANICS OR AMATEURS.

34. Eclipse Level, Nickel Plated, 6 in., with Ground Glasses, $1.50

The outer shell of the Level can be turned, so as to completely protect the glass from damage when not in use.

One in a box.

IMPROVED WOOD SCRAPER.

This Scraper can be adjusted to any desired pitch; and may be worked toward or from the person using it. The Roller acts as a support to relieve the strain on the wrist and hands of the workman. The Handle can be detached for working into corners.

No. 83. Wood Scraper, with Handles and Roller, 3 in. Blade, $1.00

☞ Please insert opposite Page 26, Illustrated Catalogue, 1892.

STANLEY'S IMPROVED TRY SQUARES AND BEVELS. 26½

TRY SQUARES, No. 20.

See Prices and Description, Page 26.

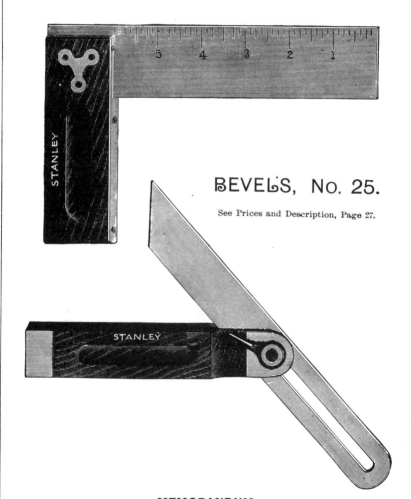

BEVELS, No. 25.

See Prices and Description, Page 27.

MEMORANDUM.

Improvements have been made in the Tools illustrated above. All sizes of both lines are now made by us with the "Hand-y" feature, first introduced by us on our Plumbs and Levels, and which has given them the preference with all first-class Mechanics.

The "Hand-y" feature will be found equally desirable on a Try Square, or Bevel, for the more convenient use of these Tools.

U.S.Patent 473,087 granted to
E.A.Schade, April 18,1892
Assigned to Stanley Rule & Level Company

☞ **Please insert opposite Page 35, Illustrated Catalogue, 1892.**

STANLEY'S PATENT
IMPROVED PLANE IRONS.

The improved form of this Plane Iron renders it unnecessary to detach the Cap-Iron, *at any time;* as the connecting screw will slide back to the extreme end of the slot in the Plane Iron, without the danger of falling out. The screw may then be tightened, by a turn with thumb and finger; and the Cap-iron will serve as a convenient handle, or rest, in whetting or sharpening the cutting edge of the Plane Iron.

Illustration of Plane Iron, with Cap-screw fastened just back of circular opening in the slot, and in position for using the Iron.

Illustration of Plane Iron, with Cap-screw moved to back end of the slot, and in position for sharpening the Iron, without removing the cap.

The greater convenience in using, and the security from loss of the different parts, both commend this Improved Plane Iron to all wood-workers.

☞ See Price List of Plane Irons at bottom of opposite page.

☞ Please insert opposite Page 36, Illustrated Catalogue, 1892.

STANLEY'S PATENT THROAT ADJUSTMENT.

US Patent No.515,063; J.A.Traut & Christian Bodner;
STANLEY'S   Feb.20,1894

PATENT THROAT ADJUSTMENT.

For opening or closing the throat of the Plane, as coarse or fine work may require.

By moving the Eccentric Plate to the right, the throat can be closed, as shown by the dotted line. A single turn of the Knob will fasten the Plate, and secure any desired width for the throat.

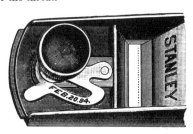

Block Planes are used largely for planing across the ends of boards, or of blocks, and in fitting close joints. The pitch of the Plane Iron is less than in ordinary Bench Planes, and the Iron is used with the bevel uppermost at the cutting edge, to further aid in cutting across the grain of wood.

A convenient adjustment for changing the width of the throat of Block Planes is especially important to wood-workers. All the Nos. (9½, 9¾, 15, 15½, 16, 17, 18 and 19) shown on page 36, opposite, are now equipped with the adjustment illustrated above.

WOODWORKER'S HANDY ROUTER PLANE.
WITH CLOSED THROAT.

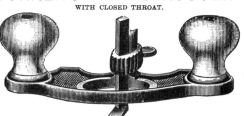

No. 71½.   Iron Stock, with Steel Bits (¼ and ½ inch)................$1.25 Each.

☞ Please insert as Page 46½ (opposite Page 46), Illustrated Catalogue, 1892.

☞ **Please insert opposite Page 40, Illustrated Catalogue, 1892.**

MISCELLANEOUS PLANES, ETC.   40½

☞ Below is an illustration of the Improved No. 45 Beading, Rabbet and Slitting Plane. An additional Plow and Dado Bit (¾ inch) and a Sash Tool, are now sent out with the Plane, making *Twenty Tools* in all. The cutters are adjusted by a screw, and also the Depth Gauge has a screw adjustment; both features adding greatly to the precision with which the Plane can be worked.

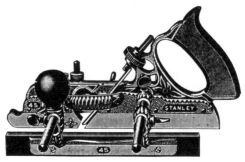

A removable Rosewood Face is attached to the Fence. The Plane is Nickel Plated; and is sent out already set up for working, packed in a neat Wooden Box.
See opposite (page 40) for general description and price.

SIDE RABBET PLANE.

A convenient tool for side-rabbeting and trimming dados, mouldings and grooves of all sorts. A reversible nose-piece will give the tool a form by which it will work close up into corners when required.

No. **98.**   Side Rabbet Plane, 4 inches in length............ **$0.90**

RABBET AND BLOCK PLANE.

A detachable side will easily change this tool from a Block Plane to a Rabbet Plane, or *vice versa*. The cutter is set on a *skew*.

No. **140.**   Rabbet and Block Plane, with detachable side,
7 inches in length, 1¾ inch cutter............ **$1.25**

40¾   CORE-BOX PLANE, ETC.

CORE-BOX PLANE.

A Tool much needed by Pattern-Makers, Wheelwrights, and others, for planing out semi-circles.

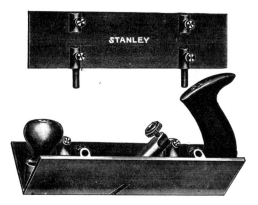

No. **57.** Core-Box Plane, for semi-circles, up to 2½ inch diameter.................................... **$3.00**

This Plane is constructed so that the sides can be extended by additional sections, 2½ inch wide, until a diameter of 10 inches can be worked, if desired.

The price for these additional Sections............per pair, **1.00**

STANLEY'S "ODD JOBS."

The Tool is now sent out with a 12 inch Graduated Ruler inserted in it; and near one end of the Ruler is an Adjustable Steel Point. This addition greatly facilitates the use of this unique tool, already favorably known to Mechanics, Amateurs and Housekeepers.

See Catalogue, page 49, for general description and price.

STANLEY'S UNIVERSAL HAND BEADER.

We are now sending out these Tools Nickel Plated, making them more than ever attractive to buyers. The Hand Beaders have an extensive sale on their merits as a practical tool.

See Catalogue, page 47, for general description and price.

☞ Please insert opposite Page 42, Illustrated Catalogue, 1892.

STANLEY'S IMPROVED VICTOR CIRCULAR PLANE, ETC.

STANLEY'S
IMPROVED VICTOR CIRCULAR PLANE.

The Flexible Steel Face of the Plane can be made concave, or convex, by turning the screw which is attached to its centre.

This Plane is an improved form of the "Victor" Adjustable Plane (page 46 of this Catalogue), and is furnished with our latest adjustments, both longitudinal and lateral, together with "Stanley's Patent Improved Plane Irons."

No. **20.**   Adjustable Circular Plane, Nickeled, 1¾ in. Cutter, **$6 00**

STANLEY'S IMPROVED SCRUB PLANE.

This Tool has a single Iron, with the cutting edge rounded; and is particularly adapted for roughing down work before using a Jack or other Plane.

No. **40.**   Iron Stock, 9½ inches in length, 1¼ in. Cutter **$1 00**

MEMORANDUM.

Improvements have been made in several of the Tools embraced in the pages of our latest Illustrated Catalogue (1892).

Planes Nos. 9½, 9¾, 15 and 15½ are now furnished with "Stanley's Patent Lateral Adjustment."

Router Plane No. 71 is accompanied by an additional appliance, which largely increases its usefulness to woodworkers.

Miscellaneous Planes Nos. 45, 46, 47, 50 and 71 are now Nickel-Plated; also, Improved Try Squares No. 12; Eureka Bevels No. 18; and our Magnetic and other Tack Hammers.

☞ Please insert opposite Page 44, Illustrated Catalogue, 1892.

STANLEY'S PATENT UNIVERSAL PLANE.

STANLEY'S PATENT UNIVERSAL PLANE.

This Tool, in the hands of an ordinary carpenter, can be used for all lines of work covered by a full assortment of so-called Fancy Planes.

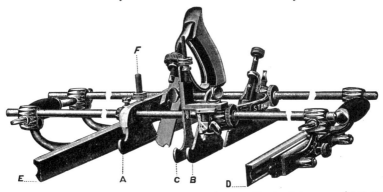

No. 55. Stanley's Universal Plane, with 52 Tools, Bits, etc. **$16.00**

☞ The Plane is Nickel Plated; the 52 Cutters are arranged in four separate cases; and the entire outfit is packed in a neat Wooden Box.

This Plane consists of :—

A MAIN STOCK (A) with two sets of transverse sliding arms, a depth gauge (F) adjusted by a screw, and a Slitting Cutter with stop.

A SLIDING SECTION (B) with a vertically adjustable bottom.

THE AUXILIARY CENTER BOTTOM (C) is to be placed in front of the Cutter, as an extra support, or stop, when needed. This bottom is adjustable both vertically and laterally.

FENCES (D) AND (E). Fence D has a lateral adjustment by means of a screw, for extra fine work. The Fences can be used on either side of the Plane, and the rosewood guides can be tilted to any desired angle, up to 45°, by loosening the screws on the face. Fence E can be reversed for Center Beading wide boards.

AN ADJUSTABLE STOP to be used in beading the edges of matched boards is inserted on left hand side of sliding section (B).

Each Plane is sold with **Fifty-two (52) Cutters,** marked ◆ on within list. The other Cutters on this list are kept in stock and will be furnished to order.

PRICES OF CUTTERS.   Cts. each.

Nos. 10, 11, 12, 13, 21, 22, 23	**15**
Nos. 14, 15, 16, 17, 18, 24, 25, 42, 43, 44, 45, 46, 47, 52, 53, 54, 55, 56, 57, 212, 222, 232	**20**
Nos. 9, 19, 26, 27	**25**
Nos. 8, 28, 29, 31, 32, 33, 34, 35, 36, 37, 38, 213, 223, 233	**30**
Nos. 214, 224, 234	**40**
Nos. 61, 62, 63, 71, 72, 73, 81, 82, 83, 91, 92, 93, 101, 102, 103, 111, 112, 113	**45**
Nos. 1, 2, 5, 6, 64, 65, 66, 74, 75, 76, 84, 85, 86, 94, 95, 96, 104, 105	**50**
Nos. 106, 114, 115, 116, 215, 225, 235	

☞ Special Cutters will be made to order; or Blanks from which the workman can file up any form he requires, may be ordered.

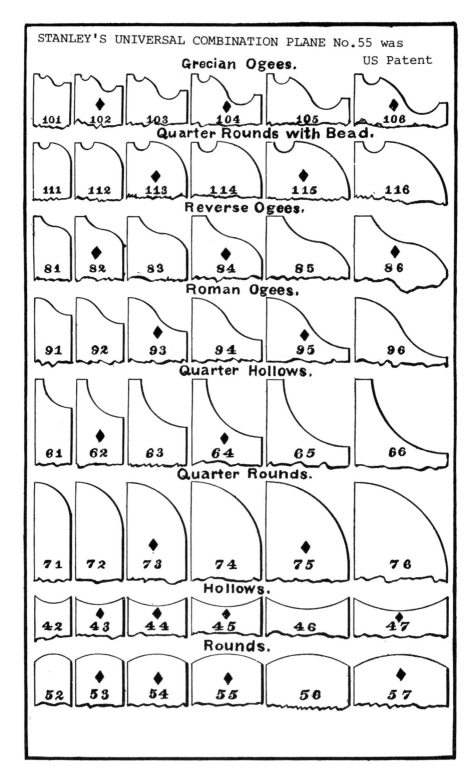

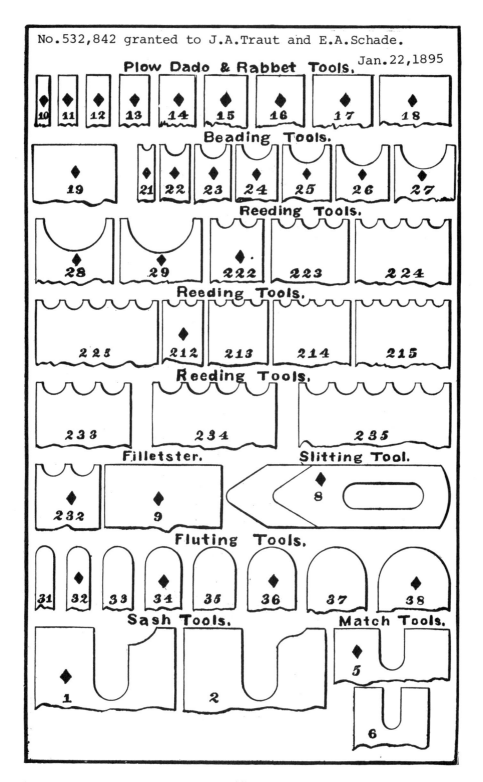

STANLEY'S PATENT UNIVERSAL PLANE.

DIRECTIONS.

MOULDING PLANE.—Insert the desired Cutter, and adjust bottom of Sliding Section (B) to conform to the shape of the Cutter; then fasten this Section firmly by means of the check-nuts on the transverse arms. When needed, adjust Auxiliary Center Bottom (C) for an additional support in front of the Cutter. By tilting the rosewood guides on Fences D and E, mouldings of various angles may be formed.

MATCH, SASH, BEADING, REEDING, FLUTING, HOLLOW, ROUND, PLOW, RABBET AND FILLETSTER PLANE.—Use in same manner as for Mouldings. In working Match and Sash Cutters, the Auxiliary Center Bottom (C) may be used as a stop.

DADO.—Remove the Fences (D and E) and set the spurs parallel with the edges of Cutter. Insert long Adjustable Stop on left-hand side of Sliding Section.

SLITTING PLANE.—Insert the Cutter and Stop on right-hand side of Main Stock and use Fence D or E for guide.

CHAMFER.—Insert the desired Cutter; fasten a Fence on each side of the Plane and tilt the rosewood guides to the desired angle.

The No.67 Universal Spokeshave illustrated below first appeared on Insert Oage 46-1/2 in 1892.

UNIVERSAL SPOKE SHAVE.

This Spoke Shave has two detachable bottoms, adapting it equally well to circular work or straight; and, by means of a movable width-gauge, the tool can be used in rabbeting.

Both Handles are detachable, and either of them can be screwed into a socket on top of the stock, thus enabling the owner to work into corners, or panels, as no other spoke shave can do.

Each.

No. **67.** Universal Spoke Shave, for curved or straight work............**$1.50**

STANLEY'S IMPROVED
MARKING AND MORTISE GAUGE.

This Gauge is made of metal, and has two graduated bars. The steel points are attached very near the ends of the bars, to admit of being used close up into a rabbet, or corner.

Each.

No. **91.** Nickel-plated Marking and Mortise Gauge, one in box, **$0 65**

STANLEY'S
IMPROVED BUTT GAUGE.

A Metallic Butt Gauge, having one bar with two steel cutters fixed upon it. When the cutter at the outer end of this bar is set for gauging on the edge of the door, the cutter at the inner end of the bar is already set for gauging from the back of the jamb. The other bar has a steel cutter to accurately gauge for the thickness of the butt.

The form of this Tool is convenient for carrying in the pocket. It is so constructed that the bars cannot fall out of the stock.

Each.

No. **95.** Nickel-plated Butt Gauge, one in a box **$0 75**

☞ **Please insert as Page 50½ (opposite Page 50), Illustrated Catalogue, 1892.**

The No. 22 Dowel Sharpener was added to the first printing of the 1892 Catalogue, page 55.

WHEELER'S PATENT COUNTERSINK.
FOR WOOD.

This Countersink works equally well for every variety of screw. The Countersink cuts rapidly, and is easily sharpened by drawing a thin file lengthwise inside of the cutter. By fastening the Gauge at a given point, any number of screws may be driven so as to leave the heads flush with the surface, or at a uniform depth below it.

No.		Per doz.
18.	Countersinks, ½ dozen in a box	$3 00
20.	Countersinks, with Gauge, ¼ dozen in a box	4 50

STANLEY'S IMPROVED
DOWEL SHARPENER.

This Tool is the reverse of the Countersink, and can be sharpened with equal ease. By its use Dowel Pins can be sharpened rapidly and uniformly.

No.		Per Doz.
22.	Dowel Sharpener, ½ dozen in a box	$3 00

The No. 92 Improved Butt and Rabbet Gauge first appeared on Page 51 of the 1892 Catalogue.

STANLEY'S BUTT AND RABBET GAUGE.
For Hanging Doors, Mortising, Marking, etc.

This Gauge has two bars, both of brass, one movable within the other; also three blades or markers. The two steel blades or markers at the extreme end of the inner bar, can be moved to any position by means of the thumb-screw at opposite end of the Gauge.

A Rosewood Head on the outer bar makes the ordinary Marking, or Slitting Gauge. At one end of the outer bar a steel plate or stop is attached; and, in connection with the Rosewood Head set at the proper distance from the plate, a double head for Mortising is made.

When the Head and Plate have been adjusted to the desired width for the mortise, both markers can still be moved by means of the thumb-screw, and the mortise lines set further in from the edge of the work in hand, or brought out nearer to the edge, without disturbing the width of the mortise.

A superior Butt Gauge, for hanging doors, is made by bringing the marker on the upper side of the bar back of, and to any required distance from the outer surface of the steel plate.

Turn the Gauge over, and place this outer surface against the back of the Rabbet in the door-casing. The owner can then gauge correctly for the back leaf of the Butt; and, without changing any of the parts, the inner surface of the steel plate will act as a guide for the other marker in gauging on the edge of the door, for the front leaf of the Butt. The thickness of the plate (1-16th inch), will set off the door sufficiently to clear the jamb, in opening and shutting.

The Rosewood Head should be secured at the end of the bar nearest the thumb-screw, when the Tool is being used as a Butt Gauge. By placing one edge of a Butt between the surface of this Head and the small blade which stands up back of it, the exact thickness of the Butt is ascertained, and can be accurately gauged on both the door-post and the door. The movable Steel Plate affords a perfect guide for gauging in, or near, Rabbets of all kinds.

No.		Per Doz.
92.	Improved Butt and Rabbet Gauge, ¼ dozen in a box	$15 00

1876 LEONARD BAILEY & Co. CATALOG and PRICE LIST

Leonard Bailey was born May 8, 1825, in Hollis, New Hampshire, the third child of Leonard and Mary Friend Bailey. He married Elizabeth Ann Wildes in 1848 in Massachusetts. (She was born in Winchester, Mass., Sept. 1827.) Bailey lived in Winchester while managing his iron plane firm, L. Bailey & Co., at 73 Haverhill Street in Boston, 1861 – 1863.

He then moved to Boston, continuing his firm there through 1867. In 1867 this became Bailey, Chany & Co. On May 19, 1869, he licensed seven of his patents for making iron planes, spokeshaves, and a veneer scraper (N°s 13,381; 20,615; 20,855; 21,311; 55,599; 67,398; and 72,443). They are described and illustrated by William B. Hilton, "A Check List of Boston Plane Makers," *The Chronicle of the EAIA,* Vol. 27, No. 2, June 1974, pp. 24–25. He thereupon moved to New Britain, Conn., to supervise the manufacture of these planes and spokeshaves at the Stanley Rule & Level Co. After his contract was suddenly terminated with SR&L Co. on September 24, 1873, he re-established himself as L. Bailey & Co. at Cushman Street in Hartford, in the factory of the Hartford Screw Co. He introduced another line of iron planes, which he called the "VICTOR" Line, Registered trade mark #3,299 granted Jan. 4, 1876.

His firm received the highest award at the 1876 Centennial Exhibition at Philadelphia, and also a Premium Award Gold Medal at the Connecticut State Fair, also held in 1876. By October 1, 1876, his products were handled by Sargent & Co. of New York as sole agent. On January 26, 1880, Stanley Rule & Level became agents for these planes and his other tools as noted in their 1879 through 1884 Catalogs. On July 16, 1884, SR&L Co. purchased the tool line of L. Bailey & Co. and removed the machinery and stock to New Britain, noting in its 1888 Catalog that these planes were "formerly manufactured by L. Bailey & Co., Hartford". Bailey remained in Hartford, continuing to manufacture and sell his patented copying press and allied line of printing materials.

In 1888 he removed his factory and residence to the adjacent village of Wethersfield. However he maintained a Post Office Box at Hartford through 1897, and opened a New York sales office in 1891. A list of his patents attest to his versatility as an inventor of both tools and printing equipment. A trade Catalog concerning his printing equipment, believed to have been published in the mid-1800s is in the collection of the Connecticut Historical Society. Leonard Bailey died while visiting in New York on February 2, 1905, months before his 80th birthday.

Ken Roberts
1981

LEONARD BAILEY & CO'S

Patent Adjustable Iron

BENCH PLANES,

Try Squares and Bevels.

Manufactured by

L. BAILEY & CO.,

HARTFORD, CONN.

SARGENT & CO., - - - Sole Agents,

37 CHAMBERS STREET, NEW YORK.

FOR SALE BY ALL HARDWARE DEALERS.

Oct. 1, 1876.

INTRODUCTORY.

The "VICTOR PLANES" are the most simple, compact, and practical adjustable Planes ever yet produced, being the natural outgrowth and result of twenty years' exclusive experience in the production of Patent Planes, and we have so much confidence in their merits, that we have no hesitation in presenting them to the public, and to trust them to find their way to public favor.

BAILEY'S PATENT HARDENED TRY SQUARES and PATENT FLUSH T BEVELS, although comparatively new in the market, have received high commendation from those who have used them. Their accuracy, style, and finish, equal, at a less cost, the best in the market.

All opportunities for the improvement of these tools have been sought, and no device that could add to their usefulness has been neglected.

We desire to call attention to the fact that these Tools are made under the direct control and supervision of Mr. Leonard Bailey, the original inventor of L. Bailey's Patent Adjustable Iron and Wood Bench Planes.

All Goods in this Catalogue are shipped from our Factory, and fully warranted.

We shall endeavor, by a rigid adherence to the policy of producing nothing but the best goods, at a low but reasonable profit, and by keeping the quality at the highest attainable point of perfection, to merit the patronage you may favor us with.

Respectfully,

L. BAILEY & CO.

Mr. R. C. Graves will mainly represent us with the trade. His practical knowledge of this branch of the business, we trust, will entitle him to your confidence.

All orders received by him, or at the Factory, will meet with prompt attention.

VICTOR PLANES.

All our Planes have our Trade Mark.   Patented Jan. 4, 1876.

PATENT IRON PLANES.

		EACH.
No. 0.	Block Plane, 7 inches in length, 1¾ inch Cutter, Japanned Finish,	$1.00
No. 00.	Block Plane, 7 inches in length, 1¾ inch Cutter, Nickel Plated Trimmings,	1.25
No. 000.	Block Plane, 7 inches in length, 1¾ inch Cutter, Nickel Plated Trimmings, with Adjustment,	1.50

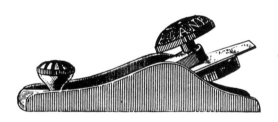

		EACH.
No. 1.	Block Plane, adjustable Mouth and Cutter, 6 inches in length, 1¾ inch Cutter, Polished Trimmings,	$2.00
No. 1¾.	Block Plane, adjustable Mouth and Cutter, 6 inches in length, 1¾ inch Cutter, Nickel Plated Trimmings,	2.25
No. 2.	Block Plane, adjustable Mouth and Cutter, 7 inches in length, 1¾ inch Cutter, Polished Trimmings,	2.25
No. 2¾.	Block Plane, adjustable Mouth and Cutter, 7 inches in length, 1¾ inch Cutter, Nickel Plated Trimmings,	2.50

☞ Complete adjustment, attachable to our No. 0 and 00 Plane, can be applied anytime, without any change of construction, at the cost of 25 cents.

VICTOR PLANES.

All our Planes have our Trade Mark. Patented Jan. 4, 1876.

PATENT IRON PLANES.

EACH.

No. 1¼. Block Plane, with Handle, adjustable Mouth and Cutter, 6 inches in length, 1¾ inch Cutter, Polished Trimmings, ... $2.25

No. 1½. Block Plane, with Handle, adjustable Mouth and Cutter, 6 inches in length, 1¾ inch cutter, Nickel Plated Trimmings, ... 2.50

No. 2¼. Block Plane, with Handle, adjustable Mouth and Cutter, 7 inches in length, 1¾ inch Cutter, Polished Trimmings, ... 2.50

No. 2½. Block Plane, with Handle, adjustable Mouth and Cutter, 7 inches in length, 1¾ inch Cutter, Nickel Plated Trimmings, ... 2.75

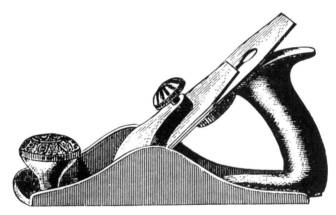

EACH.

No. 3. Smooth Plane, with Adjustment, 8½ inches in length, 1¾ inch Cutter, Polished Trimmings, . . . $4.00

No. 3½. Smooth Plane, with Adjustment, 8½ inches in length, 1¾ inch Cutter, Nickel Plated Trimmings, . . 4.50

No. 4. Smooth Plane, with Adjustment, 9 inches in length, 2 inch Cutter, Polished Trimmings, ... 4.50

No. 4½. Smooth Plane, with Adjustment, 9 inches in length, 2 inch Cutter, Nickel Plated Trimmings, . . . 5.00

VICTOR PLANES.

All our Planes have our Trade Mark. Patented Jan. 4, 1876.

PATENT IRON PLANES.

		EACH.
No. 5.	Jack Plane, with Adjustment, 14 inches in length, 2 in. Cutter, Polished Trimmings,	$5.00
No. 5½.	Jack Plane, with Adjustment, 14 inches in length, 2 in. Cutter, Nickel Plated Trimmings,	5.50
No. 6.	Fore Plane, with Adjustment, 18 inches in length, 2⅜ inch Cutter, Polished Trimmings,	6.00
No. 6½.	Fore Plane, with Adjustment, 18 inches in length, 2⅜ inch Cutter, Nickel Plated Trimmings,	6.50
No. 7.	Jointer Plane, with Adjustment, 22 inches in length, 2⅜ inch Cutter, Polished Trimmings,	7.00
No. 7½.	Jointer Plane, with Adjustment, 22 inches in length, 2⅜ inch Cutter, Nickel Plated Trimmings,	7.50
No. 8.	Jointer Plane, with Adjustment, 24 inches in length, 2⅝ inch Cutter, Polished Trimmings,	8.00
No. 8½.	Jointer Plane, with Adjustment, 24 inches in length, 2⅝ inch Cutter, Nickel Plated Trimmings,	8.50
No. 9.	Carriage Makers' Rabbet Plane, with Adjustment, 14 inches in length, 2⅛ inch Cutter, Polished Trimmings,	5.50
No. 9½.	Carriage Makers' Rabbet Plane, with Adjustment, 14 inches in length, 2⅛ inch Cutter, Nickel Plated Trimmings,	6.00

The Victor Planes are all made with solid Bed, the Cutter Seat being cast in one piece with the Stock.

VICTOR PLANES.

All our Planes have our Trade Mark. Patented Jan. 4, 1876.

PATENT IRON PLANES.

EACH.

No. 10. Circular Plane, with Adjustment, Flexible Adjustable Steel Face, 1¾ in. Cutter, Nickel Plated Trimmings, $4.75

This Plane will work a circle less in diameter than any other Flexible Face Plane.

PLANE IRONS.

SINGLE.

Inches,	1¾	2	2⅛	2⅜	2⅝
Each,	38	42	44	48	50 cts.

DOUBLE.

Inches,	1¾	2	2⅛	2⅜	2⅝
Each.	62	68	74	80	84 cts.

Block Plane Irons, 1¾ inch, 35 cts. each.

These Plane Irons are made of the best English Steel, hardened and tempered by an improved process. Each Cutter is sharpened and put in perfect working order before leaving the Factory, and fully warranted.

VICTOR PLANES.

All our Planes have our Trade Mark. Patented Jan. 4, 1876.

PATENT IRON PLANES.

Bailey's Patent Combination Plane.

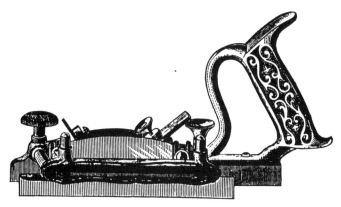

No. 14.

Plow, Filletster, Back Filletster, Dado, Rabbet Plane, and Matching Plane Combined.

Price, - - $5.50.

The main feature of this Tool is, that it has a fence or guide, which is made to change from one side to the other as the nature of the work requires, same fence or guide also in itself being vertically self-adjusting. The same cutter is used for each special tool in the combination.

We recommend this ingenious tool as most desirable and useful for all wood-working mechanics, who require in a compact form a tool designed to accomplish practically the result expected of each of the Planes named above.

As an Adjustable Dado it has no equal, as such it will cut a groove from ⅜ inch in width up to 2 inches or more, without change of cutter.

Bailey's Patent Try Squares.

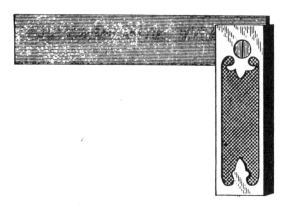

PRICES.

3 inch,	per doz., $18.00
4½ inch,	" 21.00
6 inch,	" 24.00
8 inch,	" 30.00

Half dozen in a box.

These Try Squares are what might be termed adjustable, the Handle and the Blade being connected together by means of one screw, passing through both parts; this screw is beveled or tapered where it passes through the Blade, bearing only on the lower edge of the hole in the same, which hole is also fitted to match the taper on the screw, thereby forcing the Blade, when the screw is turned home, firmly to the shoulder in the Handle.

The Handle of these Squares are made of Cast Iron, highly finished, and the Blade is made of the best Steel, hardened and tempered by an improved patented process.

Bailey's Patent Flush T Bevel.

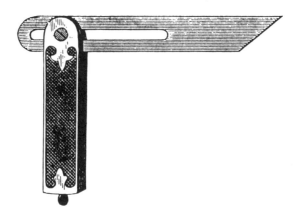

PRICES.

8 inch, . . per doz., $14.40
10 inch, . .   "   16.33

Half dozen in a box.

The method of securing the Blade at any angle desired is simple, yet effective. Fig. 2 shows the internal construction of the Handle. By moving the thumb-piece at the lower end of the Handle, the long lever acts upon the shorter one (being set upon a strong pivot). The strength of a compound lever is produced by this arrangement, and the short lever being attached to a Nut inside the upper end of the Handle, operates as a wrench to turn it upon the screw, and thus to fasten or release the Blade at the pleasure of the owner.

The Handle of this Bevel is made of Cast Iron. The Blade is of fine quality Steel, spring temper, and with perfectly parallel edges.

Fig. 2.

Automatic Boring Tool.

JOHNSON & TAINTER'S PATENT.

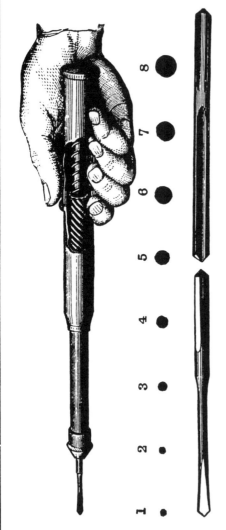

This instrument is designed for **Boring Wood** for various purposes, such as for setting BRADS, FINISHING NAILS, SCREWS, &c. Eight Bits (or Drills, sizes indicated by dots and numerals above) accompanying each Tool. It can be used in many places where the bit-brace, gimlet, or brad-awl cannot, and is superior to either for the purposes mentioned. Piano-Forte and Organ Builders, as well as Cabinet Makers and House Carpenters, who have used them, all attest their great value as a tool for doing work **rapidly** and **perfectly.**

When Nos. 1, 2, and 3 are sharpened for drilling metals, they will work with satisfaction in **brass** and **compositions;** which makes it a useful Tool for Dentists, Jewelers, and, in fact, **all Artisans.**

PRICE.
PACKED IN A TIN CASE.

Each, ...	$2.75
Extra Drills, per dozen, assorted,	1.00

STEPHENS & CO'S
BOXWOOD AND IVORY RULES.

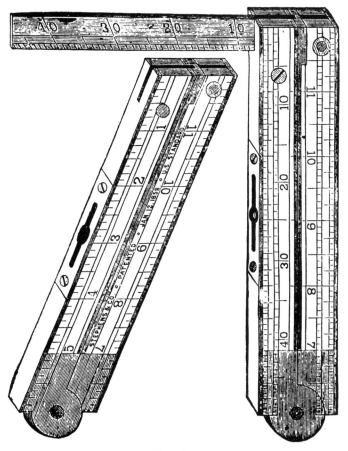

No. 36.

Boxwood Rules.

Two Foot, Two Fold.

Old Nos.		Width.	S. & Co's Nos.	Price Per Doz.
29	Round Joint,	1½ in.	1	$3.50
18	Square Joint,	"	2	5.00
—	Square Joint, Bound,	"	4	15.00
22	Square Joint, Board Measure,	"	5	8.00
26	Square Joint, Plain Slide,	"	9	9.00
2	Arch Joint,	"	13	8.00
12	Arch Joint, Gunter's Slide,	"	14	14.00
4	Arch Joint, Extra Thin,	"	15	10.00
6	Arch Joint, Gunter's Slide, Engineers,	"	16	18.00
5	Arch Joint, Bound,	"	17	16.00
—	Arch Joint, Board Measure,	"	18	10.00
—	Arch Joint, Board Measure, Bound,	"	22	18.00
—	Arch Joint, Plain Slide,	"	23	11.00
15	Arch Joint, Gunter's Slide, Bound,	"	27	24.00
16	Arch Joint, B'd, Gunter's Slide, Eng'rs,	"	28	28.00

The Gunter's and Engineer's Slide Rules are graduated with great care into 8ths, 10ths, and 16ths of inches, and are provided with Drafting and Octagonal Scales, and 100ths of a foot.

Miscellaneous Rules.

43	Ship Carpenters' Bevels, Single Tongue,	½ in.	30	6.00
42	Ship Carpenters' Bevels, Doub. Tongue,	"	31	6.00
66½	Yard Rules, Four Fold, 1-8 and 1-16 in.,	1 in.	32	8.00
58	Six Fold, Boxwood,	¾ in.	33	13.00
—	Six Fold, Ivory,	"	34	80.00
—	Six Fold, Ivory, Bound,	"	35	100.00
—	Shrinkage, Two Fold,	1½ in.	37	13.00
98	School Rules, Beveled Edge,	¾ in.	38	1.25
—	Architects' Rules, Beveled Edges,	1 in.	39	12.00

Patent Combination Rules.

Boxwood, Brass Bound, net, . . 1⅜ in. 36 20.00

This embraces a Rule, Level, Square, Plumb, Bevel, Inclinometer, &c.

See cut.

Boxwood Rules.

Two Foot, Narrow, Four Fold.

Old Nos.		Width.	S. & Co's Nos.	Price Per Doz.
68	Round Joint, ...	1 in.	41	$4.00
61	Square Joint, ...	"	42	5.00
84	Square Joint, Half Bound, . . .	"	42¼	12.00
62	Square Joint, Bound, ...	"	42½	15.00
—	Square Joint, Bound, Extra Narrow, .	¾ in.	42¾	14.00
61½	Square Joint, Extra Narrow, . .	"	44	6.00
63½	Square Joint, Ex. Narrow, Edge Plates,	"	44½	8.00
63	Square Joint, Edge Plates, . . .	1 in.	45	7.00
51	Arch Joint, ...	"	46	6.00
53	Arch Joint, Edge Plates, . . .	"	48	8.00
54	Arch Joint, Bound, ...	"	49	16.00
52	Arch Joint, Half Bound, . . .	"	49¼	13.00
59	Double Arch Joint, ...	"	50	9.00
60	Double Arch Joint, Bound, . . .	"	52	21.00

Boxwood Rules.

Two Foot, Broad, Four Fold.

67	Round Joint, ...	1⅜ in.	53	5.00
70	Square Joint, ...	"	54	7.00
—	Square Joint, Half Bound, . . .	"	54¼	15.00
72½	Square Joint, Bound, ...	"	54½	18.00
72	Square Joint, Edge Plates, . . .	"	56	9.00
73	Arch Joint, ...	"	57	9.00
75	Arch Joint, Edge Plates, . . .	"	59	11.00
76	Arch Joint, Bound, ...	"	60	20.00
—	Arch Joint, Half Bound, . . .	"	60¼	16.00
77	Double Arch Joint, ...	"	61	12.00
78½	Double Arch Joint, Bound, . . .	"	63	24.00

BOARD MEASURE.

79	Square Joint, Edge Plates, . . .	1⅜ in.	64	11.00
—	Square Joint, Bound, ...	"	65	20.00
81	Arch Joint, Edge Plates, . . .	"	66	13.00
82	Arch Joint, Bound, ...	"	67	22.00

Boxwood Rules.

One Foot—Pocket.

Old Nos.		Width.	S. & Co's Nos.	Price Per Doz.
69	Round Joint, ...	⅝ in.	70	$3.00
65	Square Joint, ...	"	71	3.50
64	Square Joint, Edge Plates, . . .	"	72	5.00
65½	Square Joint, Bound, ...	"	72½	11.00
55	Arch Joint, ...	"	73	4.00
56	Arch Joint, Edge Plates, . . .	"	74	6.00
57	Arch Joint, Bound, ...	"	75	12.00

Ivory Rules.

Two Foot, Four Fold.

85	Square Joint, Edge Plates, Ger. Silver,	¾ in.	77	54.00
—	Square Joint, Bound, German Silver, .	⅞ in.	78	72.00

Ivory Rules.

Two Foot, Four Fold.

86	Arch Joint, Edge Plates, Ger. Silver,	1 in.	83	64.00
87	Arch Joint, Bound, German Silver, .	"	84	80.00
89	Double Arch Joint, Bound, Ger. Silver,	"	86	92.00

Ivory Rules.

One Foot—Pocket.

Old Nos.		Width.	S. & Co's Nos.	Price Per Doz.
90	Round Joint, Brass,	½ in.	89	$10.00
—	Round Joint, German Silver,	"	89½	12.00
—	Square Joint, Brass,	"	90	12.00
92½	Square Joint, German Silver,	"	90½	14.00
92	Square Joint, Edge Plates, Ger. Silver,	"	91	17.00
—	Square Joint, Bound, German Silver,	⅝ in.	92	28.00
88½	Arch Joint, Edge Plates, Ger. Silver,	"	93	21.00
88	Arch Joint, Bound, German Silver,	"	94	32.00

Caliper Rules.

Boxwood.

		Width.	S. & Co's Nos.	Price Per Doz.
36	Two Fold, 6 Inch, Plain,	⅞ in.	95	7.00
—	Two Fold, 6 Inch, Brass, Case,	"	96	8.00
—	Two Fold, 6 Inch, Bound,	"	97	12.00
32	Four Fold, 12 Inch, Plain,	"	98	13.00
—	Four Fold, 12 Inch, Bound,	"	99	20.00
—	Two Fold, 6 Inch, Plain,	1⅜ in.	100	12.00

Caliper Rules.

Ivory.

		Width.	S. & Co's Nos.	Price Per Doz.
38	Two Fold, 6 Inch, Ger. Silver, Plain,	⅞ in.	95½	15.00
—	Two Fold, 6 Inch, Ger. Silver, Case,	"	96½	18.00
—	Two Fold, 6 Inch, Bound, Ger. Silver,	"	97½	30.00
39	Four Fold, 12 Inch, Ger. Silver, Plain,	"	98½	38.00
40	Four Fold, 12 Inch, Bound, Ger. Silver,	"	99½	48.00

STRATTON BROS. PATENT ADJUSTABLE PLUMBS AND LEVELS.

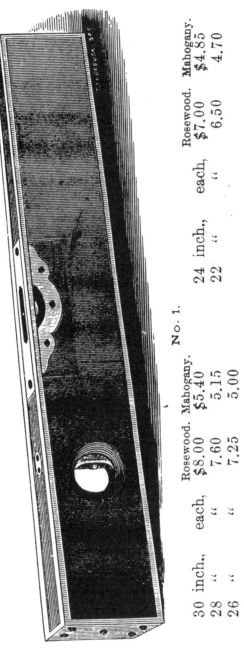

No. 1.

		Rosewood.	Mahogany.
30 inch.,	each,	$8.00	$5.40
28 "	"	7.60	5.15
26 "	"	7.25	5.00
24 inch.,	each,	$7.00	$4.85
22 "	"	6.50	4.70

The engraving represents a No. 1 Adjustable Plumb and Level. The corners are protected from injury, and preserved true and perfect by Brass Rods extending the entire length of the Level, securely attached to the wood.

These Levels have two ornamental Brass Lipped Side Views, heavy Brass Top and End Plates, and are made of the best thoroughly seasoned Mahogany and Rosewood, Polished.

Patented March 1, 1870, and July 16, 1872.

PLUMBS AND LEVELS.

Machinist.

		Rosewood.
12 inch,	each,	$2.60
10 inch,	"	2.40
8 inch,	"	2.00
6½ inch, without Plumb,	"	1.10

The Machinist Levels have Brass Side Views, heavy Brass Top and End Plates and Brass Corners. Rosewood, Polished.

Fig. 2.

The corners of the No. 1 Level are protected by Brass rods about ¼ inch square, extending the entire length, and attached to work as shown in Fig. 2. The manner of adjusting is entirely different, and considered much better than any other in use.

It is not affected by heat or cold, wet or dry weather. The wood is selected rosewood and mahogany. For workmanship they are not equaled. Mechanics who have used them are giving them the preference, and without an exception say they are the best in the market.

PLUMBS AND LEVELS.

No. 2.

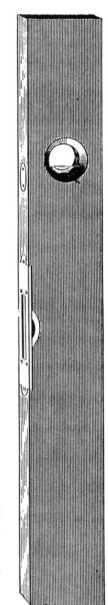

Nos. 4 and 5.

		EACH.
No. 2.	Patent Adjustable, Mahogany Plumb and Level, two ornamental Brass Lipped Side Views, heavy Circular End Top Plate, Tipped, Polished, 26, 28, and 30 inches,	$3.60
No. 3.	Patent Adjustable, Mahogany Plumb and Level, two ornamental Brass Lipped Side Views, heavy Circular End Top Plate, Tipped, Polished, 26, 28, and 30 inches.	3.00
No. 4.	Patent Adjustable, Mahogany Plumb and Level, two Side Views, heavy Circular End Top Plate, Tipped, Polished, 26, 28, and 30 inches,	2.10
No. 5.	Patent Adjustable, Cherry Plumb and Level, two Side Views, heavy Circular End Top Plate, Polished, 26, 28, and 30 inches.	1.50

PLUMBS AND LEVELS.

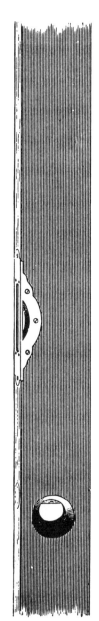

No. 6.

		EACH.
No. 6.	Patent Adjustable, MASON'S, Mahogany Plumb and Level, two ornamental Brass Lipped Side Views, heavy Circular End Top Plate, Polished 36 inches.	$3.25
No. 7.	Patent Adjustable, MASON'S, Mahogany Plumb and Level, two Side Views, heavy Circular End Top Plate, Polished, 36 inches,	2.50
No. 8.	Patent Adjustable, MASON'S, Cherry Plumb and Level, two Side Views, heavy Circular End Top Plate, Polished, 36 inches,	2.00

Double Plumb, 50 cents extra.

DIPLOMA AWARDED AT THE MECHANICS' FAIR, BOSTON, 1874.

The Glasses in all these Levels are set SOLID and IMMOVABLE in the stock, and are adjusted with a MOVEABLE BAR, which is PERFECT and RELIABLE in its operation.

H. Hammond's
Adze-Eye Cast Steel Hammers.

Adze-Eye Riveting Hammers.

Weight per single Hammer.				Price per doz.
No. 1,	0 lbs.	4 ozs.,	. . . .	$3.50
" 2,	0 "	8 "	. . . .	5.00
" 3,	0 "	12 "	. . . .	6.00
" 4,	1 "	0 "	. . . .	8.00
" 5,	1 "	4 "	. . . .	10.00
" 6,	1 "	8 "	. . . .	12.00

Adze-Eye Machinist Ball Pene.

Weight per single Hammer.				Price per doz.
No. 1,	0 lbs.	4 ozs.,	. . . .	$12.00
" 2,	0 "	8 "	. . . .	12.00
" 3,	0 "	12 "	. . . .	12.00
" 4,	1 "	0 "	. . . .	13.00
" 5,	1 "	4 "	. . . .	14.00
" 6,	1 "	8 "	. . . .	16.00

Adze-Eye Farriers' Hammers.

No. 1,   8 ozs., . . per doz., $8.00

H. HAMMOND'S ADZE-EYE CAST STEEL HAMMERS.

ADZE-EYE NAIL HAMMERS.

	Weight per Single Hammer.		Price.		
No. 1,	0 lbs.,	7½ ounces,	per dozen,	$6.50
" 2,	0 "	13 "	"	"	8.50
" 3,	1 "	0 "	"	"	10.00
" 4,	1 "	4 "	"	"	11.25

H. HAMMOND'S ADZE-EYE CAST STEEL HAMMERS.

ADZE-EYE, BELL-FACE HAMMERS.

	Weight per Single Hammer.		Price.	
No. 1,	0 lbs.	7 ounces.	per dozen,	$6.50
" 2,	0 "	12 "	"	8.50
" 3,	1 "	0 "	"	10.00
" 4,	1 "	3 "	"	11.25

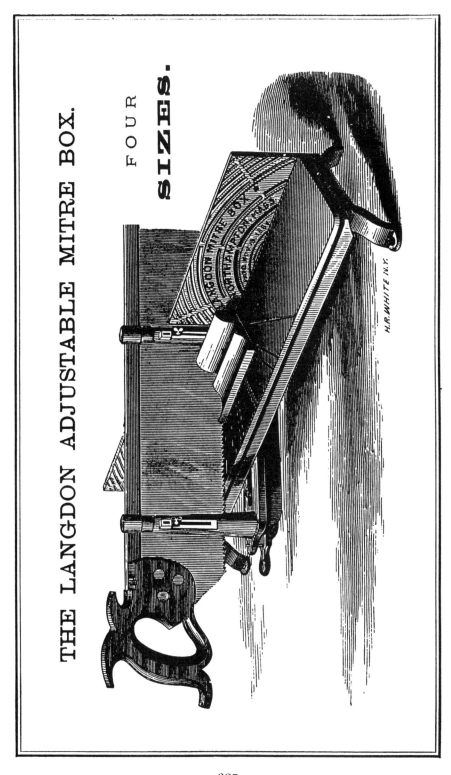

Retail Price of Mitre Boxes.

Size **A,** Designed for Printers, Paper Hangers, Small Pattern Makers, &c.

This size Box, when once used by Printers, will be found almost indispensable, as it is provided with two Saws, one for metal and one for wood; gives 3¼ inches in width at right angles, and 2¼ inches at mitre. Weight, 8 pounds, without saw.

It will be found equally useful for Paper Hangers, and Small Pattern Makers. Use no oil on saw guides.

Size A, with 14 inch Disston & Sons' Saw, 2 1-2 Inches Wide.

With 14 teeth to the inch, for sawing Wood, ...	$ 7.50
" 16 teeth for sawing Metal, ...	8.00
" both Saws, ...	10.00

Size No. 1, giving 6 inches in width at right angles, and 4 inches at Mitre. Weight, 10 1-2 pounds, without Saw.

Use no oil on saw guides, or posts in which they move.

Size No. 1, with Saw 4 Inches Wide under Back.

With 18 inch Disston & Sons' Saw, ...	$ 9.00
" 20 " " " "	9.50
" 22 " " " "	10.00
" 24 " " " "	10.50

Saws, 18, 20, 22, and 24 inches in length, are used in No. 1 size, while 22 and 24 inches give much the best satisfaction.

Retail Prices.—Continued.

Size No. 2. Weight, without Saw, 11 1-2 Pounds.

With Extension Lever, giving 9½ inches at right angles, and 6½ inches at Mitre. USE NO OIL ON SAW GUIDES.

Size No. 2. Saw 4 Inches wide under Back, with Extension Lever.

With 22 inch Disston & Sons' Saw,	$11.00
" 24 " " " "	11.50

Size No. 3.

An enlarged Box, weighing 25 pounds without saw, and giving 9½ inches at right angles, and 6½ inches at Mitre, with SAW 6 INCHES WIDE UNDER BACK. USE NO OIL ON SAW GUIDES.

Saws 24, 26, 28, and 30 inches in length are used in Size 3, but 28 and 30 inch Saws allow the longest stroke, and are much preferred.

Size No. 3, with Saw 6 Inches Wide under Back.

With 24 inch Disston & Sons' Saw,	$18.50
" 26 " " " "	19.00
" 28 " " " "	19.50
" 30 " " " "	20.00

To prevent pricking the thumb by the saw teeth in raising the Stop Lever, we have the butts of our Saws rounded. Saws are measured by the LENGTH OF THE BLADE, not of the teeth.

Mitre Box with Saw for Metal Mouldings.

With 20 inch Saws,	$9.50
" 22 " "	10.00

LIST OF PARTS,

(AS SHOWN IN CUT,)

AND PRICES OF SAME.

All parts of the Mitre Box are made to interchange, and can be sent by mail or by express, as may be desired.

In ordering parts of Machines, always state the number of the piece wanted, and the size of the machine for which it is designed.

	Part		Mitre B. Size 1.	Mitre B. Size 2.	Mitre B. Size 3.
1.	Stop Gauge,	each,	.10	.10	.20
2.	Thumb Screw,	"	.10	.10	.10
3.	Back Lever,	"	.15	.15	.30
4.	Legs,	"	.15	.15	.30
5.	Stop Lever,	"	.30		.60
6.	Stop Lever,	"		.45	
7.	Thumb Lever,	"		.20	
8.	Gibb,	"	.25	.25	.50
9.	Back,	"	.30	.30	.60
10.	Back Posts,	"	.35	.35	.70
11.	Saw Guides,	"	.55	.55	1.10
12.	Swinging Lever,	"	.90		1.80
13.	Swinging Lever,	"		1.00	
14.	Sliding Post,	"		.50	
15.	Bed,	"	1.00	1.00	2.00
	Bottom Board,	"	.10	.10	.20
	Spiral Spring,	"	.05	.05	.05
	Screws,	"	.05	.05	.05
	Lever Pins,	"	.05	.05	.05
	Set Screws,	"		.10	
	Clamp,	"			

Each Box is provided with a STOP GAUGE for holding the Saw in sawing Tenons, Rustic Frames, Scarfing, and similar work; is adjustable at any desired angle or depth of tenon within the capacity of the saw.

NUMBER OF PARTS.

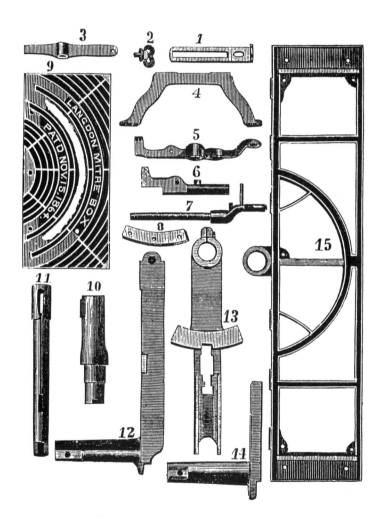

VICTOR PLANES.

All our Planes have our Trade Mark. Pat. Jan. 4, 1876.

PATENT IRON PLANES.

EACH.

No. 1 1-4. Block Plane, with Handle, adjustable Mouth and Cutter, 6 inches in length, 1 3-4 in. Cutter, Polished Trimmings, - - - - - $2.25

No. 1 1-2. Block Plane, with Handle, adjustable Mouth and Cutter, 6 inches in length, 1 3-4 in. Cutter, Nickel Plated Trimmings, - - - - - 2.50

No. 2 1-4. Block Plane, with Handle, adjustable Mouth and Cutter, 7 inches in length, 1 3-4 in. Cutter, Polished Trimmings, - - - - - - 2.50

No. 2 1-2. Block Plane, with Handle, adjustable Mouth and Cutter, 7 inches in length, 1 3-4 in. Cutter, Nickel Plated Trimmings, - - - - - 2.75

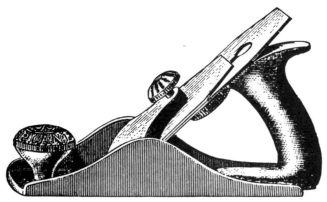

EACH.

No. 3. Smooth Plane, with Adjustment, 8 1-2 in. in length, 1 3-4 inch Cutter, Polished Trimmings, - - $4.00

No. 3 1-2. Smooth Plane, with Adjustment, 8 1-2 in. in length, 1 3-4 inch Cutter, Nickel Plated Trimmings, - 4.50

No. 4. Smooth Plane, with Adjustment, 9 inches in length, 2 inch Cutter, Polished Trimmings, - - - 4.50

No. 4 1-2. Smooth Plane, with Adjustment, 9 inches in length, 2 inch Cutter, Nickel Plated Trimmings. - - 5.00

The VICTOR PLANES

are all made with solid Bed, the Cutter Seat being cast in one piece with the stock.

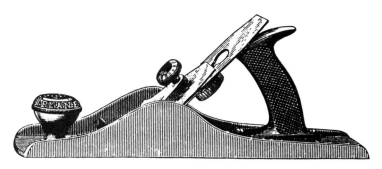

		EACH.
No. 5.	Jack Plane, with Adjustment, 14 inches in length, 2 inch Cutter, Polished Trimmings,	$5.00
No. 5 1-2.	Jack Plane, with Adjustment, 14 inches in length, 2 inch Cutter, Nickel Plated Trimmings,	5.50
No. 6.	Fore Plane, with Adjustment, 18 inches in length, 2 3-8 inch Cutter, Polished Trimmings,	6.00
No. 6 1-2.	Fore Plane, with Adjustment, 18 inches in length, 2 3-8 inch Cutter, Nickel Plated Trimmings,	6.50
No. 7.	Jointer Plane, with Adjustment, 22 in. in length, 2 3-8 inch Cutter, Polished Trimmings,	7.00
No. 7 1-2.	Jointer Plane, with Adjustment, 22 in. in length, 2 3-8 inch Cutter, Nickel Plated Trimmings,	7.50
No. 8.	Jointer Plane, with Adjustment, 24 in. in length, 2 5-8 inch Cutter, Polished Trimmings,	8.00
No. 8 1-2.	Jointer Plane, with Adjustment, 24 in. in length, 2 5-8 inch Cutter, Nickel Plated Trimmings,	8.50
No. 9.	Carriage Makers' Rabbet Plane, with Adjustment, 14 inches in length, 2 1-8 inch Cutter, Polished Trimmings,	5.50
No. 9 1-2.	Carriage Makers' Rabbet Plane, with Adjustment, 14 inches in length, 2 1-8 in. Cutter, Nickel Plated Trimmings,	6.00

The VICTOR PLANES

are the most simple, compact, and practical adjustable Planes ever yet produced, being the natural outgrowth and result of twenty years' exclusive experience in the production of Patent Planes.

All opportunities for the improvement of these tools have been sought, and no device that could add to their usefulness has been neglected.

We shall endeavor, by a rigid adherence to the policy of producing nothing but the best goods, at a low but reasonable profit, and by keeping the quality at the highest attainable point of perfection, to merit the patronage you may favor us with.

EACH.

No. 10. Circular Plane, with Adjustment, Flexible Adjustable Steel Face, 1 3-4 inch Cutter, Nickel Plated Trimmings, - - - - - - - - $4.75

This Plane will work a circle less in diameter than any other Flexible Face Plane.

PLANE IRONS

SINGLE.

Inches,	1¾	2	2⅛	2⅜	2⅝
Each,	38	42	44	48	50 cts.

DOUBLE.

Inches,	1¾	2	2⅛	2⅜	2⅝
Each,	62	68	74	80	84 cts.

Block Plane Irons, 1¾ inch, 35 cents each.

These Plane Irons are made of the best English Steel, hardened and tempered by an improved process. Each Cutter is sharpened and put in perfect working order before leaving the factory, and fully warranted.

1883 LEONARD BAILEY & Co. CATALOG and PRICE LIST

Leonard Bailey made an agreement on May 19, 1869, with the Stanley Rule & Level Co. for it to purchase certain rights to manufacture and sell Bailey's "Improved Iron & Wood Planes," Bailey patents, Aug.5, 1855, through Dec.24, 1867. He agreed to superintend the manufacture of these products at SR&L Co. About 1874 Bailey left New Britain and set up his own plane shop in Connecticut at Cushman St. in Hartford. Bailey also licensed the Bailey Wringing Machine Co. of Woonsocket, R I., to manufacture iron planes under trade marks of "Defiance" and "Battleaxe" brands. These were subsequently acquired by SR&L Co. on Jan.26, 1880. About the same time Bailey made arrangements with SR&L Co. to act as agent for his "Victor" line of planes and other products made in Hartford. They are illustrated in this Catalog, the second and only other trade catalog known to have been produced by the firm.

Other products illustrated in this Catalog under patents issued to Leonard Bailey are as follows:

- No.10 Adjustable Circular Plane, (Pat.113,003, March 28, 1871).
- No.20 Improved Adjustable Circular Plane (Pat.247,740, July 14,1881).
- No.14 Combination Plane with Charles Miller (Pat.165,356, July 6, 1875).
- No.48 Box Scraper & No.41 Spoke Shave (Pat.182,881, Oct. 3, 1876).
- Flush "T" Bevel with Samuel D.Sargent (Pat.124,779, March 19, 1872).
- Try Squares (Pat.145,715, Dec. 23, 1873).
- Hardening of Blades for Squares (Pat.157,566, Dec. 8, 1874).

Ken Roberts
1981

January, 1883.

LEONARD BAILEY & CO.'S
PATENT ADJUSTABLE IRON
BENCH PLANES,
TRY SQUARES, BEVELS,
Spoke Shaves, Box Scrapers, &c.

RECEIVED THE HIGHEST AWARD Paris Universal Exposition, 1878, Centennial Exhibition, 1876.

Also First Premium, A GOLD MEDAL, CONN. STATE FAIR, 1876. PATENTED.

Manufactured by

L. BAILEY & CO., HARTFORD, CONN.

STANLEY RULE & LEVEL COMPANY,
GENERAL AGENTS
FOR THE SALE OF "VICTOR PLANES," Etc.

NEW BRITAIN,	29 CHAMBERS ST.,
CONN.	NEW YORK.

FOR SALE BY ALL HARDWARE DEALERS.

American Bank Note Co., Type Department, 53 Broadway, N. Y.

VICTOR PLANES.

The Tools comprising this list are made under the direct supervision of LEONARD BAILEY, the original inventor of L. Bailey's Patent Adjustable Iron and Wood Bench Planes. They are genuine Bailey Tools, and are fully warranted in all their parts.

EACH.

No. 0. Block Plane, 7 inches in length, 1¾ inch Cutter, Japanned Finish, - - - - - $0.70
No. 0½. Block Plane, 7 inches in length, 1¾ inch Cutter, Japanned Finish, with Adjustment, - - - 1.00
No. 00. Block Plane, 7 inches in length, 1¾ inch Cutter, Nickel Plated Trimmings, - - - - 1.25
No. 000. Block Plane, 7 inches in length, 1¾ inch Cutter, Nickel Plated Trimmings, with Adjustment, - - 1.50

No. 1. Block Plane, adjustable Mouth and Cutter, 6 inches in length, 1¾ inch Cutter, Polished Trimmings, - $1.50
No. 1¼. Block Plane, adjustable Mouth and Cutter, 6 inches in length, 1¼ inch Cutter, Nickel Plated Trimmings, 1.75
No. 2. Block Plane, adjustable Mouth and Cutter, 7 inches in length, 1¾ inch Cutter, Polished Trimmings, - 1.75
No. 2¼. Block Plane, adjustable Mouth and Cutter, 7 inches in length, 1¾ inch Cutter, Nickel Plated Trimmings, 2.00

No. 1¼. Block Plane, with Handle, adjustable Mouth and Cutter, 6 inches in length, 1¾ inch Cutter, Polished Trimmings, - - - - - $1.75
No. 1½. Block Plane, with Handle, adjustable Mouth and Cutter, 6 inches in length, 1¾ inch Cutter, Nickel Plated Trimmings, - - - - 2.00
No. 2¼. Block Plane, with Handle, adjustable Mouth and Cutter, 7 inches in length, 1¾ inch Cutter, Polished Trimmings, - - - - - 2.00
No. 2½. Block Plane, with Handle, adjustable Mouth and Cutter, 7 inches in length, 1¾ inch Cutter, Nickel Plated Trimmings, - - - - 2.25

VICTOR PLANES.

EACH.

No. 3. Smooth Plane, with Adjustment, 8½ inches in length, 1¾ inch Cutter, Polished Trimmings, - - $3.00

No. 3½. Smooth Plane, with Adjustment, 8½ inches in length, 1¾ inch Cutter, Nickel Plated Trimmings, - - 3.75

No. 4. Smooth Plane, with Adjustment, 9 inches in length, 2 inch Cutter, Polished Trimmings, - - - 3.25

No. 4½. Smooth Plane, with Adjustment, 9 inches in length, 2 inch Cutter, Nickel Plated Trimmings, - - 4.00

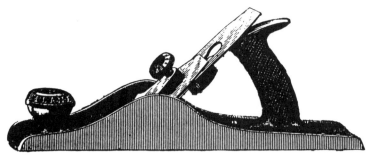

No. 5. Jack Plane, with Adjustment, 14 inches in length, 2 inch Cutter, Polished Trimmings, - - - $3.75

No. 5½. Jack Plane, with Adjustment, 14 inches in length, 2 inch Cutter, Nickel Plated Trimmings, - - 4.50

No. 6. Fore Plane, with Adjustment, 18 inches in length, 2⅜ inch Cutter, Polished Trimmings, - - 4.75

No 6½. Fore Plane, with Adjustment, 18 inches in length, 2⅜ inch Cutter, Nickel Plated Trimmings, - - 5.50

No. 7. Jointer Plane, with Adjustment, 22 inches in length, 2⅜ inch Cutter, Polished Trimmings, - - 5.50

No. 7½. Jointer Plane, with Adjustment, 22 inches in length, 2⅜ inch Cutter, Nickel Plated Trimmings, - - 6.25

No. 8. Jointer Plane, with Adjustment, 24 inches in length, 2⅝ inch Cutter, Polished Trimmings, - - 6.50

No. 8½. Jointer Plane, with Adjustment, 24 inches in length, 2⅝ inch Cutter, Nickel Plated Trimmings, - - 7.25

☞ The "Victor" Planes, on above page, have Iron Handles. They will be furnished with Wood Handles if so ordered.

VICTOR PLANES.

COMBINED
SMOOTH, RABBET & FILLETSTER PLANE.

The Tool represented below consists of an Iron Smooth Plane, (same as our No. 4 Plane,) and is so constructed that by means of the attachments which accompany it, it can also be used as a Rabbet Plane or a Filletster.

This view of the tool shows the cutter on one side to be flush with the edge of the stock to the Plane, adapting it for use as a Rabbet Plane; also, there is shown in this view the depth gauge, spur in edge of stock, and the fence, all of which belong to the tool when used as a Filletster.

This view of the tool shows the socket, which can be screwed on to the side of the stock, and through which a bar slides. The fence can thus be moved to any required distance from the edge of the stock, making an Adjustable Filletster of any desired width up to two inches.

In this tool, wood-workers of all classes, will find a practical combination of three useful tools adapted to their needs.

No. 11. Combined Smooth, Rabbet and Filletster Plane, 9 in. long, 2 inch Cutter, Polished Trimmings, Each, - $5.50

No. 11½. Combined Smooth, Rabbet and Filletster Plane, 9 in. long, 2 inch Cutter, Nickle Plated Trimmings, Each, 6.00

☞ These Tools will be furnished with Wood Handles, if so ordered.

VICTOR PLANES.

Adjustable Circular Plane

No. 10. Circular Plane, with Adjustment, Flexible Steel Face, 1¾ inch Cutter, Nickel Plated Trimmings. Each, $4.00.

IMPROVED
Adjustable Circular Plane.

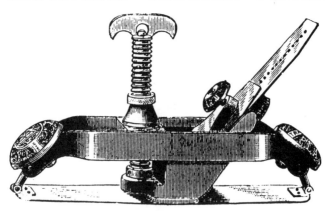

The above cut represents LEONARD BAILEY's Improved Adjustable Plane, for working Concave or Convex Surfaces.

Its smallest working capacity is 18½ inches, outside circle, and 12½ inches inside circle. Both ends of the face of the Plane are moved simultaneously and precisely alike, by means of one screw. The style, finish and general make-up of this Plane cannot fail to convince every one, at first sight, of its great superiority over all other tools of this class ever offered to mechanics. The simplicity of its operation is unequaled.

No. 20. Circular Plane, 9½ inches in length, 1¾ inch Cutter, Nickel Plated Trimmings. Each, $6.00.

VICTOR PLANES.

L. Bailey's Pocket Block Plane.

No. 12, 4½ inches in length, 1¼ inch Cutter, Japanned finish, Polished Trimmings, - - - -   EACH. $0.75

No. 12½, 4½ inches in length, 1¼ inch Cutter, Japanned finish, Nickel Plated Trimmings, - - -   1.00

No. 12¼, 4½ inches in length, 1¼ inch Cutter, Full Nickel Plated, - - - - - - - - -   1.25

☞ In removing and replacing the cutter to sharpen, take it out through the top, and put it in through the bottom or face, thus avoiding all liability of dulling by coming in contact with iron.

We desire to call special attention to our New JOINER'S POCKET BLOCK PLANE. We believe this tool when once seen will speak for itself more pointedly than anything we could possibly say, and hardly think any explanation is necessary, for it is simplicity itself both in construction and operation. It is the nicest working tool ever made, and specially recommended for amateurs, pattern makers, light scroll saw work, &c.

L. Bailey's Patent Combination Plane.

No. 14, PLOW, FILLETSTER, BACK FILLETSTER, DADO, RABBET & MATCHING PLANE COMBINED. EACH, $5.50.

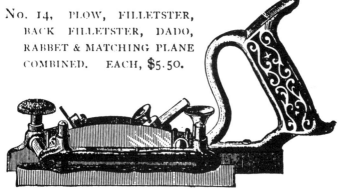

The main feature of this tool is, that it has a fence or guide, which is made to change from one side to the other as the nature of the work requires, same fence or guide also in itself being vertically self-adjusting. The same cutter is used for each special tool in the combination.

We recommend this ingenious tool as most desirable and useful for all wood-working mechanics, who require, in a compact form, a tool designed to accomplish practically the results expected of each of the planes named above.

As an Adjustable Dado it has no equal, as it will cut a groove from ⅜ in. in width up to two inches or more, without change of cutter.

THE VICTOR PLANES

are the most simple, compact and practical Adjustable Planes ever yet produced, being the natural outgrowth and result of a quarter of a century's exclusive experience in the production of Patent Planes.

All opportunities for the improvement of these tools have been sought, and no device that could add to their usefulness has been neglected.

They are manufactured and sold under various patents granted to L. BAILEY.

We shall endeavor, by rigid adherence to the policy of producing nothing but the best goods, at a low but reasonable profit, and by keeping the quality at the highest attainable point of perfection, to merit the patronage you may favor us with.

A PLANE FOR THE MILLION.

THE LITTLE VICTOR is the most perfect Toy Plane ever invented. It is not a mere toy, but will be found useful in every house, shop, factory, store, bank, insurance office, printing office, &c. It is the best pencil sharpener in the world, and especially adapted to PATTERN MAKING AND SCROLL SAW WORK. It is an article of real value and practical utility, and sells on its merits.

		EACH.
No. 50.	3¼ inches in length, 1 inch Cutter, complete Adjustment, Japanned,	$0.45
No. 50½.	3¼ inches in length, 1 inch Cutter, complete Adjustment, full Nickel Plated,	.50
No. 51.	3¼ inches in length, 1 inch Cutter, Screw Fastening, Japanned,	.30
No. 51½.	3¼ inches in length, 1 inch Cutter, Screw Fastening, full Nickel Plated,	.45
No. 52.	3¼ inches in length, 1 inch Cutter, Screw Eye Fastening, Japanned,	.25
Cutters,	per dozen,	1.25

☞ On receipt of list price, we will send to any address in the United States, by mail or express, expenses paid, any sample tool found in our list. Always order by list number.

Satisfaction Guaranteed or Money Refunded.

LEONARD BAILEY & CO.,   Hartford, Conn.

VICTOR PLANE IRONS.

On this page we illustrate a superior Compound Plane Iron, developing an improvement so simple and convenient that every man who has ever used a Double Iron Plane cannot fail to discover at sight, without any explanation, the peculiar advantages and the immense saving of *time*, with no liability of injury to *other tools* which this arrangement affords over any other double Plane Iron now in use. We offer it with full confidence that it will explain its own merits, and thereby find its way to public favor without other comments from us. We will simply say there is no necessity for ever removing or even loosening the screw which is attached to the cutting iron, until the cutter is entirely worn out. Seeing is believing in this case.

The "Victor" Patent Adjustable Planes are supplied with this SUPERIOR COMPOUND PLANE IRON.

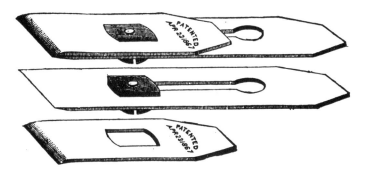

PLANE IRONS.

SINGLE,	inches,	1¾	2	2⅜	2⅝	
	each,	38	42	48	50	cents.
PATENT DOUBLE,	inches,	1¾	2	2⅜	2⅝	
	each,	62	68	80	84	cents.
" STEEL CAPS,	inches,	1¾	2	2⅜	2⅝	
	each,	25	25	35	35	cents.

All Double Irons have Patent Steel Caps.
Block Plane Irons, 1¾ inch, 35 cents each.

These Plane Irons are made of the best English Steel, hardened and tempered by a new process. Each cutter is sharpened and put in perfect working order before leaving the factory, and fully warranted.

L. BAILEY'S
VICTOR REVERSIBLE BOX SCRAPER.

Patented October 3d, 1876.

No. 48.

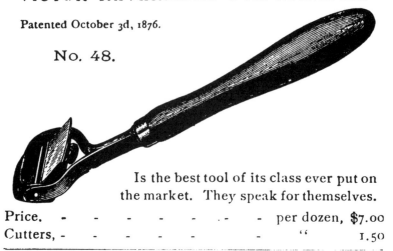

Is the best tool of its class ever put on the market. They speak for themselves.

Price. - - - - - - - per dozen, $7.00
Cutters, - - - - - - " 1.50

DOUBLE IRON SPOKE SHAVE.
No. 41.
Pat. Oct. 3d, 1876.

Are the best for all purposes, being used as either Single or Double Iron Shaves. The Cutters have no slots or holes in them. The fastenings and adjustment is new, simple and complete. The caps are cast steel.

Price per dozen, - $5.00 | Cutters, per dozen, - $1.50

ADJUSTABLE SPOKE SHAVE.
No. 43.

We warrant this to be the most practical and RELIABLE SPOKE SHAVE ever offered to mechanics. It is self-adjusting in all its parts and a perfect working tool.

Price per dozen, - $6.00 | Cutters, per dozen, - $1.50

L. Bailey's Patent Flush T Bevel.

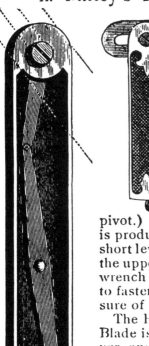

FIG. 1.

The method of securing the Blade at any angle desired is simple, yet effective. Fig. 2 shows the internal construction of the Handle. By moving the thumb-piece at the lower end of the Handle, the long lever acts upon the shorter one (being set upon a strong pivot.) The strength of a compound lever is produced by this arrangement, and the short lever being attached to the Nut inside the upper end of the Handle, operates as a wrench to turn it upon the screw, and thus to fasten or release the Blade at the pleasure of the owner.

The Handle is made of Cast Iron. The Blade is of fine quality Steel, spring temper, and with perfectly parallel edges.

Positively this is the BEST BEVEL made.

PRICES:

8 inch, - - - per dozen, $14.00
10 inch, - - -   "   16.00

FIG. 2.

L. Bailey's Patent Try Squares.

These Try Squares are what might be termed adjustable, the Handle and the Blade being connected together by

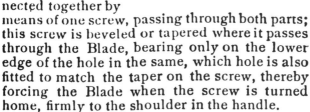

means of one screw, passing through both parts; this screw is beveled or tapered where it passes through the Blade, bearing only on the lower edge of the hole in the same, which hole is also fitted to match the taper on the screw, thereby forcing the Blade when the screw is turned home, firmly to the shoulder in the handle.

The Handles are made of Cast Iron, highly finished, and the Blade is made of the best Steel, hardened and tempered by a patented process, and warranted accurate.

3 inch, per dozen, $18.00 | 6 inch, per dozen, $24.00
4½ inch,   "   21.00 | 8 inch,   "   30.00

CONDENSED PRICE LIST.

GENTLEMEN:—We present you with a REVISED and CONDENSED PRICE LIST of "VICTOR PLANES." We will continue to keep the standard of our goods at the high point of perfection already attained; nothing but the very best materials and skilled labor will be used in their manufacture. Our facilities for the production of Iron Planes are unequaled. Our machinery is new and ingenious, and especially adapted to the business.

The Victor Planes are the product of twenty-five years' experience in manufacturing, improving and inventing Iron Planes. Each and every tool bearing the Stamp and Trade Mark of L. Bailey & Co., can be strictly relied upon for their superiority, and *warranted* perfect in all their parts.

LEONARD BAILEY & CO.

Hartford, *Conn.*, *Jan.* 1883.

VICTOR PLANES.

No. 0,	$0.70	each	No. 3,	$3.00	each	No. 11,	$5.50	each
No. 0½,	1.00	"	No. 3½,	3.75	"	No. 11½,	6.00	"
No. 00,	1.25	"	No. 4,	3.25	"	No. 12,	.75	"
No. 000,	1.50	"	No. 4½,	4.00	"	No. 12¼,	1.25	"
No. 1,	1.50	"	No. 5	3.75	"	No. 12½,	1.00	"
No. 1¼,	1.75	"	No. 5½,	4.50	"	No. 14,	5.50	"
No. 1½,	2.00	"	No. 6,	4.75	"	No. 20,	6.00	"
No. 1¾,	1.75	"	No. 6½,	5.50	"	No. 50,	.45	"
No. 2,	1.75	"	No. 7,	5.50	"	No. 50½,	.50	"
No. 2¼,	2.00	"	No. 7½,	6.25	"	No. 51,	.30	"
No. 2½,	2.25	"	No. 8,	6.50	"	No. 51½,	.45	"
No. 2¾,	2.00	"	No. 8½,	7.25	"	No. 52,	.25	"
			No. 10,	4.00	"			

PLANE IRONS.

SINGLE, - - -	inches,	1¾	2	2⅜	2⅝	
	each,	38	42	48	50	cents.
PATENT DOUBLE,	inches,	1¾	2	2⅜	2⅝	
	each,	62	68	80	84	cents.
" STEEL CAPS,	each,	25	25	35	35	cents.

Block Plane Irons, 1¾ inch, - - - each, $0.35

Little Victor Plane Irons, - - - per dozen, 1.25

TRY SQUARES.

3 inch, per dozen,	$18.00	6 inch, per dozen,	$24.00
4½ " "	21.00	8 " "	30.00

BEVELS.

8 inch, per dozen, $14.00 10 inch, per dozen, $16.00

No. 41, SPOKE SHAVE, - -	per dozen,	$5.00
No. 43, SPOKE SHAVE, - -	"	6.00
No. 48, BOX SCRAPER, - -	"	7.00
Box Scraper and Spoke Shave Cutters,	"	1.50

Milestones in the Growth of Stanley Tools

1853—Thomas S. Hall and Francis Knapp organize firm and set up production of try squares, plumbs and levels in two-story building on west side of Elm Street north of what is now Church Street.

1854—August and Timothy Stanley establish partnership with Thomas Conklin, formerly a rule manufacturer in Bristol, and begin production of rules under the firm name of A. Stanley and Company.

1857—Henry Stanley, elder brother of Augustus and Timothy Stanley, brings about consolidation of A. Stanley and Company and Hall and Knapp. Combined companies named The Stanley Rule and Level Company.

1863—The Stanley Rule and Level Company purchases rule business of E. A. Stearns and Company, Brattleboro, Vermont.

1869—The Stanley Rule and Level Company purchases Bailey, Chany and Co., Boston, manufacturer of the Bailey Plane and moves it to New Britain.

1885—The Stanley Rule and Level Company introduces variations of planes with notable success.

1892—Metal gauges and scrapers added to product line and introduced.

1900—The Stanley Rule and Level Company becomes one of the largest manufacturers of artisans' woodworking tools in the world.

1902—Company expands hand tools line through purchase of three small firms which manufacture bit braces.

1903— "Zig Zag" folding wood rules introduced.

1904—Screwdrivers added to the product line through acquisition of Hurley and Wood factory in Plantsville, Connecticut.

1906—Mitre boxes added to product line.

1907—Company enters promising Canadian market for hand tools by purchasing Roxton Tool and Mill Company, Roxton Pond, Quebec.

1908—Product line expanded by introduction of nail sets, awls and ice picks.

1912—New Britain firm of Humason and Beckley, manufacturers of hammers, purchased by the Company.

1913—To bolster hammer business, Company purchases Atha Tool Company, Newark, New Jersey. Atha presently a branch plant of Stanley Tools.

1914—Wood chisels, cold chisels and punches introduced.

1916—Eagle Square Manufacturing Company, Shaftsbury, Vermont, purchased to supplement hand tools production particularly in regard to carpenters' squares and zig zag sticks. This Company is presently a branch plant of Stanley Tools.

1918—Combination squares added to product line.

1920—Assets of The Stanley Rule and Level Company purchased by The Stanley Works.

1924—Stanley "Tools for the Handyman" introduced. Aluminum levels also added to product line.

1928—Electric drills developed and introduced by Stanley Rule and Level.

1930—Company enters coilable steel rule business through purchase of business of Hiram A. Farrand, Berlin, New Hampshire.

1934—Plastic-handle screwdrivers, utility knives and blades put on the market.

1935—Stanley Rule and Level Division renamed Stanley Tools Division.

1938—Phillips screwdrivers line introduced.

1946—Stanley purchases North Brothers Manufacturing Company, Philadelphia, manufacturer of "Yankee" tools and other products.

1956— "Steelmaster" hammer and "Surform" tools added to Division's product line.

1957—Stanley garden tools line introduced.

1963—Stanley Powerlock Tape Rules introduced. Ground breaking at the site of the new plant for Stanley Tools, April 18.

1964—Plant Dedication Day, January 17, 1968. Fiberglass hammer introduction.

1971—New Eagle Square Manufacturing plant completed and in operation.